Springer Series in Statistics

Advisors:
J. Berger, S. Fienberg, J. Gani,
K. Krickeberg, B. Singer

Springer Series in Statistics

R.-D. Reiss

Approximate Distributions of Order Statistics

With Applications to Nonparametric Statistics

With 30 Illustrations

Springer-Verlag
New York Berlin Heidelberg
London Paris Tokyo

R.-D. Reiss
Universität Gesamthochschule Siegen
Fachbereich 6, Mathematik
D-5900 Siegen
Federal Republic of Germany

Mathematics Subject Classification (1980): 62-07, 62B15, 62E20, 62G05, 62G10, 62G30

Library of Congress Cataloging-in-Publication Data
Reiss, Rolf-Dieter.
  Approximate distributions of order statistics.
  (Springer series in statistics)
  Bibliography: p.
  Includes indexes.
  1. Order statistics. 2. Asymptotic distribution
(Probability theory)   3. Nonparametric statistics.
I. Title.   II. Series.
QA278.7.R45 1989   519.5   88-24844

Printed on acid-free paper.

Typeset by Asco Trade Typesetting Ltd., Hong Kong.

9 8 7 6 5 4 3 2 1

ISBN-13:978-1-4613-9622-2   e-ISBN-13:978-1-4613-9620-8
DOI: 10.1007/978-1-4613-9620-8

To Margit, Maximilian, Cornelia, and Thomas

Preface

This book is designed as a unified and mathematically rigorous treatment of some recent developments of the asymptotic distribution theory of order statistics (including the extreme order statistics) that are relevant for statistical theory and its applications. Particular emphasis is placed on results concerning the accuracy of limit theorems, on higher order approximations, and other approximations in quite a general sense.

Contrary to the classical limit theorems that primarily concern the weak convergence of distribution functions, our main results will be formulated in terms of the variational and the Hellinger distance. These results will form the proper springboard for the investigation of parametric approximations of nonparametric models of joint distributions of order statistics. The approximating models include normal as well as extreme value models. Several applications will show the usefulness of this approach.

Other recent developments in statistics like nonparametric curve estimation and the bootstrap method will be studied as far as order statistics are concerned. In connection with this, graphical methods will, to some extent, be explored.

The prerequisite for handling the indicated problems is a profound knowledge of distributional properties of order statistics. Thus, we collect several basic tools (of finite and asymptotic nature) that are either scattered in literature or are not elaborated to such an extent that would satisfy our present requirements. For example, the Markov property of order statistics is studied in detail. This part of the book that has the characteristics of a textbook is supplemented by several well-known results.

The book is intended for students and research workers in probability and statistics, and practitioners involved in applications of mathematical results concerning order statistics and extremes. The knowledge of standard calculus and topics that are taught in introductory probability and statistics courses are necessary for the understanding of this book. To reinforce previous knowledge as well as to fill gaps, we shall frequently give a short exposition of probabilistic and statistical concepts (e.g., that of conditional distribution and approximate sufficiency).

The results are often formulated for distributions themselves (and not only for distribution functions) and so we need, as far as order statistics are concerned, the notion of Borel sets in a Euclidean space. Intervals, open sets, and closed sets are special Borel sets. Large parts of this book can be understood without prior knowledge of technical details of measure-theoretic nature.

My research work on order statistics started at the University of Cologne, where influenced by J. Pfanzagl, I became familiar with expansions and statistical problems. Lecture notes of a course on order statistics held at the University of Freiburg during the academic year 1976/77 can be regarded as an early forerunner of the book.

I would like to thank my students B. Dohmann, G. Heer, and E. Kaufmann for their programming assistance. G. Heer also skillfully read through larger parts of the manuscript. It gives me great pleasure to acknowledge the cooperation, documented by several articles, with my colleague M. Falk. The excellent atmosphere within the small statistical research group at the University of Siegen, and including A. Janssen and F. Marohn, facilitated the writing of this book. Finally, I would like to thank W. Stute, and those not mentioned individually, for their comments.

Siegen, FR Germany   Rolf-Dieter Reiss

Contents

CHAPTER 0

Introduction

Let us start with a detailed outline of the intentions and of certain characteristics of this book.

0.1. Weak and Strong Convergence

For good reasons the concept of weak convergence of random variables (in short, r.v.'s) ξ_n plays a preeminent role in literature. Whenever the distribution functions (in short, d.f.'s) F_n of the r.v.'s ξ_n are not necessarily continuous then, in general, only the weak convergence holds, that is,

$$P\{\xi_n \leq t\} \to P\{\xi_0 \leq t\}, \qquad n \to \infty, \tag{1}$$

at every point of continuity t of F_0. If F_0 is continuous then it is well known that the convergence in (1) holds uniformly in t. This may be written in terms of the Kolmogorov–Smirnov distance as

$$\sup_t |F_n(t) - F_0(t)| \to 0, \qquad n \to \infty.$$

In this sequel let us assume that F_0 is continuous. It follows from (1) that

$$P\{\xi_n \in I\} \to P\{\xi_0 \in I\}, \qquad n \to \infty, \tag{2}$$

uniformly over all intervals I. In general, (2) does not hold for every Borel set I. However, if the d.f.'s F_n have densities, say, f_n such that $f_n(t) \to f_0(t)$, $n \to \infty$, for almost all t, then it is well known that (2) is valid w.r.t. the variational distance, that is,

$$\sup_{B} |P\{\xi_n \in B\} - P\{\xi_0 \in B\}| \to 0, \qquad n \to \infty, \tag{3}$$

where the sup is taken over all Borel sets B.

Next, the remarks above will be specialized to order statistics. It is well known that central order statistics $X_{r(n):n}$ of a sample of size n are asymptotically normally distributed under weak conditions on the underlying d.f. F. In terms of weak convergence this may be written

$$P\{a_n^{-1}(X_{r(n):n} - b_n) \le t\} \to P\{\xi_0 \le t\}, \qquad n \to \infty, \tag{4}$$

for every t, with ξ_0 denoting a standard normal r.v. and a_n, b_n are normalizing constants. The two classical methods of proving (4) are

(a) an application of the central limit theorem to binomial r.v.'s,
(b) a direct proof of the pointwise convergence of the corresponding densities (e.g. H. Cramér (1946)).

However, it is clear that (b) yields the convergence in a stronger sense, namely, w.r.t. the variational distance. We have

$$\sup_{B} |P\{a_n^{-1}(X_{r(n):n} - b_n) \in B\} - P\{\xi_0 \in B\}| \to 0, \qquad n \to \infty, \tag{5}$$

where the sup is taken over all Borel sets B. A more systematic study of the strong convergence of distributions of order statistics was initiated by L. Weiss (1959, 1969a) and S. Ikeda (1963). These results particularly concern the joint asymptotic normality of an increasing number of order statistics.

The convergence of densities of central order statistics was originally studied for technical reasons; these densities are of a simpler analytical form than the corresponding d.f.'s. On the other hand, when treating weak convergence of extreme order statistics it is natural to work directly with d.f.'s. To highlight the foregoing remark the reader is reminded of the fact that F^n is the d.f. of the largest order statistic (maximum) $X_{n:n}$ of n independent and identically distributed r.v.'s with common d.f. F.

The, meanwhile, classical theory for extreme order statistics provides necessary and sufficient conditions for a d.f. F to belong to the domain of attraction of a nondegenerate d.f. G; that is, the weak convergence

$$F^n(b_n + a_n t) \to G(t), \qquad n \to \infty, \tag{6}$$

holds for some choice of constants $a_n > 0$ and reals b_n. If F has a density then one can make use of the celebrated von Mises conditions to verify (6). These conditions are also necessary for (6) under further milder conditions imposed on F. In particular, the d.f.'s treated in statistical textbooks satisfy one of the von Mises conditions. Moreover, it turns out that the convergence w.r.t. the variational distance holds. This may be written,

$$\sup_{B} |P\{a_n^{-1}(X_{n:n} - b_n) \in B\} - G(B)| \to 0, \qquad n \to \infty, \tag{7}$$

where the sup is taken over all Borel sets B. Note that the symbol G is also used for the probability measure corresponding to the d.f. G. Apparently, (7) implies (6).

The relation (7) can be generalized to the joint distribution of upper extremes $X_{n-k+1:n}, X_{n-k+2:n}, \ldots, X_{n:n}$ where $k \equiv k(n)$ is allowed to increase to infinity as the sample size n increases.

We want to give some arguments why our emphasis aims at the variational and Hellinger distance instead of the Kolmogorov–Smirnov distance:

(a) We claim mathematical reasons, namely, to formulate as strongly as possible the results. One can add that the problems involved are very challenging.
(b) Results in terms of d.f.'s look awkward if the dimension increases with the sample size. Of course, the alternative outcome is the formulation in terms of stochastic processes.
(c) It is necessary to use the variational distance (and, as an auxiliary tool, the Hellinger distance) in connection with model approximation. In other words, certain problems cannot be solved in a different way.

0.2. Approximations

The joint distributions of order statistics can explicitly be described by analytical expressions involving the underlying d.f. F and density f. However, in most cases it is extremely cumbersome to compute the exact numerical values of probabilities concerning order statistics or to find the analytical form of d.f.'s of functions of order statistics. Hence, it is desirable to find approximate distributions. In view of practical and theoretical applications these approximations should be of a simple form.

The classical approach of finding approximate distributions is given by the asymptotic theory for sequences of order statistics $X_{r(n):n}$ with the sample size n tending to infinity:

(a) If $r(n) \to \infty$ and $n - r(n) \to \infty$ as $n \to \infty$ then the order statistics are asymptotically normal under mild regularity conditions imposed on F.
(b) If $r(n) = k$ or $r(n) = n - k + 1$ for every n with k being fixed then the order statistics are asymptotically distributed according to an extreme value distribution (being unequal to the normal distribution).

In the intermediate cases—that is, $r(n) \to \infty$ and $r(n)/n \to 0$ or $n - r(n) \to \infty$ and $(n - r(n))/n \to 0$ as $n \to \infty$—one can either use the normal approximation or an approximation by means of a sequence of extreme value distributions. Thus, the problem of computing an estimate of the remainder term enters the scene; sharp estimates will make the different approximations comparable.

In the case of maxima of normal r.v.'s we shall see that a certain sequence of extreme value distributions provides a better approximation than the limit distribution.

Better insight into the problem of computing accurate approximations is obtained when higher order approximations are available. There is a trade-off between the two requirements that the higher order approximation should be of a simple form and also of a better performance than the limiting distribution.

In particular, we shall study finite expansions of length $m + 1$ which may be written

$$Q + \sum_{i=1}^{m} v_{i,n}$$

where Q is the limiting distribution and the $v_{i,n}$ are signed measures depending on the sample size n. A prominent example is provided by Edgeworth expansions. Usually, the signed measures have polynomials $f_{i,n}$ as densities w.r.t. Q. If Q has a density g then the expansion may be written

$$Q(B) + \sum_{i=1}^{m} v_{i,n}(B) = \int_{B} \left(1 + \sum_{i=1}^{m} f_{i,n}(x) \right) g(x)\, dx \qquad (8)$$

for every Borel set B. Specializing (8) to $B = (-\infty, t]$, one gets approximations to d.f.'s of order statistics.

The bound of the remainder term of an approximation will involve

(a) unknown universal constants, and
(b) some known terms which specify the dependence on the underlying d.f. and the index of the order statistic.

Since the universal constants are not explicitly stated, our considerations belong to the realm of asymptotics.

The bounds give a clear picture of the dependence on the remainder terms from the underlying distribution. Much emphasis is laid on providing numerical examples to show that the asymptotic results are relevant for small and moderate sample sizes.

0.3. The Role of Order Statistics in Nonparametric Statistics

The sample d.f. F_n is the natural, nonparametric estimator of the unknown d.f. F, and, likewise, the sample quantile function (in short, sample q.f.) F_n^{-1} may be regarded as a natural estimator of the unknown q.f. F^{-1}. For any functional $T(F^{-1})$ of F^{-1} a plausible choice of an estimator will be $T(F_n^{-1})$ if no further information is given about the underlying model.

Note that $T(F_n^{-1})$ can be expressed as $\hat{T}(X_{1:n}, \ldots, X_{n:n})$ since $F_n^{-1}(q) = X_{r(q):n}$ where $r(q) = nq$ or $r(q) = [nq] + 1$.

In many nonparametric problems one is only concerned with the local behavior of the q.f. F^{-1} so that it suffices to base a statistic on a small set of order statistics like upper extremes

$$X_{n-k+1:n} \leq \cdots \leq X_{n:n}$$

or certain central order statistics

$$X_{[nq]:n} \leq \cdots \leq X_{[np]:n} \quad \text{where } 0 < q < p < 1.$$

Thus, one is interested in the distribution of functions of order statistics of the form $T(X_{r:n}, X_{r+1:n}, \ldots, X_{s:n})$ where $1 \leq r \leq s \leq n$. This problem can be studied for a particular statistic T or within a certain class of statistics T like linear combinations of order statistics.

If the type of the statistic T is not fixed in advance, one can simplify the stochastic analysis by establishing an approximation of the joint distribution of order statistics. Upper extremes $X_{n:n}, \ldots, X_{n-k+1:n}$ may be replaced by r.v.'s $Y_1, \ldots, Y_k$ that are jointly distributed according to a multivariate extreme value distribution so that the error term

$$\sup_B |P\{(X_{n:n}, \ldots, X_{n-k+1:n}) \in B\} - P\{(Y_1, \ldots, Y_k) \in B\}| := \delta(F) \qquad (9)$$

is sufficiently small. (9) implies that for any statistic T

$$\sup_B |P\{T(X_{n:n}, \ldots, X_{n-k+1:n}) \in B\} - P\{T(Y_1, \ldots, Y_k) \in B\}| \leq \delta(F), \quad (10)$$

and hence statistical problems concerning upper extremes can approximately be solved within the parametric extreme value model. These arguments also hold for lower extremes.

A similar—yet slightly more complicated—operation is needed in the case of central order statistics. Now the joint distribution of order statistics is replaced by a multivariate normal distribution. To return from the normal model to the original model one needs a fixed Markov kernel which will be constructed by means of a conditional distribution of order statistics.

0.4. Central and Extreme Order Statistics

There are good reasons for a separate treatment of extreme order statistics and central order statistics; one can e.g. argue that the asymptotic distributions of extreme order statistics are different from those of central order statistics.

However, as already mentioned above, intermediate order statistics can be regarded as central order statistics as well as extremes so that a clear distinction between the two different classes of order statistics is not possible. The statistical extreme value theory is concerned with the evaluation of parameters of the tail of a distribution like the upper and lower endpoint. In many situations the asymptotically efficient estimator will depend on intermediate order statistics and will itself be asymptotically normal. Thus, from a certain conservative point of view statistical extreme value theory does not belong to extreme value theory.

On the other hand, some knowledge of stochastical properties of extreme order statistics is needed to examine certain aspects of the behaviour of central order statistics. To highlight this point we note that spacings $X_{r:n} - X_{r-1:n}$ of exponential r.v.'s have the same distribution as sample maxima. Another example is provided by the conditional distribution of the order statistic $X_{r:n}$ given $X_{r+1:n} = x$ that is given by distributions of sample maxima.

0.5. The Restriction to Independent and Identically Distributed Random Variables

The classical theory of extreme values deals with the weak convergence of distributions of maxima of independent and identically distributed r.v.'s. The extension of these classical results to dependent sequences was one of the celebrated achievements of the last decades. This extension was necessary to justify the applicability of classical results to many natural phenomena.

A similar development can be observed in the literature concerning the distributional properties of central order statistics, however, these results are more sporadic than systematic. In this book we shall indicate some extensions of the classical results to dependent sequences, but our attention will primarily be focused upon strengthening classical results by obtaining convergence in a stronger sense and deriving higher order approximations. Our results may also be of interest for problems which concern dependent r.v.'s like

(a) testing problems where under the null-hypothesis the r.v.'s are assumed to be independent, and
(b) cases where results for dependent random variables are formulated via a comparison with the corresponding results for independent r.v.'s.

0.6. Graphical Methods

Despite of the preference for mathematical results the author strongly believes in the usefulness of graphical methods. I have developed a very enthusiastic attitude toward graphical methods but this is only when the methods are controlled by a mathematical background.

The traditional method of visually discriminating between distributions is the use of probability papers. This method is highly successful since the eye can easily recognize whether a curve deviates from a straight line. Perhaps the disadvantages are

(a) that one can no longer see the original form of the "theoretical" d.f.,
(b) that small oscillations of the density (thus, also of probabilities) are difficult to be detected by the approach via d.f.'s.

Alternatively, one may use densities, which play a key role in our methodology. As far as visual aspects are concerned the maximum deviation of densities is more relevant than the L_1-distance (which is equivalent to the variational distance of distributions).

The problem that discrete d.f.'s (like sample d.f.'s) have no densities can be overcome by using smoothing techniques like histograms or kernel density estimates. Thus the data points can be visualized by densities. The q.f. is another useful diagnostic tool to study the tails of the distribution.

The graphical illustrations in the book were produced by means of the interactive statistical software package ADO.

0.7. A Guide to the Contents

This volume is organized in three parts, each of which is divided into chapters where univariate and multivariate order statistics are studied. The treatment of univariate order statistics is separated completely from the multivariate case.

The chapters start—as a warm-up—with an elementary treatment of the topic or with an outline of the basic ideas and concepts. In order not to overload the sections with too many details some of the results are shifted to the Problems and Supplements. The Supplements also include important theorems which are not central to this book. Historical remarks and discussions of further results in literature are collected in the Bibliographical Notes.

Given the choice between different proofs, we prefer the one which can also be made applicable within the asymptotic set-up. For example, our way of establishing the joint density of several order statistics is also applicable to derive the joint asymptotic normality of several central order statistics.

Part I lays out the basic notions and tools. In Chapter 1 we explain in detail the transformation technique, compute the densities of order statistics and study the structure of order statistics as far as representations and conditional distributions are concerned.

Chapter 2 is devoted to the multivariate case. We discuss the problem of defining order statistics in higher dimensions and study some basic properties in the special case of order statistics, these are defined componentwise.

Chapter 3 contains some simple inequalities for distributions of order statistics. Moreover, concepts and auxiliary tools are developed which are needed in Part II for the construction of approximate distributions of order statistics.

Part II provides the basic approximations of distributions of order statistics. Chapter 4 and 5 are concerned with the asymptotic normality of central order statistics and the asymptotic distributions of extreme order statistics. Both chapters start with an introduction to asymptotic theory; in a second step the accuracy of approximation is investigated. Some asymptotic properties of functionals of order statistics, the Bahadur statistic and the bootstrap method are treated in Chapter 6. Certain aspects of asymptotic theory of order statistics in the multivariate case are studied in Chapter 7.

Our own interests heavily influence the selection of statistical problems in Part III, and we believe the topics are of sufficient importance to be generally interesting.

In Chapter 8 we study the problem of estimating the q.f. and related problems within the nonparametric framework. Comparisons of semi-parametric models of actual distributions with extreme value and normal models are made in Chapters 9 and 10. The applicability of these comparisons is illustrated by several examples.

0.8. Notation and Conventions

Given some random variables (in short: r.v.'s) $\xi_1, \ldots, \xi_n$ defined on a probability space $(\Omega, \mathscr{A}, P)$ we write:

$X_{i:n}$ ith order statistic of $\xi_1, \ldots, \xi_n$,

$U_{i:n}$ ith order statistic of n independent and identically distributed (i.i.d.) r.v.'s with uniform distribution on $(0, 1)$,

F^{-1} quantile function (q.f.) corresponding to the distribution function (d.f.) F,

$\alpha(F)$ $= \inf\{x: F(x) > 0\}$ "left endpoint of d.f. F,"

$\omega(F)$ $= \sup\{x: F(x) < 1\}$ "right endpoint of d.f. F,"

1_B indicator function of a set B; thus $1_B(x) = 1$ if $x \in B$ and $1_B(x) = 0$ if $x \notin B$,

$X \overset{d}{=} Y$ equality of r.v.'s in distribution,

w.p. 1   with probability one.

We shall say, in short, density instead of Lebesgue density. In other cases, the dominating measure is stated explicitly. The family of all Borel sets is the smallest σ-field generated by intervals. When writing $\sup_B$ without any comment then it is understood that the sup ranges over all Borel sets of the respective Euclidean space. Given a d.f. F we will also use this symbol for the corresponding probability measure. Frequently, we shall use the notation TP for the distribution of T.

EXACT DISTRIBUTIONS AND BASIC TOOLS

CHAPTER 1

Distribution Functions, Densities, and Representations

After an introduction to the basic notation and elementary, important techniques which concern the distribution of order statistics we derive, in Section 1.3, the d.f. and density of a single order statistic. From this result and from the well-known fact that the spacings of exponential r.v.'s are independent (the proof is given in Section 1.6) we deduce the joint density of several order statistics in Section 1.4.

In Sections 1.3 and 1.4 we shall always assume that the underlying d.f. is absolutely continuous. Section 1.5 will provide extensions to continuous and discontinuous d.f.'s.

In Section 1.6, the independence of spacings of exponential r.v.'s and the independence of ratios of order statistics of uniform r.v.'s is treated in detail. Furthermore, we study the well-known representation of order statistics of uniform r.v.'s by means of exponential r.v.'s. This section includes extensions from the case of uniform r.v.'s to that of generalized Pareto r.v.'s.

In Section 1.7 various results are collected concerning functional parameters of order statistics—like moments, modes, and medians.

Finally, Section 1.8 provides a detailed study of the conditional distribution of one collection of order statistics conditioned on another collection of order statistics. This result which is related to the Markov property of order statistics will be one of the basic tools in this book.

1.1. Introduction to Basic Concepts

Order Statistics, Sample Maximum, Sample Minimum

Let $\xi_1, \ldots, \xi_n$ be n r.v.'s. If one is not interested in the order of the outcome of $\xi_1, \ldots, \xi_n$ but in the order of the magnitude then one has to examine the ordered sample values

$$X_{1:n} \leq X_{2:n} \leq \cdots \leq X_{n:n} \qquad (1.1.1)$$

which are the order statistics of a sample of size n.

We say that $X_{r:n}$ is the rth order statistic and the random vector $(X_{1:n}, \ldots, X_{n:n})$ is the order statistic. Note that $X_{1:n}$ is the sample minimum and $X_{n:n}$ is the sample maximum. We may write

$$X_{1:n} = \min(\xi_1, \ldots, \xi_n) \qquad (1.1.2)$$

and

$$X_{n:n} = \max(\xi_1, \ldots, \xi_n). \qquad (1.1.3)$$

When treating a sequence $X_{r(n):n}$ of order statistics, one may distinguish between the following different cases: A central sequence of order statistics is given if $r(n) \to \infty$ and $n - r(n) \to \infty$ as $n \to \infty$. A sequence of lower (upper) extremes is given if $r(n)$ (respectively, $n - r(n)$) is bounded. If $r(n) \to \infty$ and $r(n)/n \to 0$ or $n - r(n) \to \infty$ and $(n - r(n))/n \to 0$ as $n \to \infty$ then one can also speak of an intermediate sequence.

One should know that the asymptotic properties of central and extreme sequences are completely different, however, it is one of the aims of this book to show that it can be useful to combine the different results to solve certain problems.

From (1.1.2) and (1.1.3) we see that the minimum $X_{1:n}$ and the maximum $X_{n:n}$ may be written as a composition of the random vector $(\xi_1, \ldots, \xi_n)$ and the functions min and max. Sometimes it will be convenient to extend this notion to the rth order statistic. For this purpose define

$$Z_{r:n}(x_1, \ldots, x_n) = z_r \qquad (1.1.4)$$

where $z_1 \leq \cdots \leq z_n$ are the values of the reals $x_1, \ldots, x_n$ arranged in a nondecreasing order. Using this notation one may write

$$X_{r:n} = Z_{r:n}(\xi_1, \ldots, \xi_n). \qquad (1.1.5)$$

As special cases we obtain $Z_{1:n} = \min$ and $Z_{n:n} = \max$. Such a representation of order statistics is convenient when order statistics of different samples have to be dealt with simultaneously. Then, given another sequence $\xi'_1, \ldots, \xi'_n$ of r.v.'s, we can write $X'_{r:n} = Z_{r:n}(\xi'_1, \ldots, \xi'_n)$.

Sample Quantile Function, Sample Distribution Function

There is a simple device in which way we may derive results for order statistics from corresponding results concerning the frequency of r.v.'s ξ_i. Let $1_{(-\infty, t]}$ denote the indicator function of the interval $(-\infty, t]$; then the frequency of the data x_i in $(-\infty, t]$ may be written $\sum_{i=1}^{n} 1_{(-\infty, t]}(x_i)$. A moment's reflection shows that

$$z_r \le t \quad \text{iff} \quad \sum_{i=1}^{n} 1_{(-\infty, t]}(x_i) \ge r \qquad (1.1.6)$$

with $z_1 \le \cdots \le z_n$ denoting again the ordered values of $x_1, \ldots, x_n$. From (1.1.6) it is immediate that

$$\{X_{r:n} \le t\} = \left\{ \sum_{i=1}^{n} 1_{(-\infty, t]}(\xi_i) \ge r \right\}, \qquad (1.1.7)$$

and hence,

$$P\{X_{r:n} \le t\} = P\left\{ \sum_{i=1}^{n} 1_{(-\infty, t]}(\xi_i) \ge r \right\} = P\{F_n(t) \ge r/n\} \qquad (1.1.8)$$

with

$$F_n(t) = \frac{1}{n} \sum_{i=1}^{n} 1_{(-\infty, t]}(\xi_i) \qquad (1.1.9)$$

defining the sample d.f. F_n.

Given a sequence of independent and identically distributed (in short, i.i.d.) r.v.'s, the d.f. of an order statistic can easily be derived from (1.1.8) by using binomial probabilities. Keep in mind that (1.1.8) holds for every sequence $\xi_1, \ldots, \xi_n$ of r.v.'s.

Next, we turn to the basic relation between order statistics and the sample quantile function (in short, sample q.f.) F_n^{-1}. For this purpose we introduce the notion of the quantile function (in short, q.f.) of a d.f. F. Define

$$F^{-1}(q) = \inf\{t : F(t) \ge q\}, \qquad q \in (0, 1). \qquad (1.1.10)$$

Notice that the q.f. F^{-1} is a real-valued function. One could also define $F^{-1}(0) := \alpha(F) = \inf\{x : F(x) > 0\}$ and $F^{-1}(1) := \omega(F) = \sup\{x : F(x) < 1\}$; then, however, F^{-1} is no longer real-valued in general.

In Section 1.2 we shall indicate the possibility of defining a q.f. without referring to a d.f.

$F^{-1}(q)$ is the smallest q-quantile of F, that is, if ξ is a r.v. with d.f. F then $F^{-1}(q)$ is the smallest value t such that

$$P\{\xi < t\} \le q \le P\{\xi \le t\}. \qquad (1.1.11)$$

The q-quantile of F is unique if F is strictly increasing. Moreover, F^{-1} is the inverse of F in the usual sense if F is continuous and strictly increasing. As an illustration we state three simple examples.

EXAMPLES 1.1.1. (i) Let Φ denote the standard normal d.f. Then Φ^{-1} is the usual inverse of Φ.

(ii) The standard exponential d.f. is given by $F(x) = 1 - e^{-x}$, $x \ge 0$. We have $F^{-1}(q) = -\log(1 - q)$, $q \in (0, 1)$.

(iii) Let $z_1 < z_2 < \cdots < z_n$ and $F(t) = n^{-1} \sum_{i=1}^{n} 1_{(-\infty, t]}(z_i)$. Then,

$$F^{-1}(q) = z_i \text{ if } (i - 1)/n < q \le i/n, i = 1, \ldots, n.$$

From Example 1.1.1(iii), with $n = 1$, we know if F is a degenerate d.f. with jump at $z = z_1$ then F^{-1} is a constant function with value z. Notice that the converse also holds. In this case we have $F(F^{-1}(q)) = 1$ for every $q \in (0, 1)$. Thus F^{-1} is not the inverse of F in the usual sense.

If $\xi_1, \ldots, \xi_n$ are r.v.'s with continuous d.f.'s then one can ignore the possibilities of ties which occur with probability zero. Then, according to Example 1.1.1 (iii) we obtain for every $q \in (0, 1)$:

$$F_n^{-1}(q) = X_{i:n}, \qquad (i - 1)/n < q \leq i/n, \qquad i = 1, \ldots, n. \qquad (1.1.12)$$

Alternatively, we may write

$$F_n^{-1}(q) = \begin{matrix} X_{nq:n}, & nq \text{ integer}, \\ X_{[nq]+1:n}, & \text{otherwise}, \end{matrix} \qquad (1.1.13)$$

where $[nq]$ denotes the integer part of nq. Thus, we have

$$F_n^{-1}(q) = X_{\langle nq \rangle:n} \qquad (1.1.13')$$

with $\langle nq \rangle = \min\{m: m \geq nq\}$.

The r.v. $F_n^{-1}(q)$ is the smallest sample q-quantile. If $q = 1/2$ then one also speaks of the sample median.

The considerations above show that order statistics are more related to q.f.'s than to d.f.'s. Finally, we remark that according to (1.1.12), $F_n^{-1}(i/n) = X_{i:n}$ which implies $F_n(F_n^{-1}(i/n)) = i/n$ for $i = 1, \ldots, n - 1$.

1.2. The Quantile Transformation

In the finite and asymptotic treatment of order statistics we shall make use of certain special properties of order statistics of uniform and exponential r.v.'s. In a first step, one has to established the required results for these particular cases. The extension to other r.v.'s will be accomplished by a transformation technique.

Introduction and Main Results

To be more precise let us introduce i.i.d. random variables $\eta_1, \ldots, \eta_n$ and $\xi_1, \ldots, \xi_n$ where the η_i are $(0, 1)$-uniformly distributed and the ξ_i have the common d.f. F. Then, the following two relations hold:

$$(\eta_1, \ldots, \eta_n) \stackrel{d}{=} (F(\xi_1), \ldots, F(\xi_n)) \qquad (1.2.1)$$

if F is continuous, and

$$(\xi_1, \ldots, \xi_n) \stackrel{d}{=} (F^{-1}(\eta_1), \ldots, F^{-1}(\eta_n)) \qquad (1.2.2)$$

where F^{-1} is the q.f. of F.

Let $U_{1:n} \leq \cdots \leq U_{n:n}$ and, respectively, $X_{1:n} \leq \cdots \leq X_{n:n}$ be the order statistics of $\eta_1, \ldots, \eta_n$ and $\xi_1, \ldots, \xi_n$. Since an increasing order of the observations is not destroyed by a monotone (nondecreasing) transformation one obtains

$$(U_{1:n}, \ldots, U_{n:n}) \overset{\mathrm{d}}{=} (F(X_{1:n}), \ldots, F(X_{n:n})) \tag{1.2.3}$$

and

$$(X_{1:n}, \ldots, X_{n:n}) \overset{\mathrm{d}}{=} (F^{-1}(U_{1:n}), \ldots, F^{-1}(U_{n:n})). \tag{1.2.4}$$

For the details we refer to Lemma 1.2.4 and Theorem 1.2.5.

Some Preliminaries

Let us begin by noting the simple fact that given ordered values $z_1 \leq \cdots \leq z_n$ we get $\varphi(z_1) \leq \cdots \leq \varphi(z_n)$ if φ is nondecreasing [respectively, $\varphi(z_1) \geq \cdots \geq \varphi(z_n)$ if φ is nonincreasing].

Lemma 1.2.1. *Let $X_{r:n}$ be the rth order statistic of r.v.'s $\xi_1, \ldots, \xi_n$ with range R, φ a real-valued function with domain R, and $X'_{r:n}$ the rth order statistic of the r.v.'s $\varphi(\xi_1), \ldots, \varphi(\xi_n)$.*
Then,

(i) $X'_{r:n} = \varphi(X_{r:n})$ *if φ is nondecreasing,*
(ii) $X'_{r:n} = \varphi(X_{n-r+1:n})$ *if φ is nonincreasing.*

Alternatively, using the notation in (1.1.4) one can write

$$Z_{r:n}(\varphi(\xi_1), \ldots, \varphi(\xi_n)) = \varphi(Z_{r:n}(\xi_1, \ldots, \xi_n)) \tag{1.2.5}$$

if φ is nondecreasing, and

$$Z_{r:n}(\varphi(\xi_1), \ldots, \varphi(\xi_n)) = \varphi(Z_{n-r+1:n}(\xi_1, \ldots, \xi_n)) \tag{1.2.6}$$

if φ is nonincreasing.

Lemma 1.2.1(i) shows that one can interchange the nondecreasing function φ and the function $Z_{r:n}$ without changing the r.v. The main results of the present section are applications of Lemma 1.2.1(i) to $\varphi = F$ and $\varphi = F^{-1}$ where F is a d.f.

Another example is $\varphi(x) = -x$. According to (1.2.6),

$$Z_{r:n}(\xi_1, \ldots, \xi_n) = -Z_{n-r+1:n}(-\xi_1, \ldots, -\xi_n).$$

In particular, the identity

$$Z_{1:n}(\xi_1, \ldots, \xi_n) = -Z_{n:n}(-\xi_1, \ldots, -\xi_n) \tag{1.2.7}$$

indicates that results for the sample minimum can easily be deduced from those for the sample maximum.

We mention an application of Lemma 1.2.1(ii) to $\varphi(x) = 1 - x$.

EXAMPLE 1.2.2. Let $U_{1:n}, \ldots, U_{n:n}$ be the order statistics of n i.i.d. $(0, 1)$-uniformly distributed r.v.'s $\eta_1, \ldots, \eta_n$. Then,

$$(U_{1:n}, \ldots, U_{n:n}) \overset{d}{=} (1 - U_{n:n}, \ldots, 1 - U_{1:n}). \qquad (1.2.8)$$

To prove this make use of the well-known fact that

$$(1 - \eta_1, \ldots, 1 - \eta_n) \overset{d}{=} (\eta_1, \ldots, \eta_n).$$

In Lemma A.1.1 it will be proved—within a more general framework—that

$$q \le F(x) \quad \text{iff} \quad F^{-1}(q) \le x \qquad (1.2.9)$$

for every real x and $q \in (0, 1)$. Notice that (1.2.9) is equivalent to

$$q > F(x) \quad \text{iff} \quad F^{-1}(q) > x. \qquad (1.2.10)$$

Deduce from (1.2.9) that the q.f. of the d.f.

$$x \to F((x - \mu)/\sigma),$$

with location and scale parameters μ and σ, is given by $\mu + \sigma F^{-1}$.

From (1.2.9) one also obtains

$$F(F^{-1}(q)^-) \le q \le F(F^{-1}(q)), \qquad 0 < q < 1, \qquad (1.2.11)$$

[where $F(x^-)$ denotes the left-hand limit of F at x] and,

$$F^{-1}(F(x)) \le x \le F^{-1}(F(x)^+) \quad \text{if } 0 < F(x) < 1. \qquad (1.2.12)$$

Criterion 1.2.3. *A d.f. F is continuous if, and only if,*

$$F(F^{-1}(q)) = q \quad \text{for every } q \in (0, 1). \qquad (1.2.13)$$

PROOF. Obvious from (1.2.11) and the fact that every $q \in (0, 1)$ lies in the range of F. □

Notice that $F(F^{-1}(q)) = q$ if $F^{-1}(q)$ is a continuity point of F. Moreover, from (1.2.12) we get

$$F^{-1}(F(x)) = x$$

if $F(x)$ is a continuity point of F^{-1}.

Quantile and Probability Integral Transformation

Criterion 1.2.3 will be the decisive tool to prove

Lemma 1.2.4. *Let η be a $(0, 1)$-uniformly distributed r.v. Then for any d.f. F the following two results hold:*

(i) (*Quantile transformation*) $F^{-1}(\eta)$ has the d.f. F.
(ii) (*Probability integral transformation*)   Let ξ be a r.v. with d.f. F. Then,

$$F(\xi) \overset{d}{=} \eta \qquad iff \ F \ is \ continuous.$$

PROOF. (i) From (1.2.9) it is immediate that

$$P\{F^{-1}(\eta) \le x\} = P\{\eta \le F(x)\} = F(x).$$

(ii) From (i) we know that $\xi \overset{d}{=} F^{-1}(\eta)$. Thus, Criterion 1.2.3 implies that $F(\xi) \overset{d}{=} F(F^{-1}(\eta)) = \eta$ if F is continuous. Conversely, for every x,

$$P\{\xi = x\} \le P\{F(\xi) = F(x)\} = 0$$

if $F(\xi) \overset{d}{=} \eta$, and hence the d.f. F of ξ is continuous.   □

Let us note a direct consequence of the quantile transformation and the transformation theorem for integrals. Apparently,

$$\int g \, dF = \int_0^1 g(F^{-1}(x)) \, dx \qquad (1.2.14)$$

provided one of the integrals exists.

For independent r.v.'s ξ_1, ξ_2 with common continuous d.f. F we deduce from (1.2.9), (1.2.10), (1.2.13) and Lemma 1.2.4(i) that

$$P\{\xi_1 \overset{\le}{_<} \xi_2\} = P\{F^{-1}(\eta_1) \overset{\le}{_<} F^{-1}(\eta_2)\}$$

$$= P\begin{Bmatrix} \eta_1 \le F(F^{-1}(\eta_2)) \\ F(F^{-1}(\eta_1)) < \eta_2 \end{Bmatrix} = P\{\eta_1 \overset{\le}{_<} \eta_2\}$$

where η_1, η_2 are independent $(0, 1)$-uniformly distributed r.v.'s. Thus, the probability

$$P\{\xi_1 \overset{\le}{_<} \xi_2\}$$

is independent of the continuous d.f. F.

We remark that the probability integral transformation in case of not necessarily continuous d.f.'s will be given in Section 1.5.

The Quantile Transformation of Order Statistics

Combining Lemma 1.2.1 and Lemma 1.2.4 we obtain the main result of this section [as already formulated in (1.2.3) and (1.2.4)].

Theorem 1.2.5. Let $X_{1:n}, \ldots, X_{n:n}$ be the order statistics of n i.i.d. random variables with common d.f. F. Then,

(i) $(F^{-1}(U_{1:n}), \ldots, F^{-1}(U_{n:n})) \overset{d}{=} (X_{1:n}, \ldots, X_{n:n}),$

and if, in addition, F is continuous, then

(ii) $(U_{1:n}, \ldots, U_{n:n}) \overset{\mathrm{d}}{=} (F(X_{1:n}), \ldots, F(X_{n:n})).$

PROOF. (i) Using the quantile transformation we obtain

$$(\xi_1, \ldots, \xi_n) \overset{\mathrm{d}}{=} (F^{-1}(\eta_1), \ldots, F^{-1}(\eta_n))$$

where $\xi_1, \ldots, \xi_n$ are i.i.d. random variables with common d.f. F and $\eta_1, \ldots, \eta_n$ are i.i.d. random variables with common uniform distribution on $(0, 1)$. Moreover, w.l.g. the r.v.'s η_i are $(0, 1)$-valued. Since F^{-1} is a nondecreasing function it is immediate from Lemma 1.2.1 that

$$\begin{aligned}
(X_{1:n}, \ldots, X_{n:n}) &\overset{\mathrm{d}}{=} (Z_{1:n}(F^{-1}(\eta_1), \ldots, F^{-1}(\eta_n)), \ldots, Z_{n:n}(F^{-1}(\eta_1), \ldots, F^{-1}(\eta_n))) \\
&= (F^{-1}(Z_{1:n}(\eta_1, \ldots, \eta_n)), \ldots, F^{-1}(Z_{n:n}(\eta_1, \ldots, \eta_n))) \\
&\overset{\mathrm{d}}{=} (F^{-1}(U_{1:n}), \ldots, F^{-1}(U_{n:n})).
\end{aligned}$$

(ii) From (i) it is obvious that

$$\begin{aligned}
(F(X_{1:n}), \ldots, F(X_{n:n})) &\overset{\mathrm{d}}{=} (F(F^{-1}(U_{1:n})), \ldots, F(F^{-1}(U_{n:n}))) \\
&= (U_{1:n}, \ldots, U_{n:n})
\end{aligned}$$

where the second identity follows from Criterion 1.2.3.   □

Combining the two results of Theorem 1.2.5 we obtain

Corollary 1.2.6. *Suppose that* $X_{1:n}, \ldots, X_{n:n}$ *are the order statistics of n i.i.d. random variables with common continuous d.f. F and* $X'_{1:n}, \ldots, X'_{n:n}$ *are the order statistics of n i.i.d. random variables with common d.f. G. Then,*

$$(X'_{1:n}, \ldots, X'_{n:n}) \overset{\mathrm{d}}{=} (G^{-1}(F(X_{1:n})), \ldots, G^{-1}(F(X_{n:n}))). \tag{1.2.15}$$

Since G^{-1} is defined on $(0, 1)$ it may happen that the right-hand side of (1.2.15) is only defined on a set with probability one. This, however, creates no difficulties under the convention that the right-hand side is equal to some fixed constant on the set $\bigcup_{i=1}^n \{F(X_{i:n}) \in \{0, 1\}\}$ which has probability zero.

Corollary 1.2.7. *Let* $U_{r:n}$ *and* $X_{r:n}$ *be as in Theorem 1.2.5(i). Then, for reals* $t_1, \ldots, t_k$ *and integers* $1 \le r_1 < r_2 < \cdots < r_k \le n$ *we obtain,*

$$P\{X_{r_1:n} \le t_1, \ldots, X_{r_k:n} \le t_k\} = P\{U_{r_1:n} \le F(t_1), \ldots, U_{r_k:n} \le F(t_k)\}.$$

PROOF. Theorem 1.2.5 and (1.2.9) yield

$$\begin{aligned}
P\{X_{r_1:n} \le t_1, \ldots, X_{r_k:n} \le t_k\} &= P\{F^{-1}(U_{r_1:n}) \le t_1, \ldots, F^{-1}(U_{r_k:n}) \le t_k\} \\
&= P\{U_{r_1:n} \le F(t_1), \ldots, U_{r_k:n} \le F(t_k)\}. \quad □
\end{aligned}$$

An Alternative Approach to Q.F.'s

Next, we investigate the question whether it makes sense to speak of a q.f. without referring to a d.f. In order to treat this question in a satisfactory way it is useful to study the inverse of a nondecreasing function in a greater generality. The proof of Theorem 1.2.8 and further technical details are postponed until Appendix 1.

Theorem 1.2.8. (i) *The q.f.* F^{-1} *of a d.f.* F *is nondecreasing and left continuous.* (ii) *For every real-valued, nondecreasing and left continuous function* G *with domain* $(0, 1)$ *there exists a unique d.f.* F *such that* $G = F^{-1}$.

We remark that the d.f. can be regained from its q.f. by

$$F(x) = \sup\{q \in (0, 1): F^{-1}(q) \le x\}.$$

From Theorem 1.2.8 we know that it makes sense to say that a real-valued function G with domain $(0, 1)$ is a q.f. if G is nondecreasing and left continuous.

Since order statistics are more related to q.f.'s than to d.f.'s it is tempting to formulate assumptions via conditions imposed on q.f.'s instead of d.f.'s. However, we shall not follow this advice because of the dominant role of d.f.'s in literature.

Weak Convergence of Q.F.'s

Finally, we treat the well-known result that the weak convergence of d.f.'s is equivalent to the "weak convergence" of q.f.'s.

Lemma 1.2.9. *A sequence of d.f.'s* F_n *converges weakly to a d.f.* F_0 *if, and only if,*

$$F_n^{-1}(q) \to F_0^{-1}(q), \qquad n \to \infty,$$

at every continuity point q *of* F_0^{-1}.

PROOF. First let us assume that F_n weakly converges to F_0. Let q be a continuity point of F_0^{-1}. Since the set of all discontinuity points of F_0 is finite or countable it is obvious that for every $\varepsilon > 0$ we find continuity points y_1, y_2 of F_0 such that $y_1 < F_0^{-1}(q) < y_2$ and $|y_1 - y_2| \le \varepsilon$.

From (1.2.10) we conclude that $F_0(y_1) < q \le F_0(y_2)$. Moreover, $q < F_0(y_2)$ because $q = F_0(y_2)$ implies $F_0^{-1}(q) = F_0^{-1}(F_0(y_2)) = y_2$ since y_2 is a continuity point of F_0 [compare with (1.2.12)] which is a contradiction.

Thus, $F_n(y_1) < q < F_n(y_2)$ for all sufficiently large n because $F_n(y_i) \to F_0(y_i)$, $n \to \infty$. Now it is immediate from (1.2.9) that $y_1 \le F_n^{-1}(q) \le y_2$ and hence $|F_n^{-1}(q) - F_0^{-1}(q)| \le \varepsilon$ for all sufficiently large n. Since $\varepsilon > 0$ is arbitrary we know that $|F_n^{-1}(q) - F_0^{-1}(q)| \to 0$, $n \to \infty$.

To prove the converse conclusion repeat the argument above with (1.2.9) and (1.2.12) replaced by Lemma A.1.3 and (1.2.11). $\square$

Let F_n denote again the sample d.f. According to the Glivenko–Cantelli theorem, $\sup_t |F_n(t) - F(t)| \to 0$, $n \to \infty$, w.p. 1. Thus one obtains as an immediate consequence of Lemma 1.2.9 that, w.p. 1, the sample q.f. F_n^{-1} converges to the underlying q.f. F^{-1} at every continuity point of F^{-1}.

1.3. Single Order Statistic, Extremes

In this section we derive the explicit form of the d.f. and the density of a single order statistic.

The D.F. of a Single Order Statistic

Let us start with the most simple result.

Lemma 1.3.1. *Let $X_{r:n}$ be the rth order statistic of n i.i.d. random variables $\xi_1, \ldots, \xi_n$ with common d.f. F. Then, for every t,*

$$P\{X_{r:n} \le t\} = \sum_{i=r}^{n} \binom{n}{i} F(t)^i (1 - F(t))^{n-i}. \tag{1.3.1}$$

PROOF. Obvious from (1.1.8) by noting that $\sum_{i=1}^{n} 1_{(-\infty, t]}(\xi_i)$ is a binomial r.v. $\square$

Lemma 1.3.1 proves once more the special case of $k = 1$ in Corollary 1.2.7. It is obvious from (1.3.1) that

$$P\{X_{r:n} \le t\} = P\{U_{r:n} \le F(t)\},$$

where $U_{r:n}$ is again the rth order statistic of n i.i.d. random variables with common uniform d.f. on $(0, 1)$.

As special cases of Lemma 1.3.1 we note the d.f. of the maximum $X_{n:n}$ and the minimum $X_{1:n}$. We have

$$P\{X_{n:n} \le t\} = F(t)^n, \tag{1.3.2}$$

and

$$P\{X_{1:n} \le t\} = 1 - (1 - F(t))^n. \tag{1.3.3}$$

Notice that (1.3.2) can easily be proved in a direct way since for i.i.d. random variables $\xi_1, \ldots, \xi_n$ we have

$$P\{X_{n:n} \le t\} = P\{\xi_1 \le t, \ldots, \xi_n \le t\} = F(t)^n.$$

It is apparent that if $\xi_1, \ldots, \xi_n$ are independent and not necessarily identically distributed (in short, i.n.n.i.d.) r.v.'s then

$$P\{X_{n:n} \le t\} = \prod_{i=1}^{n} F_i(t) \tag{1.3.4}$$

with F_i denoting the d.f. of ξ_i.

The Density of a Single Order Statistic

It is easily seen from Lemma 1.3.1 that the d.f. of the rth order statistic is absolutely continuous if F is absolutely continuous. To prove this recall that the composition of monotone absolutely continuous functions is absolutely continuous (see e.g. Hewitt–Stromberg, Exercise (18.37)) or use the argument as given at the beginning of Section 1.5. Hence, the density of $X_{r:n}$ can easily be established as the derivative of its d.f. (compare e.g. with Hewitt–Stromberg, Theorem (18.3)).

Theorem 1.3.2. Let $X_{r:n}$ be the rth order statistic of n i.i.d. random variables with common d.f. F and density f. Then, $X_{r:n}$ has the density

$$f_{r:n} = n! f \frac{F^{r-1}(1 - F)^{n-r}}{(r - 1)!(n - r)!}. \tag{1.3.5}$$

PROOF. From Lemma 1.3.1 we know that the d.f. of $X_{r:n}$, say, G can be written as the composition $G = H \circ F$ where the function H is defined by $H(t) = \sum_{i=r}^{n} \binom{n}{i} t^i (1 - t)^{n-i}$. For every t where $f(t)$ is the derivative of F at t we know that the derivative of G at t exists and $G'(t) = f(t)H'(F(t))$; it suffices to prove that $G'(t) = f_{r:n}(t)$. The derivative of H is given by

$$H'(t) = n! \frac{t^{r-1}(1 - t)^{n-r}}{(r - 1)!(n - r)!} \tag{1.3.6}$$

and hence the assertion of the theorem holds. For proving (1.3.6) check that

$$H'(t) = \sum_{i=r}^{n} i \binom{n}{i} t^{i-1}(1 - t)^{n-i} - \sum_{i=r}^{n-1} (n - i) \binom{n}{i} t^i (1 - t)^{n-i-1}$$

$$= r \binom{n}{r} t^{r-1}(1 - t)^{n-r} + \sum_{i=r+1}^{n} i \binom{n}{i} t^{i-1}(1 - t)^{n-i}$$

$$- \sum_{i=r+1}^{n} i \binom{n}{i} t^{i-1}(1 - t)^{n-i}$$

$$= n! \frac{t^{r-1}(1 - t)^{n-r}}{(r - 1)!(n - r)!}$$

where the final step is obvious from the identities

$$i\binom{n}{i} = (n - i + 1)\binom{n}{i-1} = n!/((i-1)!(n-i)!). \qquad \square$$

An alternative, more elegant proof of Theorem 1.3.2 will be given in Section 1.5. This proof will enable us to replace the condition that F is absolutely continuous by the weaker condition that F is continuous.

We note simple special cases of (1.3.5). The densities of the sample maximum and the sample minimum are given by

$$f_{n:n} = nfF^{n-1}$$

and $\qquad\qquad\qquad\qquad\qquad\qquad\qquad\qquad\qquad\qquad\qquad$ (1.3.7)

$$f_{1:n} = nf(1 - F)^{n-1}.$$

Moreover, observe that $U_{r:n}$ is a beta r.v. with parameters r and $n - r + 1$. This becomes obvious by noting that a beta r.v. with parameters r and s has the density

$$x \to x^{r-1}(1 - x)^{s-1}/b(r, s), \qquad 0 < x < 1, \qquad (1.3.8)$$

where $b(r, s) = \int_0^1 x^{r-1}(1 - x)^{s-1} dx$ is the beta function. Recall that $b(r, s) = \Gamma(r)\Gamma(s)/\Gamma(r + s)$ where Γ is the gamma function [with $\Gamma(r) = (r - 1)!$ for positive integers r].

The following example, concerning sample medians, gives a flavor of the asymptotic treatment of central order statistics. It indicates that central order statistics are asymptotically normal.

Let φ denote the standard normal density given by

$$\varphi(x) = (2\pi)^{-1/2} \exp(-x^2/2).$$

Deduce from (1.3.8) that the density h_m of the normalized sample median

$$2(2m)^{1/2}(U_{m+1:2m+1} - 1/2)$$

is given by

$$h_m(x) = c(m)^{-1}(1 - x^2/2m)^m, \qquad 1 - x^2/2m > 0,$$

and $= 0$, otherwise, where $c(m)$ is a constant. Since

$$(1 - x^2/2m)^m \to \exp(-x^2/2), \qquad m \to \infty,$$

and $0 \le (1 - x^2/2m)^m \le \exp(-x^2/2)$ it follows from Lebesgue's dominated convergence theorem that

$$c(m) \to \int \exp(-x^2/2) \, dx = (2\pi)^{1/2}, \qquad m \to \infty,$$

and hence

$$h_m(x) \to \varphi(x), \qquad m \to \infty, \qquad (1.3.9)$$

for every x. The Scheffé lemma 3.3.2 yields that the distribution of the normalized sample median converges to the standard normal distribution w.r.t. the variational distance as the sample size goes to infinity.

Extreme Value D.F.'s

Next (1.3.2) and (1.3.3) will be examined in the special case of limiting d.f.'s of sample maxima or sample minima (in other words: extreme value d.f.'s).

The nondegenerate limiting d.f.'s of sample maxima are of the type

$$G_{1,\alpha}(x) = \begin{cases} 0 \\ \exp(-x^{-\alpha}) \end{cases} \quad \text{if} \quad \begin{array}{l} x \le 0 \\ x > 0, \end{array} \quad \text{"Fréchet"}$$

$$G_{2,\alpha}(x) = \begin{cases} \exp(-(-x)^{\alpha}) \\ 1 \end{cases} \quad \text{if} \quad \begin{array}{l} x \le 0 \\ x > 0, \end{array} \quad \text{"Weibull"} \quad (1.3.10)$$

$$G_3(x) = \exp(-e^{-x}) \quad \text{for every } x. \quad \text{"Gumbel"}$$

where $\alpha > 0$ is a shape parameter. We say that two d.f.'s G_1 and G_2 are of the same type if $G_1(b + ax) = G_2(x)$ for some $a > 0$ and real b.

Frequently, it will be convenient to write $G_{3,\alpha}$ in place of G_3 where α is always understood to be equal to 1. The following identities show that the d.f.'s $G_{i,\alpha}$ are in fact limiting d.f.'s of sample maxima. We have

$$G_{1,\alpha}^n(n^{1/\alpha}x) = G_{1,\alpha}(x),$$

$$G_{2,\alpha}^n(n^{-1/\alpha}x) = G_{2,\alpha}(x), \quad (1.3.11)$$

$$G_3^n(x + \log n) = G_3(x).$$

Every limiting d.f. has to be max-stable in the sense of (1.3.11). It is one of the admirable achievements of the classical extreme value theory that one can show that the d.f.'s in (1.3.10) are the possible nondegenerate limiting d.f.'s of sample maxima [see e.g. Galambos (1987), Theorems 2.4.1 and 2.4.2, or Leadbetter et al. (1983), Theorem 1.4.2]. It is understood that the nondegenerate limiting d.f.'s have to be of the same type as $G_{1,\alpha}$, $G_{2,\alpha}$, G_3.

Frequently, (1.3.11) will be summarized by

$$G_{i,\alpha}^n(c_n x + d_n) = G_{i,\alpha}(x) \quad (1.3.12)$$

where

$$c_n = n^{1/\alpha}, \qquad d_n = 0 \qquad i = 1$$

$$c_n = n^{-1/\alpha}, \qquad d_n = 0 \qquad \text{if} \quad i = 2 \quad (1.3.13)$$

$$c_n = 1, \qquad d_n = \log n \qquad i = 3.$$

Notice that $G_{2,1}(x) = e^x$, $x < 0$, defines the "negative" standard exponential d.f. The d.f. $G_{2,1}$ will usually be taken as a starting point of our investigations. This is partly due to the fact that $G_{2,1}$ is the limiting d.f. of the maximum $U_{n:n}$ of $(0, 1)$-uniformly distributed r.v.'s. To prove this, notice that

$$P\{n(U_{n:n} - 1) \le x\} = (1 + x/n)^n, \qquad -n \le x \le 0,$$

and   (1.3.14)

$$(1 + x/n)^n \to e^x = G_{2,1}(x), \qquad n \to \infty, x \le 0.$$

It is obvious that the pertaining densities $(1 + x/n)^{n-1} 1_{[-n, 0]}(x)$ converge to the density e^x, $x \le 0$, of $G_{2,1}$ which again yields the convergence w.r.t. the variational distance. We remark that (1.3.14) will be extended from the special case of uniform r.v.'s to generalized Pareto r.v.'s in Section 1.6. A detailed study of the asymptotic behavior of extremes will be made in Chapter 5.

Lemma 1.3.1 may be applied to show that a stability relation corresponding to (1.3.12) does not hold for the kth largest order statistic if $k > 1$.

For the sake of completeness we also state the nondegenerate limiting d.f.'s of sample minima (again with parameters $\alpha > 0$):

$$F_{1,\alpha}(x) = 1 - G_{1,\alpha}(-x), x < 0,$$
$$F_{2,\alpha}(x) = 1 - G_{2,\alpha}(-x), x > 0,$$ (1.3.15)
$$F_3(x) = 1 - G_3(-x).$$

The pertaining stability relations may be summarized by

$$1 - (1 - F_{i,\alpha}(c_n x + d_n))^n = F_{i,\alpha}(x)$$ (1.3.16)

where c_n and d_n are the constants in (1.3.13).

Von Mises Parametrization

In the statistical context, one includes a location parameter μ and a scale parameter $\sigma > 0$ into the considerations. Starting with the standard Fréchet, Weibull, and Gumbel d.f.'s as given in (1.3.10) we obtain d.f.'s of the form

$$x \to G_{i,\alpha}((x - \mu)/\sigma).$$

If the index i is unknown then these d.f.'s should be unified to a 3-parameter family by using the von Mises parametrization: For $\beta \neq 0$ define

$$H_\beta(x) = \exp[-(1 + \beta x)^{-1/\beta}], \qquad 1 + \beta x > 0.$$ (1.3.17)

Moreover,

$$H_0 = G_3.$$ (1.3.18)

Since $(1 + \beta x)^{1/\beta} \to e^x$, $\beta \to 0$, it is clear that $H_\beta(x) \to H_0(x)$, $\beta \to 0$. The Fréchet and Weibull d.f.'s can be regained from H_β by the identities

$$G_{1,1/\beta}(x) = H_\beta((x-1)/\beta) \qquad\qquad \beta > 0$$

and $\qquad\qquad\qquad\qquad\qquad\qquad$ if $\qquad\qquad\qquad\qquad$ (1.3.19)

$$G_{2,-1/\beta}(x) = H_\beta(-(x+1)/\beta) \qquad \beta < 0.$$

Graphical Representation of von Mises Densities

To get a visual impression of the "von Mises densities" we include their graphs for special parameters. We shall concentrate our attention on the behavior of the densities with parameter β close to zero. The explicit form of the densities $h_\beta = H'_\beta$ is given by

$$h_0(x) = e^{-x}\exp(-e^{-x}) \quad \text{for every } x$$

if $\beta = 0$, and

$$h_\beta(x) = (1 + \beta x)^{-(1+1/\beta)}\exp(-(1+\beta x)^{-1/\beta}) \quad \text{if} \quad \begin{array}{l} x > -1/\beta, \beta > 0 \\ x < -1/\beta, \beta < 0, \end{array}$$

and $=0$, otherwise.

Figure 1.3.1 shows the standard Gumbel density h_0. Notice that the mode of the standard Gumbel density is equal to zero.

Figure 1.3.2 indicates the convergence of the rescaled Fréchet densities to the Gumbel density as $\beta \downarrow 0$. Figure 1.3.3 concerns the convergence of the rescaled Weibull densities to the Gumbel density as $\beta \uparrow 0$.

The illustrations indicate that extreme value densities—in their von Mises parametrization—form a nice, smooth family of densities. Fréchet densities (recall that this is the case of $\beta > 0$ in the von Mises parametrization) are skewed to the right. This property is shared by the Gumbel density and

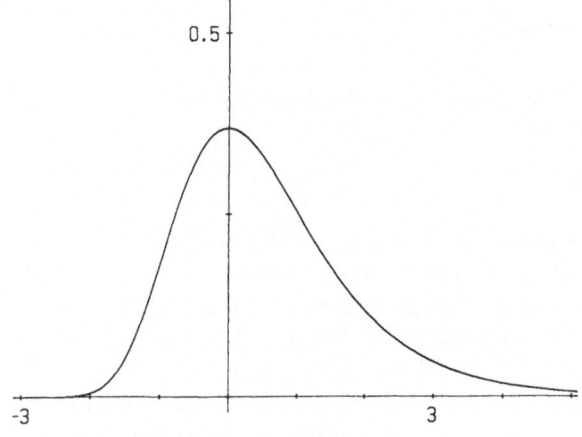

Figure 1.3.1. Gumbel density h_0.

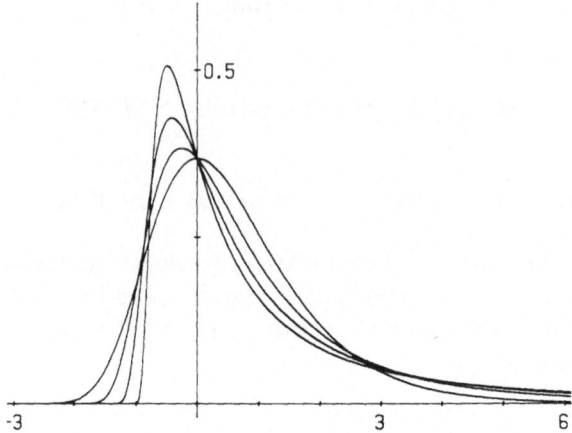

Figure 1.3.2. Gumbel density h_0 and Fréchet densities h_β (von Mises parametrization) with parameters $\beta = 0.3, 0.6, 0.9$.

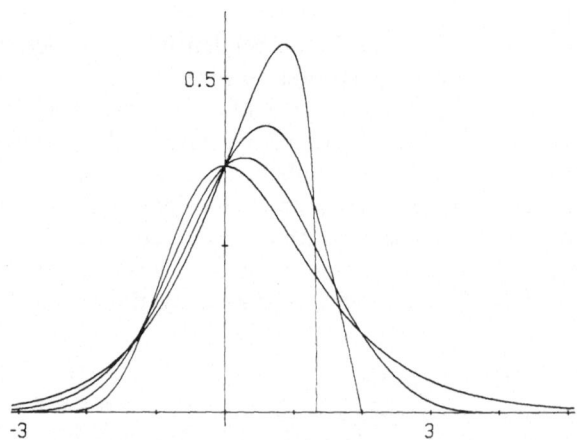

Figure 1.3.3. Gumbel density h_0 and Weibull densities h_β (von Mises parametrization) with parameters $\beta = -0.75, -0.5, -0.25$.

Weibull densities for $\beta \equiv -1/\alpha$ larger than $-1/3.6$. For parameters β close to $-1/3.6$ (that is, α close to 3.6) the Weibull densities look symmetrical. Finally, for parameters β smaller than $-1/3.6$ the Weibull densities are skewed to the left. For illustrations of Fréchet and Weibull densities, with large parameters $|\beta|$, we refer to Figures 5.1.1 and 5.1.2.

In Figure 1.3.4 we demonstrate that for certain location, scale and shape parameters μ, σ and $\alpha = -1/\beta$ it is difficult to distinguish visually the Weibull density from a normal density. Those readers having good eyes will recognize

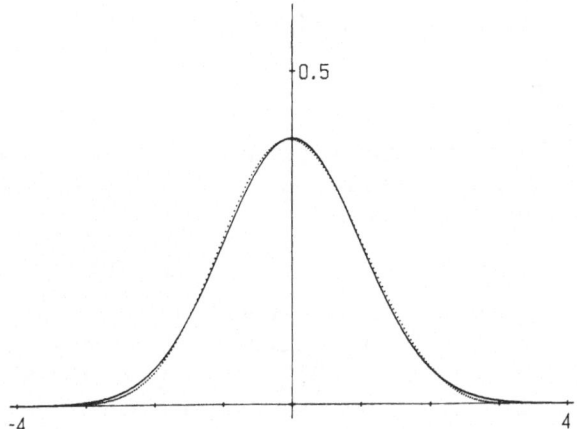

Figure 1.3.4. Standard normal density and Weibull density (dotted line) with parameters $\mu = 3.14$, $\sigma = 3.48$, and $\alpha = 3.6$.

a difference at the tails of the densities (with the dotted line indicating the Weibull density).

1.4. Joint Distribution of Several Order Statistics

In analogy to the proof of Lemma 1.3.1 which led to the explicit form of the d.f. of a single order statistic one can find the joint d.f. of several order statistics $X_{r_1:n}, \ldots, X_{r_k:n}$ by using multinomial probabilities. The resulting expression looks even more complicated than that in the case of a single order statistic.

Thus, we prefer to work with densities instead of d.f.'s. The basic results that will enable us to derive the joint density of several order statistics are (a) Theorem 1.3.2 that provides the explicit form of the density of a single order statistic in the special case of exponential r.v.'s and (b) Theorem 1.4.1 which concerns the density of the order statistic

$$\mathbf{X}_n := (X_{1:n}, \ldots, X_{n:n}).$$

Density of the Order Statistic

The density of the order statistic $\mathbf{X}_n$ can be established by some straight-forward arguments.

Theorem 1.4.1. *Suppose that $\xi_1, \ldots, \xi_n$ are i.i.d. random variables having the common density f. Then, the order statistic $\mathbf{X}_n$ has the density $f_{1,2,\ldots,n:n}$ given by*

$$f_{1,2,\ldots,n:n}(x_1,\ldots,x_n) = n! \prod_{i=1}^{n} f(x_i), \qquad x_1 < x_2 < \cdots < x_n, \qquad (1.4.1)$$

and $= 0$, *otherwise.*

PROOF. Let S_n be the permutation group on $\{1,\ldots,n\}$; thus, $(\tau(1),\ldots,\tau(n))$ is a permutation of $(1,\ldots,n)$ for every $\tau \in S_n$. Define $B_\tau = \{\xi_{\tau(1)} < \xi_{\tau(2)} < \cdots < \xi_{\tau(n)}\}$ for every $\tau \in S_n$. Note that

$$(X_{1:n},\ldots,X_{n:n}) = (\xi_{\tau(1)},\ldots,\xi_{\tau(n)}) \text{ on } B_\tau,$$

and $(\xi_{\tau(1)},\ldots,\xi_{\tau(n)})$ has the same distribution as $(\xi_1,\ldots,\xi_n)$.

Moreover, since the r.v.'s ξ_i have a continuous d.f. we know that ξ_i and ξ_j have no ties for $i \neq j$ (that is, $P\{\xi_i = \xi_j\} = 0$) so that $P(\sum_{\tau \in S_n} B_\tau) = 1$. Finally, notice that the sets B_τ, $\tau \in S_n$, are mutually disjoint. Let $A_0 = \{(x_1,\ldots,x_n): x_1 < x_2 < \cdots < x_n\}$, and let A be any Borel set. We obtain

$$P\{\mathbf{X}_n \in A\} = \sum_{\tau \in S_n} P(\{\mathbf{X}_n \in A\} \cap B_\tau) = \sum_{\tau \in S_n} P(\{(\xi_{\tau(1)},\ldots,\xi_{\tau(n)}) \in A\} \cap B_\tau)$$

$$= \sum_{\tau \in S_n} P\{(\xi_{\tau(1)},\ldots,\xi_{\tau(n)}) \in A \cap A_0\} = n! P\{(\xi_1,\ldots,\xi_n) \in A \cap A_0\}$$

$$= \int_A f_{1,2,\ldots,n:n}(x_1,\ldots,x_n)\,dx_1 \ldots dx_n$$

which is the desired representation. $\square$

Theorem 1.4.1 will be specialized to the order statistic of exponential and uniform r.v.'s.

EXAMPLES 1.4.2. (i) If $\xi_1,\ldots,\xi_n$ are i.i.d. standard exponential r.v.'s then

$$f_{1,2,\ldots,n:n}(x_1,\ldots,x_n) = n! \exp\left[-\sum_{i=1}^{n} x_i\right], \qquad 0 < x_1 < \cdots < x_n, \qquad (1.4.2)$$

and $= 0$, otherwise.
(ii) If $\xi_1,\ldots,\xi_n$ are i.i.d. random variables with uniform distribution on $(0,1)$ then

$$f_{1,2,\ldots,n:n}(x_1,\ldots,x_n) = n!, \qquad 0 < x_1 < \cdots < x_n < 1, \qquad (1.4.3)$$

and $= 0$, otherwise.

Using Example 1.4.2(i) we shall prove that spacings $X_{r:n} - X_{r-1:n}$ of exponential r.v.'s are independent (see Theorem 1.6.1). As an application one obtains the following lemma which will be the decisive tool to establish the joint density of several (in other words, sparse) order statistics $X_{r_1:n},\ldots,X_{r_k:n}$.

Lemma 1.4.3. *Let* $X_{i:n}$ *be the ith order statistic of n i.i.d. standard exponential r.v.'s. Then, for* $1 \leq r_1 < \cdots < r_k \leq n$, *the following two results hold:*

(i) *The spacings $X_{r_1:n}$, $X_{r_2:n} - X_{r_1:n}$, ..., $X_{r_k:n} - X_{r_{k-1}:n}$ are independent, and*

(ii) $$X_{r_i:n} - X_{r_{i-1}:n} \stackrel{d}{=} X_{r_i - r_{i-1}:n - r_{i-1}}$$

for $i = 1, \ldots, k$ (where $r_0 = 0$ and $X_{0:n} = 0$).

PROOF. (i) follows from Theorem 1.6.1 since $X_{1:n}$, $X_{2:n} - X_{1:n}$, ..., $X_{n:n} - X_{n-1:n}$ are independent.

(ii) From Theorem 1.6.1 we also know that $(n - r + 1)(X_{r:n} - X_{r-1:n})$ is a standard exponential r.v. Hence, using an appropriate representation of $X_{s:n} - X_{r:n}$ by means of spacings we obtain for $0 \leq r < s \leq n$,

$$X_{s:n} - X_{r:n} = \sum_{i=1}^{s-r} \frac{(n - (r + i) + 1)(X_{r+i:n} - X_{r+i-1:n})}{n - (r + i) + 1}$$

$$\stackrel{d}{=} \sum_{i=1}^{s-r} \frac{((n - r) - i + 1)(X_{i:n-r} - X_{i-1:n-r})}{(n - r) - i + 1} = X_{s-r:n-r}. \qquad \square$$

From Lemma 1.4.3 and Theorem 1.3.2 we shall deduce the density of $X_{r_i:n} - X_{r_{i-1}:n}$, and at the next step the joint density of

$$X_{r_1:n}, X_{r_2:n} - X_{r_1:n}, \ldots, X_{r_k:n} - X_{r_{k-1}:n}$$

in the special case of exponential r.v.'s. Therefore, the joint density of order statistics $X_{r_1:n}, \ldots, X_{r_k:n}$ of exponential r.v.'s can easily be established by means of a simple application of the transformation theorem for densities.

Transformation Theorem for Densities

The following version of the well-known transformation theorem for densities will frequently be used in this sequel.

Let ξ be a random vector with density f and range B where B is an open set in the Euclidean k-space $\mathbb{R}^k$. Moreover, let $T = (T_1, \ldots, T_k)$ be an $\mathbb{R}^k$-valued, injective map with domain B such that all partial derivatives $\partial T_i / \partial x_j$ are continuous. Denote by $(\partial T / \partial x)$ the matrix $(\partial T_i / \partial x_j)_{i,j}$ of all partial derivatives. Assume that $\det(\partial T / \partial x)$ is unequal to zero on B. Then, the density of $T(\xi)$ is given by

$$(f \circ T^{-1}) |\det(\partial T^{-1} / \partial x)| 1_{T(B)} \qquad (1.4.4)$$

where T^{-1} denotes the inverse of T. It is well-known that

$$\det(\partial T^{-1} / \partial x) = 1/\det(\partial T / \partial x) \circ T^{-1} \qquad (1.4.5)$$

under the conditions imposed on T.

EXAMPLE 1.4.4. Let $\xi_1, \ldots, \xi_k$ be i.i.d. standard exponential r.v.'s. Put $x = (x_1, \ldots, x_k)$. The joint distribution of the partial sums $\xi_1, \xi_1 + \xi_2, \ldots, \sum_{i=1}^{k} \xi_i$

has the density

$$\mathbf{x} \to e^{-x_k} 1_D(\mathbf{x}) \tag{1.4.6}$$

where $D = \{\mathbf{y}: 0 < y_1 < \cdots < y_k\}$. This is immediate from (1.4.4) applied to $B = (0, \infty)^k$ and $T_i(\mathbf{x}) = \sum_{j=1}^{i} x_j$. Notice that $T(B) = D$, $T^{-1}(\mathbf{x}) = (x_1, x_2 - x_1, \ldots, x_k - x_{k-1})$ and $\det(\partial T/\partial \mathbf{x}) = 1$ since $(\partial T/\partial \mathbf{x})$ is a triangle matrix with $\partial T_i/\partial x_i = 1$ for $i = 1, \ldots, k$.

The reader is reminded of the fact that $\sum_{i=1}^{k} \xi_i$ is a *gamma r.v.* with parameter k (see also Lemma 1.6.6(ii)).

The Joint Density of Several Order Statistics

To establish the joint density of $X_{r_1:n}, \ldots, X_{r_k:n}$ we shall first examine the special cases of exponential and uniform r.v.'s. Part III of the proof of Theorem 1.4.5 will concern the general case. The proof looks a little bit technical, however, it can be developed step by step without much effort or imagination. Another advantage of this method is that it is applicable to r.v.'s with continuous d.f.'s (see Theorem 1.5.2).

Theorem 1.4.5. *Let* $1 \le k \le n$ *and* $0 = r_0 < r_1 < \cdots < r_k < r_{k+1} = n + 1$. *Suppose that the common d.f. F of the i.i.d. random variables* $\xi_1, \ldots, \xi_n$ *is absolutely continuous and has the density f.*

Then, $X_{r_1:n}, \ldots, X_{r_k:n}$ *have the joint density* $f_{r_1, r_2, \ldots, r_k:n}$ *given by*

$$f_{r_1, r_2, \ldots, r_k:n}(\mathbf{x}) = n! \left(\prod_{i=1}^{k} f(x_i) \right) \prod_{i=1}^{k+1} \frac{(F(x_i) - F(x_{i-1}))^{r_i - r_{i-1} - 1}}{(r_i - r_{i-1} - 1)!}$$

if $0 < F(x_1) < F(x_2) < \cdots < F(x_k) < 1$, *and* $= 0$, *otherwise.* [*We use the convention that* $F(x_0) = 0$ *and* $F(x_{k+1}) = 1$.]

PROOF. (I) First assume that $\xi_1, \ldots, \xi_n$ are standard exponential r.v.'s. Lemma 1.4.3 and Theorem 1.3.2 imply that the joint density g of

$$X_{r_1:n}, X_{r_2:n} - X_{r_1:n}, \ldots, X_{r_k:n} - X_{r_{k-1}:n}$$

is given by

$$g(\mathbf{x}) = \prod_{i=1}^{k} \left[(n - r_{i-1})! e^{-x_i} \frac{(1 - e^{-x_i})^{r_i - r_{i-1} - 1} (e^{-x_i})^{n - r_i}}{(r_i - r_{i-1} - 1)!(n - r_i)!} \right],$$

$$x_i \ge 0, i = 1, \ldots, n,$$

and $= 0$, otherwise.

From (1.4.4) and Example 1.4.4 we get, writing in short $f_{\mathbf{r}:n}$ instead of $f_{r_1, \ldots, r_k:n}$, that for $0 = x_0 < x_1 < \cdots < x_k$,

$$f_{\mathbf{r}:n}(\mathbf{x})\left[\prod_{i=1}^{k+1}(r_i - r_{i-1} - 1)!\right]\bigg/n!$$

$$= \prod_{i=1}^{k} e^{-(n-r_i+1)(x_i-x_{i-1})}[1 - e^{-(x_i-x_{i-1})}]^{r_i-r_{i-1}-1}$$

$$= \prod_{i=1}^{k} e^{-(n-r_i+1)(x_i-x_{i-1})}e^{(r_i-r_{i-1}-1)x_{i-1}}[e^{-x_{i-1}} - e^{-x_i}]^{r_i-r_{i-1}-1}$$

$$= \left(\prod_{i=1}^{k} e^{-x_i}\right)e^{-\sum_{i=1}^{k}[(n-r_i)x_i-(n-r_{i-1})x_{i-1}]}\prod_{i=1}^{k}[e^{-x_{i-1}} - e^{-x_i}]^{r_i-r_{i-1}-1},$$

and $f_{\mathbf{r}:n} = 0$, otherwise. The proof for the exponential case is complete.

(II) For $X_{i:n}$ as in part I we obtain, according to Theorem 1.2.5(ii) that

$$(U_{r_1:n}, \ldots, U_{r_k:n}) \stackrel{d}{=} (G(X_{r_1:n}), \ldots, G(X_{r_k:n}))$$

where $G(x) = 1 - e^{-x}$, $x \geq 0$. Using this representation, the assertion in the uniform case is immediate from part I and (1.4.4) applied to

$$B = \{\mathbf{x}: 0 < x_1 < \cdots < x_k\} \quad \text{and} \quad T(\mathbf{x}) = (G(x_1), \ldots, G(x_k)).$$

(III) Denote by Q the probability measure pertaining to F, and by $g_{\mathbf{r}:n}$ the density of $(U_{r_1:n}, \ldots, U_{r_k:n})$. It suffices to prove that for $t_1, \ldots, t_k$ the identity

$$P\{X_{r_1:n} \leq t_1, \ldots, X_{r_k:n} \leq t_k\}$$

$$= \int_{\mathsf{X}_{i=1}^{k}(-\infty, t_i]} g_{\mathbf{r}:n}(F(x_1), \ldots, F(x_k))dQ^k(x_1, \ldots, x_k)$$

holds since Q^k has the density $\mathbf{x} \to \prod_{i=1}^{k} f(x_i)$. From Corollary 1.2.7 and part II we get

$$P\{X_{r_1:n} \leq t_1, \ldots, X_{r_k:n} \leq t_k\} = P\{U_{r_1:n} \leq F(t_1), \ldots, U_{r_k:n} \leq F(t_k)\}$$

$$= \int_{\mathsf{X}_{i=1}^{k}(-\infty, F(t_i)]} g_{\mathbf{r}:n}(x_1, \ldots, x_k)\,dx_1 \ldots dx_k$$

$$= \int 1_{\mathsf{X}_{i=1}^{k}(-\infty, F(t_i)]}(F(x_1), \ldots, F(x_k))g_{\mathbf{r}:n}(F(x_1), \ldots, F(x_k))dQ^k(x_1, \ldots, x_k)$$

where the 3rd identity follows by means of the probability integral transformation (Lemma 1.2.4(ii)). This lemma is applicable since F is continuous. The proof is complete if

$$1_{(-\infty, F(t)]}(F(x)) = 1_{(-\infty, t]}(x) \quad \text{for } Q \text{ almost all } x.$$

This, however, is obvious from the fact that $(-\infty, t] \subset \{y: F(y) \leq F(t)\}$ and that both sets have equal probability w.r.t. Q (prove this by applying the probability integral transformation). $\qquad\square$

Remark 1.4.6. The condition $0 < F(x_1) < \cdots < F(x_k) < 1$ in Theorem 1.4.5 can be replaced by the condition $x_1 < \cdots < x_k$. To prove this notice that

$\{0 < F(\xi_1) < \cdots < F(\xi_k) < 1\} \subset \{\xi_1 < \cdots < \xi_k\}$ and show that both sets have the same probability.

We mention some special cases. For $k = 1$ and $k = n$ we obtain again Theorem 1.3.2 and Theorem 1.4.1. Moreover, we note the joint density of the k smallest and k largest order statistics. We have

$$f_{1,2,\ldots,k:n}(\mathbf{x}) = n! \left[\prod_{i=1}^{k} f(x_i) \right] \frac{(1 - F(x_k))^{n-k}}{(n-k)!}, \qquad x_1 < \cdots < x_k, \qquad (1.4.7)$$

and $= 0$, otherwise. Moreover,

$$f_{n-k+1,\ldots,n:n}(\mathbf{x}) = n! \left[\prod_{i=1}^{k} f(x_i) \right] \frac{F(x_1)^{n-k}}{(n-k)!}, \qquad x_1 < \cdots < x_k, \qquad (1.4.8)$$

and $= 0$, otherwise. The joint density of $(X_{1:n}, X_{n:n})$ is given by

$$f_{1,n:n}(x_1, x_2) = n(n-1)f(x_1)f(x_2)(F(x_2) - F(x_1))^{n-2}, \qquad x_1 < x_2, \qquad (1.4.9)$$

and $= 0$, otherwise.

A slight modification of the proof of Theorem 1.4.5 will enable us to establish the corresponding result for continuous d.f.'s.

1.5. Extensions to Continuous and Discontinuous Distribution Functions

The results of this section are not required for the understanding of the main ideas of this book and can be omitted at the first reading.

Let $\xi_1, \ldots, \xi_n$ be again i.i.d. random variables with common distribution Q and d.f. F. It is easy to check that the joint distribution of k order statistics possesses a Q^k-density. To simplify the arguments let us treat the case of a single order statistic $X_{r:n}$. Since $\{X_{r:n} \in B\} \subset \bigcup_{i=1}^{n} \{\xi_i \in B\}$ we have $P\{X_{r:n} \in B\} \leq n P\{\xi_1 \in B\}$, thus, $P\{\xi_1 \in B\} = 0$ implies $P\{X_{r:n} \in B\} = 0$ for every Borel set B. Therefore, the distribution of $X_{r:n}$ is absolutely continuous w.r.t. Q, and hence the Radon–Nikodym theorem implies that $X_{r:n}$ has a Q-density.

The knowledge of the existence of the density stimulates the interest in its explicit form. One can argue that Theorem 1.5.1 is highly sophisticated, however in many cases one would otherwise just be able to prove less elegant results (see e.g. P.1.31).

Density of a Single Order Statistic under a Continuous D.F.

First we give an alternative proof to Theorem 1.3.2. This proof enables us to weaken the condition that F is absolutely continuous to the condition that F is continuous.

Theorem 1.5.1. *Let $X_{r:n}$ be the rth order statistic of n i.i.d. random variables with common continuous d.f. F. Then, $X_{r:n}$ has the F-density*

$$n! \frac{F^{r-1}(1 - F)^{n-r}}{(r - 1)!(n - r)!}. \tag{1.5.1}$$

PROOF. It suffices to prove that

$$P\{X_{r:n} \leq x\} = \int_{-\infty}^{x} H'(F)\,dF$$

with H' as in (1.3.6). According to (1.2.4), Criterion 1.2.3 and (1.2.9), the right-hand side above is equal to $\int_{0}^{F(x)} H'(x)\,dx$. Moreover,

$$\int_{0}^{F(x)} H'(x)\,dx = H(F(x)) = P\{X_{r:n} \leq x\}. \qquad \square$$

Notice that Theorem 1.3.2 is immediate from Theorem 1.5.1 under the condition that F is absolutely continuous.

Joint Density of Several Order Statistics under a Continuous D.F.

Another look at the proof of Theorem 1.4.5 reveals that the essential condition adopted in the proof was the continuity of the d.f. F. In a second step we also made use of the density $\mathbf{x} \to \prod_{i=1}^{k} f(x_i)$. When omitting the second step in the proof one gets the following theorem for continuous d.f.'s which is an extension of Theorem 1.4.5.

Theorem 1.5.2. *Let $1 \leq k \leq n$ and $0 = r_0 < r_1 < \cdots < r_k < r_{k+1} = n + 1$. Let $\xi_1, \ldots, \xi_n$ be i.i.d. random variables with common distribution Q and d.f. F. If F is continuous then the order statistics $X_{r_1:n}, \ldots, X_{r_k:n}$ have the joint Q^k-density $g_{r_1,\ldots,r_k:n}$ given by*

$$g_{r_1,\ldots,r_k:n}(\mathbf{x}) = n! \prod_{i=1}^{k+1} \frac{(F(x_i) - F(x_{i-1}))^{r_i - r_{i-1} - 1}}{(r_i - r_{i-1} - 1)!}, \tag{1.5.2}$$

if $x_1 < x_2 < \cdots < x_k$, and $= 0$, otherwise (where again $F(x_0) = 0$ and $F(x_{k+1}) = 1$).

Note that Theorem 1.4.5 is immediate from Theorem 1.5.2 since Q^k has the Lebesgue density $\mathbf{x} \to \prod_{i=1}^{k} f(x_i)$ if Q has the Lebesgue density f.

Remark 1.5.3. Part III of the proof of Theorem 1.4.5 shows that the following result holds true: Let Q_0 be the uniform distribution on $(0, 1)$ and let Q_1 be a probability measure with continuous d.f. F.

If $(\xi_1, \ldots, \xi_k)$ is a random vector with Q_0^k-density g then the random vector $(F^{-1}(\xi_1), \ldots, F^{-1}(\xi_k))$ has the Q_1^k-density

$$\mathbf{x} \to g(F(x_1), \ldots, F(x_k)).$$

Probability Integral Transformation for Discontinuous D.F.'s

Let ξ be a r.v. with distribution Q having a continuous d.f. F. The uniformly distributed r.v. $F(\xi)$, as studied in Lemma 1.2.4(ii), corresponds to the following experiment: If x is a realization of ξ then in a second step the realization $F(x)$ will be observed.

Next, let F be discontinuous at x. Consider a 2-stage random experiment where we include a further r.v. which is uniformly distributed on the interval $(F(x^-), F(x))$. Here, $F(x^-)$ denotes again the left-hand limit of F at x. For example, we may take the r.v. $F(x^-) + \eta(F(x) - F(x^-))$ where η is uniformly distributed on $(0, 1)$.

If x is a realization of ξ, and y is a realization of η then the final outcome of the experiment will be $F(x^-) + y(F(x) - F(x^-))$. This 2-stage random experiment is also governed by the uniform distribution. This idea will be made rigorous in the following lemma.

Lemma 1.5.4. *Suppose that ξ is a r.v. with d.f. F, and that η is a r.v. with uniform distribution on $(0, 1)$. Moreover, ξ and η are assumed to be independent. Define*

$$H(y, x) = F(x^-) + y(F(x) - F(x^-)). \tag{1.5.3}$$

Then, $H(\eta, \xi)$ is uniformly distributed on $(0, 1)$.

PROOF. It suffices to prove that $P\{H(\eta, \xi) < q\} = q$ for every $q \in (0, 1)$. From (1.2.9) we know that $\xi < F^{-1}(q)$ implies $F(\xi) < q$ and $\xi > F^{-1}(q)$ implies $F(\xi) \geq q$. Therefore, by setting $x = F^{-1}(q)$, we have

$$P\{H(\eta, \xi) < q\} = P\{\xi < x\} + P\{H(\eta, \xi) < q, \xi = x\}$$
$$= F(x^-) + P\{F(x^-) + \eta(F(x) - F(x^-)) < q\}P\{\xi = x\} = q. \qquad \square$$

Lemma 1.5.4 will be reformulated by using a Markov kernel K. Note that inducing with the d.f. F is equivalent to inducing with the Markov kernel $(B, x) \to 1_B(F(x))$.

Corollary 1.5.5. *Let Q be a probability measure with d.f. F. Define $K(B|x) = 1_B(F(x))$ for every Borel set B if x is a continuity point of the d.f. F, and $K(\cdot|x)$ is the uniform distribution on $(F(x^-), F(x))$ if F is discontinuous at x. Then,*

$$KQ = \int K(\cdot|x) dF(x)$$

is the uniform distribution on $(0, 1)$.

PROOF. Let ξ and η be as in Lemma 1.5.4. Thus, $K(\cdot|x)$ is the distribution of $F(x^-) + \eta(F(x) - F(x^-))$. By Fubini's theorem we obtain for every t,

$$\int K((-\infty, t]|x)dF(x) = \int P\{F(x^-) + \eta(F(x) - F(x^-)) \le t\}dF(x)$$

$$= P\{F(\xi^-) + \eta(F(\xi) - F(\xi^-)) \le t\} = t$$

where the final identity is obvious from Lemma 1.5.4. $\qquad\square$

Joint Density of Order Statistics under a Discontinuous D.F.

Hereafter, let $\xi_1, \ldots, \xi_n$ be i.i.d. random variables with common distribution Q and d.f. F. For example, F is allowed to be a discrete d.f. Let again $H(y, x) = F(x^-) + y(F(x) - F(x^-))$.

Theorem 1.5.6. *For $1 \le k \le n$ and $0 = r_0 < r_1 < \cdots < r_k < r_{k+1} = n + 1$ the Q^k-density of $(X_{r_1:n}, \ldots, X_{r_k:n})$, say, $f_{r_1, \ldots, r_k:n}$ is given by*

$$f_{r_1, \ldots, r_k:n}(x_1, \ldots, x_k) = \int_{(0,1)^k} g_{r_1, \ldots, r_k:n}(H(y_1, x_1), \ldots, H(y_k, x_k)) \, dy_1 \ldots dy_k$$

where $g_{r_1, \ldots, r_k:n}$ is the joint density of $U_{r_1:n}, \ldots, U_{r_k:n}$.

PROOF. The proof runs along the lines of part (III) in the proof of Theorem 1.4.5. Instead of Lemma 1.2.4(ii) apply its extension Lemma 1.5.4 to discontinuous d.f.'s. We have

$$P\{X_{r_1:n} \le t_1, \ldots, X_{r_k:n} \le t_k\}$$

$$= E[1_{\chi_{i=1}^k(-\infty, F(t_i)]}(H(\eta_1, \xi_1), \ldots, H(\eta_k, \xi_k))g_{r_1, \ldots, r_k:n}(H(\eta_1, \xi_1), \ldots, H(\eta_k, \xi_k))]$$

$$= E[1_{\chi_{i=1}^k(-\infty, t_i]}(\xi_1, \ldots, \xi_k)g_{r_1, \ldots, r_k:n}(H(\eta_1, \xi_1), \ldots, H(\eta_k, \xi_k))]$$

where $\eta_1, \xi_1, \ldots, \eta_k, \xi_k$ are independent r.v.'s such that $\xi_1, \ldots, \xi_n$ possess the common d.f. F, and $\eta_1, \ldots, \eta_k$ are uniformly distributed on $(0, 1)$. The second identity is established in the same way as the corresponding step in the proof of Theorem 1.4.5 by applying Lemma 1.5.4 instead of Lemma 1.2.4(ii). Now the assertion is immediate by applying Fubini's theorem. $\qquad\square$

Notice that $H(y_1, x_1) < H(y_2, x_2)$ if and only if either $x_1 < x_2$, or $x_1 = x_2$ and $y_1 < y_2$. Hence, by using the lexicographical ordering one may write Theorem 1.5.6 in a different way:

Corollary 1.5.7. *Define B_k as the set of all vectors $(x_1, y_1, \ldots, x_k, y_k)$ with $0 < y_i < 1$, $i = 1, \ldots, k$, and $x_i < x_{i+1}$ or $x_i = x_{i+1}$ and $y_i < y_{i+1}$ for $i = 1, \ldots, k - 1$.*

Then, the density $f_{r_1, \ldots, r_k:n}$, given in Theorem 1.5.6, is of the following form:

$$f_{r_1,\ldots,r_k:n}(\mathbf{x}) = n! \int_{B_k} \prod_{i=1}^{k+1} \frac{[H(y_i, x_i) - H(y_{i-1}, x_{i-1})]^{r_i - r_{i-1} - 1}}{(r_i - r_{i-1} - 1)!} dy_1 \ldots dy_k$$

(with the convention that $H(y_0, x_0) = 0$ and $H(y_{k+1}, x_{k+1}) = 1$).

I.N.N.I.D. Random Variables

This is perhaps the proper place to mention an interesting result due to Guilbaud (1982). This result connects the distribution of order statistics of i.n.n.i.d. (independent not necessarily identically distributed) random variables to that of order statistics of i.i.d. random variables.

Theorem 1.5.8. Let $X_{1:n} \leq \cdots \leq X_{n:n}$ be the order statistics of i.n.n.i.d. random variables $\xi_1, \ldots, \xi_n$. Denote by F_i the d.f. of ξ_i.
   Then, for every Borel set B,

$$P\{(X_{1:n}, \ldots, X_{n:n}) \in B\} = \sum_{m=1}^{n} (-1)^{n-m} \frac{m^n}{n!} \sum_{|S|=m} P\{(X_{1:n}^S, \ldots, X_{n:n}^S) \in B\}$$

where the summation runs over all subsets S of $\{1, \ldots, n\}$ with m elements. Moreover, $X_{1:n}^S \leq \cdots \leq X_{n:n}^S$ are the order statistics of n i.i.d. random variables with common d.f.

$$F^S = |S|^{-1} \sum_{i \in S} F_i.$$

We do not know whether Theorem 1.5.8 is of any practical relevance.

1.6. Spacings, Representations, Generalized Pareto Distribution Functions

In this section we collect some results concerning spacings (and thus also for order statistics) of generalized Pareto r.v.'s. We start with the particular cases of exponential and uniform r.v.'s.

Spacings of Exponential R.V.'s

The independence of spacings of exponential r.v.'s was already applied to establish the joint density of several order statistics. The following well-known result is due to Sukhatme (1937) and Rényi (1953).

Theorem 1.6.1. If $X_{1:n}, \ldots, X_{n:n}$ are the order statistics of i.i.d. standard exponential r.v.'s $\eta_1, \ldots, \eta_n$ then

(i) *the spacings* $X_{1:n}, X_{2:n} - X_{1:n}, \ldots, X_{n:n} - X_{n-1:n}$ *are independent, and*

(ii) $(n - r + 1)(X_{r:n} - X_{r-1:n})$ *is again a standard exponential r.v. for each* $r = 1, \ldots, n$ *(with the convention that* $X_{0:n} = 0$*).*

PROOF. Put $\mathbf{x} = (x_1, \ldots, x_n)$. It suffices to prove that the function

$$\mathbf{x} \to \prod_{i=1}^{n} \exp(-x_i) 1_{(0, \infty)}(x_i)$$

is a joint density of

$$nX_{1:n}, (n - 1)(X_{2:n} - X_{1:n}), \ldots, (X_{n:n} - X_{n-1:n}).$$

From Example 1.4.2(i), where the density of the order statistic of exponential r.v.'s was established, the desired result is immediate by applying the transformation theorem for densities to the map $T = (T_1, \ldots, T_n)$ defined by $T_i(\mathbf{x}) = (n - i + 1)(x_i - x_{i-1})$, $i = 1, \ldots, n$.

Notice that $\det(\partial T/\partial \mathbf{x}) = n!$ and $T^{-1}(\mathbf{x}) = (\sum_{j=1}^{i} x_j/(n - j + 1))_{i=1}^{n}$. Moreover, use the fact that $\sum_{i=1}^{n} x_i = \sum_{i=1}^{n} \sum_{j=1}^{i} x_j/(n - j + 1)$. $\square$

From Theorem 1.6.1 the following representation for order statistics $X_{r:n}$ of exponential r.v.'s is immediate:

$$(X_{1:n}, \ldots, X_{n:n}) \overset{d}{=} \left(\sum_{i=1}^{r} \eta_i/(n - i + 1) \right)_{r=1}^{n}. \tag{1.6.1}$$

Note that spacings of independent r.v.'s $\eta_1, \ldots, \eta_n$ with common d.f. $F(x) = 1 - \exp[-a(x - b)]$, $x \geq b$, are also independent. It is well known (see e.g. Galambos (1987), Theorem 1.6.3) that these d.f.'s are the only continuous d.f.'s so that spacings are independent.

Ratios of Order Statistics of Uniform R.V.'s

Spacings of uniform r.v.'s cannot be independent. However it was shown by Malmquist (1950) that certain ratios of order statistics $U_{i:n}$ of uniform r.v.'s are independent. This will be immediate from Theorem 1.6.1. A simple generalization may be found at the end of the section.

Corollary 1.6.2.

(i) $1 - U_{1:n}, (1 - U_{2:n})/(1 - U_{1:n}), \ldots, (1 - U_{n:n})/(1 - U_{n-1:n})$

are independent r.v.'s, and

(ii) $(1 - U_{r:n})/(1 - U_{r-1:n}) \overset{d}{=} U_{n-r+1:n-r+1}$, $r = 1, \ldots, n$,

(with the convention that $U_{0:n} = 0$*).*

PROOF. Let $X_{r:n}$ be as in Theorem 1.6.1 and let F be the standard exponential d.f. Since $U_{r:n} \overset{d}{=} F(X_{r:n})$ we get

$$[(1 - U_{r:n})/(1 - U_{r-1:n})]_{r=1}^n \overset{d}{=} [(1 - F(X_{r:n}))/(1 - F(X_{r-1:n}))]_{r=1}^n$$
$$\overset{d}{=} [\exp(-(X_{r:n} - X_{r-1:n}))]_{r=1}^n$$

which yields (i) according to Theorem 1.6.1.

Moreover, by Lemma 1.4.3(ii) and Example 1.2.2 we obtain

$$\exp(-(X_{r:n} - X_{r-1:n})) = 1 - F(X_{1:n-r+1}) \overset{d}{=} 1 - U_{1:n-r+1} \overset{d}{=} U_{n-r+1:n-r+1}.$$

The proof of (ii) is complete.   □

The original result of Malmquist is a slight modification of Corollary 1.6.2.

Corollary 1.6.3.

(i) $U_{1:n}/U_{2:n}, \ldots, U_{n-1:n}/U_{n:n}, U_{n:n}$ *are independent r.v.'s, and*

(ii) $U_{r:n}/U_{r+1:n} \overset{d}{=} U_{r:r}$ *for* $r = 1, \ldots, n$

(*with the convention that* $U_{n+1:n} = 1$).

PROOF. Immediate from Corollary 1.6.2 since by Example 1.2.2

$$(U_{r:n}/U_{r+1:n})_{r=1}^n \overset{d}{=} [(1 - U_{n-r+1:n})/(1 - U_{n-r:n})]_{r=1}^n.\qquad □$$

Since $U_{r:n}/U_{r+1:n}, U_{r+1:n}$ are independent one could have the idea that also $U_{r:n}, U_{r:n}/U_{r+1:n}$ are independent which however is wrong. This becomes obvious by noting that $0 \le U_{r:n} \le U_{r:n}/U_{r+1:n} \le 1$.

Representations of Order Statistics of Uniform R.V.'s

One purpose of the following lines will be to establish a representation of the order statistics $U_{1:n}, \ldots, U_{n:n}$ related to that in (1.6.1). In a preparatory step we prove the following.

Lemma 1.6.4. *Let* $\eta_1, \ldots, \eta_{n+1}$ *be independent exponential r.v.'s with* η_i *having the d.f.* $F_i(x) = 1 - \exp(-\alpha_i x)$ *for* $x \ge 0$ *where* $\alpha_i > 0$. *Put* $\zeta_i = \eta_i/(\sum_{j=1}^{n+1} \eta_j)$, $i = 1, \ldots, n$, *and* $\zeta_{n+1} = \sum_{j=1}^{n+1} \eta_j$. *Then, the joint density of* $\zeta_1, \ldots, \zeta_{n+1}$, *say* g_{n+1}, *is given by*

$$g_{n+1}(x_{n+1}) = \left(\prod_{i=1}^{n+1} \alpha_i\right) x_{n+1}^n \exp\left[-x_{n+1}\left(\alpha_{n+1} + \sum_{i=1}^n (\alpha_i - \alpha_{n+1})x_i\right)\right]$$

if $x_i > 0$ *for* $i = 1, \ldots, n+1$, $\sum_{i=1}^n x_i < 1$, *and* $g_{n+1} = 0$, *otherwise.*

PROOF. The transformation theorem for densities (see (1.4.4)) is applicable to $B = (0, \infty)^{n+1}$ and $T = (T_1, \ldots, T_{n+1})$ where $T_{n+1}(\mathbf{x}_{n+1}) = \sum_{j=1}^{n+1} x_j$ and $T_i(\mathbf{x}_{n+1}) = x_i / \sum_{j=1}^{n+1} x_j$ for $i = 1, \ldots, n$. The range of T is given by

$$T(B) = \left\{ \mathbf{x}_{n+1} : x_i > 0 \text{ for } i = 1, \ldots, n+1 \text{ and } \sum_{j=1}^{n} x_j < 1 \right\}.$$

The inverse function $S = (S_1, \ldots, S_{n+1})$ of T is given by $S_i(\mathbf{x}_{n+1}) = x_i x_{n+1}$ for $i = 1, \ldots, n$ and $S_{n+1}(\mathbf{x}_{n+1}) = (1 - \sum_{j=1}^{n} x_j) x_{n+1}$. Since the joint density of $\eta_1, \ldots, \eta_{n+1}$ is given by

$$\mathbf{x}_{n+1} \rightarrow \left(\prod_{i=1}^{n+1} \alpha_i \right) \exp\left(-\sum_{i=1}^{n+1} \alpha_i x_i \right) 1_B(\mathbf{x}_{n+1})$$

the asserted form of g_{n+1} is immediate from (1.4.4) if $\det(\partial S/\partial \mathbf{x}) = x_{n+1}^n$ (where $(\partial S/\partial \mathbf{x})$ is the matrix of partial derivatives). This, however, follows at once from the equation

$$
\begin{bmatrix} x_{n+1} & & & x_1 \\ & 0 & & \vdots \\ & & \ddots & \\ 0 & & x_{n+1} & x_n \\ & & & 1 \end{bmatrix}
=
\begin{bmatrix} 1 & & & \\ & \ddots & 0 & \\ & & \ddots & \\ 0 & & \ddots & \\ 1 & \cdots & \cdots & 1 \end{bmatrix}
\begin{bmatrix} x_{n+1} & & & x_1 \\ & \ddots & 0 & \vdots \\ & & \ddots & \\ 0 & & x_{n+1} & x_n \\ -x_{n+1} & \cdots & -x_{n+1} & (1 - \sum_{i=1}^{n} x_i) \end{bmatrix}
$$

since $\det(AB) = \det(A)\det(B)$. Notice that the 3rd matrix is $(\partial S/\partial \mathbf{x})$. $\qquad \square$

The joint density of the r.v.'s $\zeta_i = \eta_i / (\sum_{j=1}^{n+1} \eta_j)$, $i = 1, \ldots, n$, was computed in a more direct way by Weiss (1965).

Corollary 1.6.5. *The r.v.'s ζ_i, $i = 1, \ldots, n$, above have the joint density h_n given by*

$$h_n(\mathbf{x}_n) = n! \left(\prod_{i=1}^{n+1} \alpha_i \right) \left[\alpha_{n+1} + \sum_{i=1}^{n} (\alpha_i - \alpha_{n+1}) x_i \right]^{-(n+1)}$$

if $x_i > 0$ for $i = 1, \ldots, n$ and $\sum_{i=1}^{n} x_i < 1$, and $h_n = 0$, otherwise.

PROOF. Straightforward by applying Lemma 1.6.4 and by computing the density of the marginal distribution in the first n coordinates. $\qquad \square$

Lemma 1.6.4 will only be applied in the special case of i.i.d. random variables. We specialize Lemma 1.6.4 to the case of $\alpha_1 = \alpha_2 = \cdots = \alpha_{n+1} = 1$.

Lemma 1.6.6. *Let $\eta_1, \ldots, \eta_{n+1}$ be i.i.d. standard exponential r.v.'s. Then,*

(i) $(\eta_r / (\sum_{j=1}^{n+1} \eta_j))_{r=1}^{n}$, $\sum_{j=1}^{n+1} \eta_j$ *are independent,*

(ii) $\sum_{j=1}^{n+1} \eta_j$ *is a gamma r.v. with parameter $n + 1$ (thus having the density $x \rightarrow e^{-x} x^n / n!$, $x \geq 0$),*

(iii) $\eta_1, \eta_1 + \eta_2, \ldots, \sum_{j=1}^{n} \eta_j$ *have the joint density*

$$\mathbf{x}_n \to \exp(-x_n) \quad \text{if } 0 < x_1 < \cdots < x_n,$$

and the density is zero, otherwise.

PROOF. (i) and (ii) are obvious since the density g_{n+1} in Lemma 1.6.4 is of the form

$$g_{n+1}(\mathbf{x}_{n+1}) = n! \exp(-x_{n+1}) x_{n+1}^n / n!$$

if $0 < \sum_{i=1}^{n} x_i < 1$ and $x_{n+1} > 0$.
(iii) Standard calculations! See Example 1.4.4. $\qquad\square$

We prove that spacings of $(0, 1)$-uniformly distributed r.v.'s have the same joint distribution as the r.v.'s $\eta_r / (\sum_{j=1}^{n+1} \eta_j)$ above by comparing the densities of the distributions.

Theorem 1.6.7. *If $\eta_1, \ldots, \eta_{n+1}$ are i.i.d. standard exponential r.v.'s, then*

$$(U_{1:n}, U_{2:n} - U_{1:n}, \ldots, U_{n:n} - U_{n-1:n}, 1 - U_{n:n}) \stackrel{d}{=} \left(\eta_r \Big/ \sum_{j=1}^{n+1} \eta_j \right)_{r=1}^{n+1}. \quad (1.6.2)$$

PROOF. It suffices to prove that

$$(U_{r:n} - U_{r-1:n})_{r=1}^{n} \stackrel{d}{=} \left(\eta_r \Big/ \sum_{j=1}^{n+1} \eta_j \right)_{r=1}^{n}$$

(where $U_{0:n} = 0$) because the random vectors with $n + 1$ components are induced by those above and the map $\mathbf{x}_n \to (x_1, \ldots, x_n, 1 - \sum_{i=1}^{n} x_i)$.
From Corollary 1.6.5 we know that $(\eta_r / \sum_{j=1}^{n+1} \eta_j)_{r=1}^{n}$ has the density $h_n(\mathbf{x}_n) = n!$ if $x_i > 0$, $i = 1, \ldots, n$, and $\sum_{i=1}^{n} x_i < 1$. Starting with the density of $(U_{r:n})_{r=1}^{n}$ (see Example 1.4.2(ii)) it is immediate from (1.4.4) and Example 1.4.4 that h_n is also the density of $(U_{r:n} - U_{r-1:n})_{r=1}^{n}$. $\qquad\square$

Since i.i.d. random variables are exchangeable it is obvious that the r.v.'s $\eta_1 / (\sum_{j=1}^{n+1} \eta_j), \ldots, \eta_{n+1} / (\sum_{j=1}^{n+1} \eta_j)$ are also exchangeable. Thus, Theorem 1.6.7 yields that the distribution of $(U_{r:n} - U_{r-1:n})_{r=1}^{n+1}$ (where $U_{n+1:n} = 1$) is invariant under the permutation of its components. This implies, in particular, that all marginal distributions of $(U_{r:n} - U_{r-1:n})_{r=1}^{n+1}$ of equal dimension are equal.

Corollary 1.6.8. *For every permutation τ on $\{1, \ldots, n + 1\}$,*

$$(U_{\tau(r):n} - U_{\tau(r)-1:n})_{r=1}^{n+1} \stackrel{d}{=} (U_{r:n} - U_{r-1:n})_{r=1}^{n+1}. \quad (1.6.3)$$

Let us also formulate Theorem 1.6.7 in terms of the order statistics $U_{r:n}$ themselves. Since $U_{r:n} = \sum_{i=1}^{r} (U_{i:n} - U_{i-1:n})$ we obtain

Corollary 1.6.9. *If $\eta_1, \ldots, \eta_{n+1}$ are i.i.d. standard exponential r.v.'s, then*

$$(U_{1:n}, U_{2:n}, \ldots, U_{n:n}) \stackrel{\mathrm{d}}{=} \left(\sum_{i=1}^{r} \eta_i \middle/ \sum_{j=1}^{n+1} \eta_j \right)_{r=1}^{n}. \tag{1.6.4}$$

Reformulation of Results

At a first step, the results above will be reformulated to order statistics $V_{i:n}$ of n i.i.d. random variables uniformly distributed on $(-1, 0)$. From Section 1.2 we know that

$$(V_{i:n})_{i=1}^{n} \stackrel{\mathrm{d}}{=} (U_{i:n} - 1)_{i=1}^{n} \quad \text{and} \quad (V_{n-i+1:n})_{i=1}^{n} \stackrel{\mathrm{d}}{=} (-U_{i:n})_{i=1}^{n}. \tag{1.6.5}$$

In this sequel, we shall deal with "negative" standard exponential r.v.'s $\xi_i = -\eta_i$ in place of standard exponential r.v.'s η_i. Thus, $\xi_1, \ldots, \xi_{n+1}$ are i.i.d. random variables with common d.f. $G_{2,1}$ (compare with (1.3.10)). We introduce the partial sums

$$S_k = \sum_{i=1}^{k} \xi_i. \tag{1.6.6}$$

From Lemma 1.6.6(ii) it is obvious that S_k is a "negative" gamma r.v. with parameter k having density $x \to e^x(-x)^{k-1}/(k-1)!$, $x < 0$. Corollary 1.6.9 is equivalent to

Corollary 1.6.10.

$$(V_{n:n}, V_{n-1:n}, \ldots, V_{1:n}) \stackrel{\mathrm{d}}{=} \left(\frac{S_1}{-S_{n+1}}, \frac{S_2}{-S_{n+1}}, \ldots, \frac{S_n}{-S_{n+1}} \right). \tag{1.6.7}$$

Notice that $-S_{n+1}/n \to 1$, $n \to \infty$, w.p. 1, which in conjunction with (1.6.7) indicates that, for every fixed k, asymptotically in distribution,

$$(nV_{n:n}, nV_{n-1:n}, \ldots, nV_{n-k+1:n}) \stackrel{\mathrm{d}}{=} (S_1, S_2, \ldots, S_k). \tag{1.6.8}$$

Recall that for $k = 1$ such a relation was proved in (1.3.14). For further details see Section 5.3.

Next, we reformulate Malmquist's result.

Corollary 1.6.11. *We have*

(i) $$\left(\frac{V_{n:n}}{V_{n-1:n}}, \ldots, \frac{V_{2:n}}{V_{1:n}}, -V_{1:n} \right) \stackrel{\mathrm{d}}{=} \left(\frac{S_1}{S_2}, \frac{S_2}{S_3}, \ldots, \frac{S_{n-1}}{S_n}, \frac{S_n}{S_{n+1}} \right), \tag{1.6.9}$$

(ii) $$V_{n-r+1:n}/V_{n-r:n} \stackrel{\mathrm{d}}{=} -V_{1:r} \quad for \ r = 1, \ldots, n-1,$$

(iii) $$\frac{S_1}{S_2}, \frac{S_2}{S_3}, \ldots, \frac{S_{n-1}}{S_n}, \frac{S_n}{S_{n+1}}, S_{n+1} \quad are \ independent \ r.v.'s.$$

PROOF. (i) is obvious from (1.6.7). (ii) is immediate from Corollary 1.6.2(ii).

Ad (iii): From Corollary 1.6.2(i) we know that the first n components of the vectors in (1.6.9) are independent. Moreover, it is immediate from Lemma 1.6.6(i) that $(S_r/S_{n+1})_{r=1}^n, S_{n+1}$ are independent and this property also holds for $(S_r/S_{r+1})_{r=1}^n, S_{n+1}$. Thus, (iii) holds.   □

Generalized Pareto D.F.'s

The uniform distribution on $(-1,0)$ is the generalized Pareto d.f. $W_{2,1}$. We introduce the class $\{W_{1,\alpha}, W_{2,\alpha}, W_3: \alpha > 0\}$ of generalized Pareto d.f.'s and extend the results above to this class.

Associated with the extreme value d.f. $G_{i,\alpha}$ is the generalized Pareto d.f. $W_{i,\alpha}$ that will be introduced by means of the map

$$T_{i,\alpha} = G_{i,\alpha}^{-1} \circ G_{2,1}$$

defined on the support of $G_{2,1}$. Explicitly, we have for $x \in (-\infty, 0)$,

$$T_{i,\alpha}(x) = \begin{cases} (-x)^{-1/\alpha} & i = 1 \\ -(-x)^{1/\alpha} & \text{if} \quad i = 2 \\ -\log(-x) & i = 3. \end{cases} \tag{1.6.10}$$

with the convention that $T_{3,\alpha} \equiv T_{3,1} \equiv T_3$.

If ξ is a r.v. with "negative" exponential d.f. $G_{2,1}$ then we know (see (1.2.15)) that

$$T_{i,\alpha}(\xi) \text{ is distributed according to } G_{i,\alpha}.$$

In analogy to this construction we get for a $(-1,0)$-uniformly distributed r.v. η that

$$T_{i,\alpha}(\eta) \text{ is distributed according to } W_{i,\alpha}$$

with

$$W_{i,\alpha} = 1 + \log G_{i,\alpha}$$

whenever $-1 < \log G_{i,\alpha} < 0$. Thus, the class of generalized Pareto d.f.'s arises out of $W_{2,1}$ in the same way as the extreme value d.f.'s out of $G_{2,1}$.

For $\alpha > 0$ we have

$$W_{1,\alpha}(x) = \begin{cases} 0 \\ 1 - x^{-\alpha} \end{cases} \text{if} \quad \begin{matrix} x \le 1 \\ x > 1, \end{matrix} \qquad \text{"Pareto"}$$

$$W_{2,\alpha}(x) = \begin{cases} 0 & x \le -1 \\ 1 - (-x)^\alpha & \text{if} \quad x \in (-1,0) \\ 1 & x \ge 0, \end{cases} \qquad \text{"Uniform etc."} \tag{1.6.11}$$

$$W_3(x) = \begin{cases} 0 \\ 1 - e^{-x} \end{cases} \text{if} \quad \begin{matrix} x \le 0, \\ x > 0. \end{matrix} \qquad \text{"Exponential"}$$

This class of d.f.'s was introduced by J. Pickands (1975) in extreme value theory. The importance of the generalized Pareto d.f.'s will become apparent later.

Order Statistics of Generalized Pareto R.V.'s

For the rth order statistic $X_{r:n}$ of n i.i.d. random variables with common generalized Pareto d.f. $W_{i,\alpha}$ we obtain the representation

$$X_{r:n} \stackrel{d}{=} T_{i,\alpha}(V_{r:n}). \tag{1.6.12}$$

The use of the transformation $T_{i,\alpha}$ automatically leads to the proper normalization. Check that

$$T_{i,\alpha}(nV_{r:n}) = c_n^{-1}(T_{i,\alpha}(V_{r:n}) - d_n) \stackrel{d}{=} c_n^{-1}(X_{r:n} - d_n) \tag{1.6.13}$$

where c_n and d_n are the normalizing constants as defined in (1.3.13). By combining (1.6.13) and (1.6.8) one finds that

$$(c_n^{-1}(X_{n:n} - d_n), \ldots, c_n^{-1}(X_{n-k+1:n} - d_n)) \stackrel{d}{=} (T_{i,\alpha}(S_1), \ldots, T_{i,\alpha}(S_k)),$$

asymptotically in distribution, for every fixed $k \geq 1$.

Next, Malmquist's result will be extended to generalized Pareto r.v.'s. Here the cases $i = 1, 2$ are relevant. Check that for negative reals a, b,

$$T_{1,\alpha}(a)/T_{1,\alpha}(b) = T_{1,\alpha}(-a/b)$$

and

$$\tag{1.6.14}$$

$$T_{2,\alpha}(a)/T_{2,\alpha}(b) = -T_{2,\alpha}(-a/b).$$

Combining (1.6.12) and Corollary 1.6.11 one obtains

Corollary 1.6.12. *Let $X_{r:n}$ be the rth order statistic of n i.i.d. random variables with common d.f. $W_{i,\alpha}$ for $i \in \{1, 2\}$ and $\alpha > 0$. Then,*

(i) $\dfrac{X_{n:n}}{X_{n-1:n}}, \cdots, \dfrac{X_{2:n}}{X_{1:n}}, X_{1:n}$ *are independent r.v.'s,*

(ii) $X_{n-r+1:n}/X_{n-r:n} \stackrel{d}{=} \begin{array}{l} X_{1:r} \\ -X_{1:r} \end{array}$ *if* $\begin{array}{l} i = 1 \\ i = 2 \end{array}$ *for $r = 1, \ldots, n - 1$.*

It can easily be seen that the independence of ratios of consecutive order statistics still holds if we include a scale parameter into our considerations.

As mentioned above, spacings of i.i.d. random variables with common continuous d.f. are independent if, and only if, F is an exponential d.f. As a consequence of this result one obtains (see Rossberg (1972) or Galambos (1987), Corollary 1.6.2) that the ratios of consecutive order statistics of positive or negative i.i.d. random variables with common continuous d.f. F are independent if, and only if, F is of the type $W_{1,\alpha}$ or $W_{2,\alpha}$ (where a scale parameter has to be included).

1.7. Moments, Modes, and Medians

The calculation of the exact values of moments of order statistics has received much attention in literature. Since this aspect will not be central to our investigations we shall only touch on moments of order statistics of uniform and exponential r.v.'s.

Two results are included concerning conditions which ensure that moments of order statistics exist and are finite. This topic will further be pursued in Section 3.1 where some inequalities for moments of order statistics will be established. The section concludes with a short summary of results concerning modes and medians of distributions of order statistics.

Exact Moments

Let $U_{1:n}, \ldots, U_{n:n}$ again denote the order statistics of n i.i.d. random variables with common uniform distribution on $(0, 1)$. The first result is a nice application of Malmquist's lemma (see Corollary 1.6.3).

Lemma 1.7.1. *Let $0 < r_1 < \cdots < r_k < r_{k+1} = n + 1$, and let $m_1, \ldots, m_k$ be integers such that $r_i + \sum_{j=1}^{i} m_j \geq 1$ for $i = 1, \ldots, k$. Then,*

$$E \prod_{i=1}^{k} U_{r_i:n}^{m_i} = \prod_{i=1}^{k} b\left(r_i + \sum_{j=1}^{i} m_j, r_{i+1} - r_i\right) \Big/ b(r_i, r_{i+1} - r_i) \qquad (1.7.1)$$

where $b(r, s) = (r - 1)!(s - 1)!/(r + s - 1)!$ is the beta function.

PROOF. Put $U_{n+1:n} = 1$ and $s_i = \sum_{j=1}^{i} m_j$. By Corollary 1.6.3 (see also P.1.16) and by inserting the explicit form of the density of $U_{r_i:r_{i+1}-1}$ we obtain

$$E \prod_{i=1}^{k} U_{r_i:n}^{m_i} = E \prod_{i=1}^{k} (U_{r_i:n}/U_{r_{i+1}:n})^{s_i}$$

$$= \prod_{i=1}^{k} E U_{r_i:r_{i+1}-1}^{s_i}$$

$$= \prod_{i=1}^{k} \int_0^1 x^{r_i-1+s_i}(1 - x)^{r_{i+1}-r_i-1} \, dx/b(r_i, r_{i+1} - r_i)$$

which easily leads to the right-hand side of (1.7.1). $\qquad \square$

(1.7.1) may alternatively be written in the following form.

$$E \prod_{i=1}^{k} U_{r_i:n}^{m_i} = \left[n! \Big/ \left(n + \sum_{j=1}^{k} m_j\right)! \right] \prod_{i=1}^{k} \left[\left(r_i - 1 + \sum_{j=1}^{i} m_j\right)! \Big/ \left(r_i - 1 + \sum_{j=1}^{i-1} m_j\right)! \right].$$

$$(1.7.2)$$

From (1.7.2) we obtain as special cases:

$$EU_{r:n} = r/(n + 1) = \mu_{r,n},$$ (1.7.3)

and, more generally,

$$EU_{r:n}^m = \frac{r(r + 1)\dots(r + m - 1)}{(n + 1)(n + 2)\dots(n + m)}.$$ (1.7.4)

After some busy calculations one also gets, for $r \leq s$,

$$E[(U_{r:n} - \mu_{r,n})(U_{s:n} - \mu_{s,n})] = \frac{\mu_{r,n}(1 - \mu_{s,n})}{n + 2}$$ (1.7.5)

and, for $r \leq s \leq t$,

$$E[(U_{r:n} - \mu_{r,n})(U_{s:n} - \mu_{s,n})(U_{t:n} - \mu_{t,n})] = \frac{2\mu_{r,n}(1 - 2\mu_{s,n})(1 - \mu_{t,n})}{(n + 2)(n + 3)}.$$ (1.7.6)

For $r = s$ we obtain in (1.7.5) that

$$E[U_{r:n} - \mu_{r,n}]^2 = \frac{r(n - r + 1)}{(n + 1)^2(n + 2)}.$$

Next we state the expectation and the variance of the rth order statistic $X_{r:n}$ of i.i.d. standard exponential r.v.'s. From Theorem 1.6.1 we know that $X_{r:n} \overset{d}{=} \sum_{i=1}^{r} \eta_i/(n - i + 1)$ where $\eta_1, \dots, \eta_r$ are standard exponential r.v.'s (thus, having common expectation and variance equal to 1). This implies immediately that

$$EX_{r:n} = \sum_{i=1}^{r} (n - i + 1)^{-1} =: \mu_{r,n}$$ (1.7.7)

and

$$E(X_{r:n} - \mu_{r,n})^2 = \sum_{i=1}^{r} (n - i + 1)^{-2}.$$ (1.7.8)

Inequalities for Moments

The first result yields that the mth moment of any order statistic $X_{r:n}$ exists and is finite if the mth absolute moment of the underlying distribution is finite.

Lemma 1.7.2. Let $0 = r_0 < r_1 < \dots < r_k < r_{k+1} = n + 1$. Let $X_{r_1:n}, \dots, X_{r_k:n}$ be order statistics of i.i.d. random variables $\xi_1, \dots, \xi_n$.
Then for every non-negative, measurable function g on the Euclidean k-space we have

$$E(g(X_{r_1:n}, \dots, X_{r_k:n})) \leq \left[n! \middle/ \prod_{i=1}^{k+1} (r_i - r_{i-1} - 1)! \right] E(g(\xi_1, \dots, \xi_k)).$$

PROOF. Let F be the d.f. of ξ_1. Put $C = n!/\prod_{i=1}^{k+1}(r_i - r_{i-1} - 1)!$, $B = \{(x_1, \ldots, x_k): 0 < x_1 < \cdots < x_k < 1\}$, $x_0 = 0$ and $x_{k+1} = 1$. From Theorem 1.2.5(i) and Theorem 1.4.5 we get

$$Eg(X_{r_1:n}, \ldots, X_{r_k:n})$$

$$= C \int_B g(F^{-1}(x_1), \ldots, F^{-1}(x_k)) \prod_{i=1}^{k+1} (x_i - x_{i-1})^{r_i - r_{i-1} - 1} dx_1 \ldots dx_k$$

$$\leq C \int_{(0,1)^k} g(F^{-1}(x_1), \ldots, F^{-1}(x_k)) dx_1 \ldots dx_k = CEg(\xi_1, \ldots, \xi_k)$$

where the final identity becomes obvious by using the quantile transformation.

□

For $g(x) = |x|^m$ we obtain as a special case

$$E|X_{r:n}|^m \leq \frac{n!}{(r-1)!(n-r)!} E|\xi_1|^m. \tag{1.7.9}$$

Next, we find some necessary and sufficient conditions which ensure that moments of central order statistics exist and are finite if the sample size n is sufficiently large.

Lemma 1.7.3. *Let* $X_{i:j}$ *be the ith order statistic of* j *i.i.d. random variables* $\xi_1, \ldots, \xi_j$ *with common d.f. F. Assume that*

$$E|X_{s:j}|^m < \infty \tag{1.7.10}$$

for some positive integers j, m *and* $s \in \{1, \ldots, j\}$. *Then there exists a constant* $C > 0$ *such that*

$$|F^{-1}(x)|^m x^s (1-x)^{j-s+1} \leq C, \qquad x \in (0,1). \tag{1.7.11}$$

Conversely, (1.7.11) *implies that*

$$E|X_{r:n}|^k < \infty \tag{1.7.12}$$

whenever $1 + ks/m \leq r \leq n - (j - s + 1)k/m$.

PROOF. By the same arguments as in the proof of Lemma 1.7.2 we get

$$E|X_{s:j}|^m = \frac{j!}{(s-1)!(j-s)!} \int_0^1 |F^{-1}(x)|^m x^s (1-x)^{j-s+1}/(x(1-x)) dx$$

and hence, (1.7.11) holds under condition (1.7.10) since

$$\int_0^1 (1/x) dx = \int_0^1 (1/(1-x)) dx = \infty.$$

Moreover, (1.7.11) implies (1.7.12) since

$$E|X_{r:n}|^k = \frac{n!}{(r-1)!(n-r)!} \int_0^1 |F^{-1}(x)|^k x^{r-1}(1-x)^{n-r}\,dx$$

$$= \frac{n!}{(r-1)!(n-r)!} C^{k/m} \int_0^1 x^{r-1-ks/m}(1-x)^{n-r-(j-s+1)k/m}\,dx < \infty,$$

and $r - 1 - ks/m \geq 0$ as well as $n - r - (j - s + 1)k/m \geq 0$. $\square$

We formulate a slightly weaker version of Lemma 1.7.3.

Corollary 1.7.4. *For every positive integer k and $0 < \alpha < 1/2$ the following three conditions are equivalent:*

(i) $E|X_{r:n}|^k < \infty$

for all sufficiently large n and $n\alpha \leq r \leq (1 - \alpha)n$.
(ii) *There exists $\delta > 0$ such that*

$$\sup_{q \in (0,1)} |F^{-1}(q)|^\delta q(1-q) < \infty. \tag{1.7.13}$$

(iii) *There exists $\rho > 0$ such that*

$$\sup_x |x|^\rho F(x)(1 - F(x)) < \infty. \tag{1.7.14}$$

PROOF. If (i) holds for all $n \geq n_0$, say, then the implication $(1.7.10) \Rightarrow (1.7.11)$ yields (ii) with $\delta = k/(n_0\alpha + 2)$.

Moreover, if (ii) holds then $(1.7.11) \Rightarrow (1.7.12)$ yields (i) for $n_0 = [(1 + k(1 + 1/\delta))/\alpha]$. Thus (i) and (ii) are equivalent.

To prove the equivalence of (ii) and (iii) notice that (1.7.13) holds iff there exists $\delta > 0$ such that

$$\text{(a) } |F^{-1}(q)|^\delta q < 1 \quad \text{and (b) } |F^{-1}(q)|^\delta(1-q) \leq 1 \tag{1.7.13'}$$

for sufficiently small values of q in (a) and $(1 - q)$ in (b).

Moreover, (iii) holds iff there exists $\delta > 0$ such that

$$\text{(a) } |y|^\delta F(y) < 1 \quad \text{and (b) } |y|^\delta(1 - F(y)) < 1 \tag{1.7.14'}$$

for sufficiently small values of $F(y)$ in (a) and $(1 - F(y))$ in (b).

We are going to prove the equivalence of (1.7.13')(a) and (1.7.14')(a): For sufficiently small q, the inequality $|F^{-1}(q)|^\delta q < 1$ is equivalent to $F^{-1}(q) > -q^{-1/\delta}$ which holds, according to (1.2.10), iff $q > F(-q^{-1/\delta})$. Setting $y = -q^{-1/\delta}$ we see that (1.7.13')(a) holds iff for all sufficiently small y we have $|y|^{-\delta} > F(y)$ which is equivalent to (1.7.14')(a).

In a similar manner one can prove that (1.7.13')(b) is equivalent to (1.7.14')(b) which completes the proof. $\square$

Unimodality of D.F.'s of Order Statistics

In this part of the section we find conditions which imply the unimodality of the d.f. of an order statistic.

A d.f. F is unimodal if there exists a number u such that the restriction $F|(-\infty, u)$ of F to the interval $(-\infty, u)$ is convex and $F|(u, \infty)$ is concave. Every u with this property is a mode of F. If u is a mode of F and F is continuous at u then F possesses a density, say f, where f is nondecreasing on $(-\infty, u]$ and nonincreasing on $[u, \infty)$. We also say that a density f is unimodal if it has these properties.

Hereafter let $X_{i:n}$ be the order statistic of n i.i.d. random variables with common d.f. F and density f. Moreover, assume that f is differentiable and strictly positive on $(\alpha(F), \omega(F))$. Denote by $f_{r:n}$ again the density of $X_{r:n}$. Given a real number u we write $I(u) = (\alpha(F), \omega(F)) \cap (-\infty, u)$ and $J(u) = (\alpha(F), \omega(F)) \cap (u, \infty)$. The following results are essentially due to Alam (1972).

Standard calculations yield that $f_{r:n}$ is unimodal if, and only if, there exists some u such that

$$f'_{r:n}|I(u) \geq 0 \quad \text{and} \quad f'_{r:n}|J(u) \leq 0. \qquad (1.7.15)$$

Check that $f'_{r:n} = b(r, n-r+1)^{-1} f^2 F^{r-1}(1-F)^{n-r} g_{r,n}$ on $(\alpha(F), \omega(F))$ where

$$g_{r,n} = \frac{f'}{f^2} + \frac{r-1}{F} - \frac{n-r}{1-F} \quad \text{on } (\alpha(F), \omega(F)). \qquad (1.7.16)$$

The unimodality of $f_{r:n}$ will be characterized by means of the function $g_{r,n}$.

Lemma 1.7.5. *The density $f_{r:n}$ of $X_{r:n}$ is unimodal if, and only if, there exists u such that $g_{r,n}|I(u) \geq 0$ and $g_{r,n}|J(u) \leq 0$.*

PROOF. Immediate from (1.7.15) and (1.7.16). Define $u := \sup\{x: \alpha(F) < x < \omega(F)$ and $g_{r,n}(x) \geq 0\}$ if $\{x: \alpha(F) < x < \omega(F), g_{r,n}(x) \geq 0\} \neq \varnothing$, and $u = \inf\{x: \alpha(F) < x < \omega(F), g_{r,n}(x) < 0\}$, otherwise. $\qquad \square$

The density $f_{r:n}$ is not unimodal, in general, if the underlying density f is unimodal. We mention the following counterexample due to Huang and Gosh (1982): Consider the density f defined by

$$f(x) = \begin{cases} 1 \\ \frac{1}{2} \end{cases} \quad \text{if} \quad \begin{matrix} -\frac{1}{2} < x < 0 \\ 0 \leq x < 1 \end{matrix}$$

that is zero otherwise. Obviously, f is unimodal. However, it can be shown that the density of the kth order statistic of a sample of size n is not unimodal for $k > (n+1)/2$.

However, if f is strongly unimodal [that is, $\log f$ is concave on the support $(\alpha(F), \omega(F))$] then it can be shown that $f_{r:n}$ is unimodal. Notice that the strong unimodality of f implies that f'/f^2 is nonincreasing on $(\alpha(F), \omega(F))$. This follows at once from the fact that $1/f$ is convex if f is strongly unimodal.

Corollary 1.7.6. (i) *If f'/f^2 is nonincreasing on $(\alpha(F), \omega(F))$ then $f_{r:n}$ is unimodal.*
(ii) *If, in addition, $g_{r,n}(u) = 0$ for some $u \in (\alpha(F), \omega(F))$ and $n \geq 2$ then u is the unique mode of $f_{r:n}$.*

PROOF. (i) Obvious from Lemma 1.7.5 since F is nondecreasing.
(ii) Since F is strictly increasing on $(\alpha(F), \omega(F))$ we know that $g_{r,n}$ is strictly decreasing. This implies that the solution of the equation $g_{r,n}(u) = 0$ (that is necessarily a mode of $f_{r:n}$) is unique.   □

The Cauchy distribution provides an example of a unimodal density which is not strongly unimodal, however, f'/f^2 is nonincreasing.

EXAMPLES 1.7.7. (i) The normal, exponential and uniform densities are strongly unimodal.
(ii) If $f = 1_{[0,1]}$ and $n \geq 2$ then $(r - 1)/(n - 1)$ is the unique mode of $f_{r:n}$.
(iii) The condition that f'/f^2 is nonincreasing is not necessary for the uni-modality of $f_{r:n}$: Let $F(x) = x^\alpha$ for $x \in (0, 1)$ and some $\alpha \in (0, 1)$. Then $f_{r:n}$ is unimodal, however, f'/f^2 is strictly increasing on $(0, 1)$.

It follows from Corollary 1.7.6 and P.3.4 that the weak convergence of distributions of order statistics is equivalent to the convergence w.r.t. the variational distance if the underlying density is strongly unimodal (or if f'/f^2 is nonincreasing).

Medians

As a third functional parameter of order statistics we consider the median of the distribution of an order statistic. Again we are interested in the relationship between the underlying distribution and the distributions of order statistics.
Recall that a median u of a r.v. ξ is defined by the property that

$$P\{\xi < u\} \leq \tfrac{1}{2} \leq P\{\xi \leq u\}. \tag{1.7.17}$$

(1.7.17) holds if $F(u) = \tfrac{1}{2}$. Moreover, if the d.f. F of ξ is continuous, then (1.7.17) is equivalent to the condition $F(u) = \tfrac{1}{2}$.

Lemma 1.7.8. *Let $X_{i:2m+1}$ be the ith order statistic of i.i.d. random variables $\xi_1, \ldots, \xi_{2m+1}$ with common d.f. F where m is a positive integer. Then, every median of ξ_1 is a median of $X_{m+1:2m+1}$.*

PROOF. Let u be a median of ξ_1. Since $F(u) \geq \tfrac{1}{2}$ we obtain from Corollary 1.2.7 that

$$P\{X_{m+1:2m+1} \leq u\} = P\{U_{m+1:2m+1} \leq F(u)\} \geq P\{U_{m+1:2m+1} \leq \tfrac{1}{2}\}.$$

Example 1.2.2 implies that $P\{U_{m+1:2m+1} \leq \tfrac{1}{2}\} = P\{U_{m+1:2m+1} \geq \tfrac{1}{2}\}$. Hence $P\{U_{m+1:2m+1} \leq \tfrac{1}{2}\} = \tfrac{1}{2}$ and, thus, $P\{X_{m+1:2m+1} \leq u\} \geq \tfrac{1}{2}$.

Since $P\{X_{m+1:2m+1} \le v\} \uparrow P\{X_{m+1:2m+1} < u\}$ as $v \uparrow u$ it remains to prove that $P\{X_{m+1:2m+1} \le v\} \le \frac{1}{2}$ for every $v < u$. This follows by the same arguments as in the first part of the proof by using the fact that $F(v) \le \frac{1}{2}$.   □

Lemma 1.7.8 reveals that the sample medians for odd sample sizes are median unbiased estimators of the underlying (unknown) median. However, this is an exceptional case. For even sample sizes $2m$ it is impossible, in general, to find some $r \in \{1, \ldots, 2m\}$ such that the underlying median is the median of $X_{r:2m}$.

EXAMPLE 1.7.9. For every positive integer m and $r \in \{1, \ldots, 2m\}$ we have

$$P\{U_{r:2m} \le \tfrac{1}{2}\} \ne \tfrac{1}{2}. \tag{1.7.18}$$

To prove this notice that for $r \ne 2m - r + 1$ we have $P\{U_{r:2m} \le \frac{1}{2}\} \ne P\{U_{2m-r+1:2m} \le \frac{1}{2}\}$ and hence by Example 1.2.2

$$P\{U_{r:2m} \le \tfrac{1}{2}\} = 1 - P\{U_{2m-r+1:2m} \le \tfrac{1}{2}\} \ne 1 - P\{U_{r:2m} \le \tfrac{1}{2}\}.$$

This implies (1.7.18).

The discussion above can be extended to the question whether the q-quantile $F^{-1}(q)$ is a median of the distribution of the sample q-quantile $F_n^{-1}(q)$; in other words, whether the sample q-quantile is a median unbiased estimator of the underlying q-quantile. Clearly, the answer is negative in general, however, as pointed out in (8.1.9), randomized sample q-quantiles have this property. In the present section we shall only examine randomized sample medians. The reader not familiar with Markov kernels and their interpretation is adviced first to read Section 10.1.

Denote by ε_x the Dirac measure with mass 1 at x; thus, we have $\varepsilon_x(B) = 1_B(x)$. Define the Markov kernel $M_{r,n}$ by

$$M_{r,n}(B|\cdot) = (\varepsilon_{X_{r:n}}(B) + \varepsilon_{X_{n-r+1:n}}(B))/2 \tag{1.7.19}$$

which is a randomized sample median if $r = [(n+1)/2]$. Thus, $X_{r:n}$ as well as $X_{n-r+1:n}$ are chosen with probability $\frac{1}{2}$. Notice that if $n = 2m + 1$ and $r = m + 1$ then the (non-randomized) sample median $X_{m+1:2m+1}$ is taken.

Denote by $M_{r,n}P$ the distribution of the Markov kernel $M_{r,n}$ (compare with (10.1.2)). We have $(M_{r,n}P)(B) = EM_{r,n}(B|\cdot)$.

Lemma 1.7.10. *Let $X_{i:n}$ denote the ith order statistic of n i.i.d. random variables $\xi_1, \ldots, \xi_n$ with continuous d.f. F. Then every median of ξ_1 is a median of $M_{r,n}$.*

PROOF. Since F is continuous we have $F(u) = 1/2$ for every median u of ξ_1. We will prove that $(M_{r,n}P)(-\infty, u] = 1/2$ and hence u is a median of $M_{r,n}$. From Corollary 1.2.7 and Example 1.2.2 we get

$$(M_{r,n}P)(-\infty, u] = \tfrac{1}{2}[P\{X_{r:n} \leq u\} + P\{X_{n-r+1:n} \leq u\}]$$
$$= \tfrac{1}{2}[P\{U_{r:n} \leq \tfrac{1}{2}\} + P\{U_{n-r+1:n} \leq \tfrac{1}{2}\}]$$
$$= \tfrac{1}{2}[P\{U_{r:n} \leq \tfrac{1}{2}\} + P\{U_{r:n} > \tfrac{1}{2}\}] \qquad = \tfrac{1}{2}. \qquad \square$$

Lemma 1.7.10 shows that $M_{r,n}$ is a median unbiased estimator of the underlying median.

1.8. Conditional Distributions of Order Statistics

Throughout this section, we shall assume that $X_{1:n}, \ldots, X_{n:n}$ are the order statistics of n i.i.d. random variables with common continuous d.f. F. The aim of the following lines will be to establish the conditional distribution of $(X_{s_1:n}, \ldots, X_{s_m:n})$ conditioned on $(X_{r_1:n}, \ldots, X_{r_k:n})$.

Introductionary Remarks

At the beginning let us touch on some essential definitions and properties concerning the conditional distribution

$$P(Y \in \cdot | X) \quad \text{of } Y \text{ given } X.$$

In the present context it is always possible to factorize the conditional distribution $P(Y \in \cdot | X)$ by means of the conditional distribution $P(Y \in \cdot | X = x)$ of Y given $X = x$. Moreover, $P(Y \in B | X)$ is the composition of $P(Y \in B | X = \cdot)$ and X. By writing, in short, $P(Y \in B | \cdot)$ in place of $P(Y \in B | X = \cdot)$ we have $P(Y \in B | X) = P(Y \in B | \cdot) \circ X$.

Apart from a measurability condition and the fact that $P(Y \in \cdot | X = x)$ is a probability measure the defining property of $P(Y \in \cdot | X)$ is

$$E(1_A(X)P(Y \in B | X)) = P\{X \in A, Y \in B\} \qquad (1.8.1)$$

for all Borel sets (in general, measurable sets) A and B.

From (1.8.1) we see that $P(Y \in \cdot | X = x)$ has only to be defined for elements x in a set having probability 1 w.r.t. the distribution of X. For x in the complement of the this set, $P(Y \in \cdot | X = x)$ may e.g. be defined as the distribution of Y.

In the statistical context, one is primarily interested in the consequence that the distribution of Y can rebuilt by means of the conditional distribution $P(Y \in \cdot | X = \cdot)$ and the distribution of X. Obviously,

$$EP(Y \in B | X) = P\{Y \in B\}. \qquad (1.8.2)$$

Assume that the joint distribution of X and Y has a density, say, f w.r.t. some product measure $\mu_1 \times \mu_2$. Then we know that the conditional distribution $P(Y \in \cdot | X = x)$ has a μ_2-density, say, $f_2(\cdot | x)$ which, by the definition of a density, has the property

$$P(Y \in B | X = x) = \int_B f_2(\cdot | x) \, d\mu_2.$$

The density $f_2(\cdot | x)$ is the conditional density of Y given $X = x$. It is well known that $f_2(\cdot | x) = f(x, \cdot)/f_1(x)$ if $f_1(x) > 0$ where f_1 is a μ_1-density of the distribution of X.

We mention another simple consequence of (1.8.1). The conditional distribution

$$P((X, Y) \in \cdot | X = x) \quad \text{of} \quad (X, Y) \quad \text{given } X = x$$

is the product of $P(Y \in \cdot | X = x)$ and the Dirac-measure δ_x at x defined by $\delta_x(B) = 1_B(x)$. This becomes obvious by noting that

$$E[1_A(X) P(Y \in B_2 | X) \delta_x(B_1)] = P\{X \in A, (X, Y) \in B_1 \times B_2\}. \quad (1.8.3)$$

The Basic Theorem

Starting with the joint density of order statistics it is straightforward to deduce the desired conditional distributions. A detailed proof of this result is justified because of its importance. We remark that the proof can slightly be clarified (however not shortened) if P.1.32, which concerns conditional independence under the Markov property, is utilized.

Let $r_1 < \cdots < r_k$. The conditional distribution of the order statistic $Y :=(X_{1:n}, \ldots, X_{n:n})$ given

$$X := (X_{r_1:}, \ldots, X_{r_k:n}) = (x_{r_1}, \ldots, x_{r_k}) =: x$$

has only to be computed for vectors x with $\alpha(F) < x_{r_1} < \cdots < x_{r_k} < \omega(F)$ (compare with Theorem 1.5.2). We shall prove that $P(Y \in \cdot | X = x)$ is the joint distribution of certain independent order statistics W_i and degenerated r.v.'s Y_{r_j}. More precisely, W_i is the order statistic of i.i.d. random variables with common d.f. $F_{i,x}$ which is F truncated on the left of $x_{r_{i-1}}$ and on the right of x_{r_i} (where $x_{r_0} = \alpha(F)$ and $x_{r_{k+1}} = \omega(F)$). Thus,

$$F_{i,x}(y) = [F(y) - F(x_{r_{i-1}})]/[F(x_{r_i}) - F(x_{r_{i-1}})], \qquad x_{r_{i-1}} < y < x_{r_i},$$

and $i = 1, \ldots, k + 1$.

Theorem 1.8.1. *Let F be a continuous d.f., and let $0 = r_0 < r_1 < \cdots < r_k < r_{k+1} = n + 1$. If $\alpha(F) = x_{r_0} < x_{r_1} < \cdots < x_{r_k} < x_{r_{k+1}} = \omega(F)$ then the conditional distribution of $(X_{1:n}, \ldots, X_{n:n})$ given $(X_{r_1:n}, \ldots, X_{r_k:n}) = (x_{r_1}, \ldots, x_{r_k})$ is the joint distribution of the r.v.'s $Y_1, \ldots, Y_n$ which are characterized by the following three properties:*

(a) *For every* $i \in I := \{j: 1 \leq j \leq k + 1, r_j - r_{j-1} > 1\}$ *the random vector*

$$W_i = (Y_{r_{i-1}+1}, \ldots, Y_{r_i-1})$$

is the order statistic of $r_i - r_{i-1} - 1$ *i.i.d. random variables with common d.f.*
$F_{i,x}$.
(b) Y_{r_i} *is a degenerate r.v. with fixed value* x_{r_i} *for* $i = 1, \ldots, k$.
(c) $W_i, i \in I$, *are independent.*

PROOF. Put $M := \{1, \ldots, n\} \setminus \{r_1, \ldots, r_k\}$. In view of (1.8.3) it suffices to show that the conditional distribution of the order statistics $X_{i:n}, i \in M$, given $X =: (X_{r_1:n}, \ldots, X_{r_k:n}) = (x_{r_1}, \ldots, x_{r_k}) =: x$ is equal to the joint distribution of the r.v.'s $Y_j, j \in M$. This will be verified by constructing the conditional density in the way as described above.

Denote by Q the probability measure corresponding to the d.f. F. Let f be the Q^n-density of the order statistic $(X_{1:n}, \ldots, X_{n:n})$ and g the Q^k-density of X (as computed in Theorem 1.5.2). Then, the conditional Q^{n-k}-density, say, $f(\cdot|x)$ of $X_{i:n}, i \in M$, given $X = x$ has the representation

$$f(z|x) = f(x_1, \ldots, x_n)/g(x),$$

if $g(x) > 0$ where z denotes the vector $(x_i)_{i \in M}$. Notice that the condition $g(x) > 0$ is equivalent to $\alpha(F) < x_{r_1} < \cdots < x_{r_k} < \omega(F)$. Check that $f(z|x)$ may be written

$$f(z|x) = \prod_{i \in I} h_i(x_{r_{i-1}+1}, \ldots, x_{r_i-1})/(F(x_{r_i}) - F(x_{r_{i-1}}))^{r_i-r_{i-1}-1}$$

where h_i is the $Q_{i,x}^{r_i-r_{i-1}-1}$-density of W_i and $Q_{i,x}$ is the probability measure corresponding to the truncated d.f. $F_{i,x}$.

Since $1/[F(x_{r_i}) - F(x_{r_{i-1}})]$ defines a Q-density of $Q_{i,x}$ it follows that $f(\cdot|x)$ is the Q^{n-k}-density of $Y_j, j \in M$. The particular structure of $f(\cdot|x)$ shows that the random vectors $W_i, i \in I$, are independent and W_i is the asserted order statistic.   □

Theorem 1.8.1 shows that the following two random experiments are equivalent as far as their distributions are concerned. First, generate the ordered values $x_1 < \cdots < x_n$ according to the d.f. F. Then, take $x_{r_1} < \cdots < x_{r_k}$ and replace the ordered values $x_{r_{i-1}+1} < \cdots < x_{r_i-1}$ by the ordered values $y_{r_{i-1}+1} < \cdots < y_{r_i-1}$ which are generated according to the truncated d.f. $F_{i,x}$ as defined above. Then, in view of Theorem 1.8.1 the final outcomes

$$y_1 < \cdots < y_{r_1-1} < x_{r_1} < y_{r_1+1} < \cdots < y_{r_2-1} < x_{r_2} < \cdots$$

$$< x_{r_k} < y_{r_k+1} < \cdots < y_n$$

as well as $x_1 < \cdots < x_n$ are governed by the same distribution.

In Corollary 1.8.2 we shall consider the conditional distribution of $(X_{s_1:n}, \ldots, X_{s_m:n})$ given $(X_{r_1:n}, \ldots, X_{r_k:n}) = (x_{r_1}, \ldots, x_{r_k})$ instead of the conditional distribution of the order statistic $(X_{1:n}, \ldots, X_{n:n})$. This corollary will be an immediate consequence of Theorem 1.8.1 and the following trivial remarks.

Let X and Y be r.v.'s, and g a measurable map defined on the range of Y. Then,

$$P(Y \in g^{-1}(\cdot)|X) \qquad (1.8.4)$$

is the conditional distribution of $g(Y)$ given X. This becomes obvious by noting that as a consequence of (1.8.1) for measurable sets A,

$$E[1_A(X)P(Y \in g^{-1}(C)|X)] = P\{X \in A, g(Y) \in C\}. \qquad (1.8.5)$$

An application of (1.8.4), with g being the projection $(x_1, \ldots, x_n) \to (x_{s_1}, \ldots, x_{s_m})$ yields

Corollary 1.8.2. *Let* $1 \le s_1 < \cdots < s_m \le n$. *The conditional distribution of* $(X_{s_1:n}, \ldots, X_{s_m:n})$ *given* $(X_{r_1:n}, \ldots, X_{r_k:n}) = (x_{r_1}, \ldots, x_{r_k})$ *is the joint distribution of the r.v.'s* $Y_{s_1}, \ldots, Y_{s_m}$ *with* Y_i *defined as in Theorem 1.8.1.*

As an illustration to Theorem 1.8.1 and Corollary 1.8.2 we note several special cases.

EXAMPLES 1.8.3. (i) The conditional distribution of $X_{s:n}$ given $X_{r:n} = x$ is the distribution of

(a) the $(s - r)$th order statistic $Y_{s-r:n-r}$ of $n - r$ i.i.d. random variables with d.f. $F_{(x,\infty)}$ (the truncation of F of the left of x) if $1 \le r < s \le n$,
(b) the $(r - s)$th order statistic $Y_{r-s:n-s}$ of $n - s$ i.i.d. random variables with d.f. $F_{(-\infty,x)}$ (the truncation of F on the right of x) if $1 \le s < r \le n$,
(c) a degenerate r.v. with fixed value x if $r = s$.

(ii) More generally, if in (i) $X_{s:n}$ is replaced by

(a) $X_{s:n}$, $r < s \le n$, then in (i)(a) $Y_{s-r:n-r}$ has to be replaced by $(Y_{1:n-r}, \ldots, Y_{n-r:n-r})$,
(b) $X_{s:n}$, $1 \le s < r$, then in (i)(a) $Y_{r-s:n-s}$ has to be replaced by $(Y_{1:n-s}, \ldots, Y_{n-s:n-s})$.

(iii) The conditional distribution of $X_{r+1:n}, \ldots, X_{s-1:n}$ given $X_{r:n} = x$ and $X_{s:n} = y$ is the distribution of the order statistic $(Y_{1:s-r+1}, \ldots, Y_{s-r+1:s-r+1})$ of $s - r + 1$ i.i.d. random variables with d.f. $F_{(x,y)}$ (the truncation of F on the left of x and on the right of y).

(iv) (Markov property) The conditional distribution of $X_{s:n}$ given $X_{1:n} = x_1, \ldots, X_{s-1:n} = x_{s-1}$ is the conditional distribution of $X_{s:n}$ given $X_{s-1:n} = x_{s-1}$. Hence, the sequence $X_{1:n}, \ldots, X_{n:n}$ has the Markov property.

The Conditional Distribution of Exceedances

Let again $X_{i:n}$ be the ith order statistic of n i.i.d. random variables $\xi_1, \ldots, \xi_n$ with common continuous d.f. F. As a special case of Example 1.8.3(ii) we obtain the following result concerning the k largest order statistics: The conditional distribution of $(X_{n-k+1:n}, \ldots, X_{n:n})$ given $X_{n-k:n} = x$ is the distribution of the order statistic $(Y_{1:k}, \ldots, Y_{k:k})$ of k i.i.d. random variables $\eta_1, \ldots, \eta_k$ with common d.f. $F_{(x,\infty)}$.

By rearranging $X_{n-k+1:n}, \ldots, X_{n:n}$ in the original order of their outcome we obtain the k exceedances, say, $\zeta_1, \ldots, \zeta_k$ of the r.v.'s $\xi_1, \ldots, \xi_n$ over the "random threshold" $X_{n-k:n}$.

We have $(\zeta_1, \ldots, \zeta_k) = (\xi_{i(1)}, \ldots, \xi_{i(k)})$ whenever $1 \leq i(1) < \cdots < i(k) \leq n$ and $\min(\xi_{i(1)}, \ldots, \xi_{i(k)}) > X_{n-k:n}$. This defines the exceedances ζ_i with probability one because F is assumed to be continuous.

Corollary 1.8.4. *Let $\alpha(F) < x < \omega(F)$. The conditional distribution of the exceedances $\zeta_1, \ldots, \zeta_k$ given $X_{n-k:n} = x$ is the joint distribution of k i.i.d. random variables $\eta_1, \ldots, \eta_k$ with common d.f. $F_{(x,\infty)}$ (the truncation of the d.f. F on the left of x).*

PROOF. Let S_k be the permutation group on $\{1, \ldots, k\}$. For every permutation $\tau \in S_k$ we get the representation

$$(\zeta_1, \ldots, \zeta_k) = (X_{n-\tau(1)+1:n}, \ldots, X_{n-\tau(k)+1:n})$$

on the set A_τ where

$$A_\tau = \{(R_{i(1)}, \ldots, R_{i(k)}) = \tau \quad \text{for some } 1 \leq i(1) < \cdots < i(k) \leq n\}$$

and $(R_1, \ldots, R_n)$ is the rank statistic (see P.1.30). Check that $P(A_\tau) = 1/k!$ for every $\tau \in S_k$. Using the fact that the order statistic and the rank statistic are independent we obtain for every Borel set B

$$P((\zeta_1, \ldots, \zeta_k) \in B | X_{n-k:n} = x)$$

$$= \sum_{\tau \in S_k} P(A_\tau \cap \{(X_{n-\tau(1)+1:n}, \ldots, X_{n-\tau(k)+1:n}) \in B\} | X_{n-k:n} = x)$$

$$= (1/k!) \sum_{\tau \in S_k} P((X_{n-\tau(1)+1:n}, \ldots, X_{n-\tau(k)+1:n}) \in B | X_{n-k:n} = x)$$

$$= (1/k!) \sum_{\tau \in S_k} P\{(Y_{\tau(1):k}, \ldots, Y_{\tau(k):k}) \in B\}$$

where the $Y_{i:k}$ are the order statistics of the r.v.'s η_j. The last step follows from Example 1.8.3(ii). By P.1.30,

$$P((\zeta_1, \ldots, \zeta_k) \in B | X_{n-k:n} = x) = P\{(\eta_1, \ldots, \eta_k) \in B\}.$$

The proof is complete. $\qquad\qquad\square$

Extensions of Corollary 1.8.4 can be found in P.1.33 and P.2.1.

Convex Combination of Two Order Statistics

From Example 1.8.3(i) we deduce the following result which will further be pursued in Section 6.2.

Corollary 1.8.5. *Let F be a continuous d.f., and let* $1 \le r < s \le n$.
Then, for every ρ and t,

$$P\{(1 - \rho)X_{r:n} + \rho X_{s:n} \le t\}$$

$$= F_{r,n}(t) - \int_{-\infty}^{t} P\{\rho(Y_{s-r:n-r} - x) > t - x\}\, dF_{r,n}(x)$$

where $F_{r,n}$ is the d.f. of $X_{r:n}$, and $Y_{s-r:n-r}$ is the $(s - r)$th order statistic of $n - r$ i.i.d. random variables with common d.f. $F_{(x,\infty)}$ [the truncation of F on the left of x].

This identity shows that it is possible to get an approximation to the d.f. of the convex combination of two order statistics by using approximations to distributions of single order statistics.

In Section 6.2 we shall study the special case of the convex combination of consecutive order statistics $X_{r:n}$ and $X_{r+1:n}$ where $X_{r:n}$ is a central order statistic and, thus, $Y_{s-r:n-r}$ is a sample minimum.

PROOF OF COROLLARY 1.8.5. Example 1.8.3(i) implies that

$$P\{(1 - \rho)X_{r:n} + \rho X_{s:n} \le t\} = \int P\{(1 - \rho)x + \rho Y_{s-r:n-r} \le t\}\, dF_{r,n}(x)$$

$$= \int_{-\infty}^{t} P\{\rho(Y_{s-r:n-r} - x) \le t - x\}\, dF_{r,n}(x)$$

since $P\{Y_{s-r:n-r} \le x\} = 0$. This implies the assertion.   □

P.1. Problems and Supplements

Let $\xi_1, \ldots, \xi_n$ be i.i.d. random variables with common d.f. F, and let $X_{r:n}$ denote the rth order statistic.

1. Prove that the order statistic is measurable.

2. Denote by $I(q)$ the set of all q-quantiles of F. If $r(n)/n \to q$ as $n \to \infty$ then $X_{r(n):n} \in U$, eventually, w.p. 1 for every open interval U containing $I(q)$.

3. Denote by S_n the group of permutations on $\{1, \ldots, n\}$.
   (i) For every function f,

   $$\sum_{\tau \in S_n} f(X_{\tau(1):n}, \ldots, X_{\tau(n):n}) = \sum_{\tau \in S_n} f(\xi_{\tau(1)}, \ldots, \xi_{\tau(n)}).$$

   (ii) Using the notation of (1.1.4),

   $$Z_{r:n}(\xi_1, \ldots, \xi_n) = Z_{r:n}(\xi_{\tau(1)}, \ldots, \xi_{\tau(n)})$$

   (that is, the order statistic is invariant w.r.t the permutation of the given r.v.'s).

4.   (i)   A d.f. F is continuous if F^{-1} is strictly increasing.
     (ii) F^{-1} is continuous if F is strictly increasing on $(\alpha(F), \omega(F))$.
     (iii) Denote by F_z the truncation of the d.f. F on the left of z. Prove that

$$F_z^{-1}(q) = F^{-1}[(1 - F(z))q + F(z)].$$

5. Let η be a $(0, 1)$-valued r.v. with d.f. F. Then, $G^{-1}(\eta)$ has the d.f. $F \circ G$ for every d.f. G.

6. Let η be a r.v. with uniform distribution on the interval (u_1, u_2) where $0 \le u_1 < u_2 \le 1$. Let F be a d.f. and put $v_i = F^{-1}(u_i)$ [with the convention that $F^{-1}(0) = \alpha(F)$ and $F^{-1}(1) = \omega(F)$]. Then, $F^{-1}(\eta)$ has the d.f.

$$G(x) = (F(x) - F(v_1))/(F(v_2) - F(v_1)), \qquad v_1 < x < v_2.$$

7. Let F and G be d.f.'s. If $F(x) \le G(x)$ for every $x \ge u$ then $F^{-1}(q) \ge G^{-1}(q)$ for every $q > G(u)$.

8. Let $\xi_i, i = 1, 2, 3, \ldots$ be r.v.'s which weakly converge to ξ_0. Then, there exist r.v.'s ξ_i' such that $\xi_i \stackrel{d}{=} \xi_i'$ and $\xi_i', i = 1, 2, 3, \ldots$ converge pointwise to ξ_0' w.p. 1. [Hint: Use Lemma 1.2.9.]

9. For the beta d.f. $I_{r,s}$ with parameters r and s [compare with (1.3.8)] the following recurrence relation holds:

$$(r + s)I_{r,s} = rI_{r+1,s} + I_{r,s+1}.$$

10. (Joint d.f. of two order statistic)
    Let $X_{i:n}$ be the ith order statistic of n i.i.d. random variables with common d.f. F.
    (i) If $1 \le r < s \le n$ then for $u < v$,

$$P\{X_{r:n} \le u, X_{s:n} \le v\}$$

$$= \sum_{i=r}^{n} \sum_{j=\max(0, s-i)}^{n-i} \frac{n!}{i!j!(n - i - j)!} F(u)^i (F(v) - F(u))^j (1 - F(v))^{n-i-j}$$

and for $u \ge v$,

$$P\{X_{r:n} \le u, X_{s:n} \le v\} = P\{X_{s:n} \le v\}.$$

[Hint: Use the fact that $\sum_{k=1}^{n} [1_{(-\infty, u]}(\xi_k), 1_{(u, v]}(\xi_k), 1_{(v, \infty)}(\xi_k)]$ is a multinomial random vector.]
    (ii) Denote again by $I_{r,s}$ the beta d.f. Then for $u < v$,

$$P\{X_{r:n} \le u, X_{s:n} \le v\}$$

$$= I_{r,n-r+1}(F(u)) - \frac{n!}{(r - 1)!} \sum_{i=0}^{s-r-1} (-1)^i F(u)^{r+i} \frac{I_{n-s+1, s-r-i}(1 - F(v))}{n!(n - r - i)!(r + i)}.$$

(Wilks, 1962)

11. (Transformation theorem)
    Let v be a finite signed measure with density f. Let T be a strictly monotone, real-valued function defined on an open interval J. Assume that $I = T(J)$ is an open interval and that the inverse $S: I \to J$ of T is absolutely continuous. Then $|S'|(f \circ S)1_I$ is a density of Tv (the measure induced by v and T).
    [Hint: Apply Hewitt & Stromberg, 1975, Corollary (20.5).]

12. Derive Theorem 1.3.2 from Theorem 1.4.1 by computing the density of the rth marginal distribution in the usual way by integration.

    (Hájek & Sidák, 1967, pages 39, 78)

13. Extension to Theorem 1.4.1: Suppose that the random vector $(\xi_1, \ldots, \xi_n)$ has the (Lebesgue) density g. Then, the order statistic $(X_{1:n}, \ldots, X_{n:n})$ has the density $f_{1,\ldots,n:n}$ given by

$$f_{1,\ldots,n:n}(\mathbf{x}) = \sum_{\tau \in S_n} g(x_{\tau(1)}, \ldots, x_{\tau(n)}), \qquad x_1 < \cdots < x_n,$$

    and $= 0$, otherwise (here S_n again denotes the permutation group).

    (Hájek & Sidák, 1967, page 36)

14. For $i = 1, 2$ let $X_{1:n}^{(i)}, \ldots, X_{n:n}^{(i)}$ be the order statistics of n i.i.d. random variables with common continuous d.f. F_i. If the restrictions $F_1|B_j$ and $F_2|B_j$ are equal on the fixed measurable sets B_j, $j = 1, \ldots, k$, then for every measurable set $B \subset B_1 \times \cdots \times B_k$ and $1 \le r_1 < \cdots < r_k \le n$:

$$P\{(X_{r_1:n}^{(1)}, \ldots, X_{r_k:n}^{(1)}) \in B\} = P\{(X_{r_1:n}^{(2)}, \ldots, X_{r_k:n}^{(2)}) \in B\}.$$

15. If the continuity condition in P.1.14 is omitted then the result remains to hold if the sets B_j are open.

16. (Modifications of Malmquist's result)

    Let $1 \le r_1 < \cdots < r_k \le n$.

    (i) Prove that the following r.v.'s are independent:

$$1 - U_{r_1:n}, (1 - U_{r_2:n})/(1 - U_{r_1:n}), \ldots, (1 - U_{r_k:n})/(1 - U_{r_{k-1}:n}).$$

    Moreover,

$$(1 - U_{r_i:n})/(1 - U_{r_{i-1}:n}) \stackrel{d}{=} U_{n-r_i+1:n-r_{i-1}}$$

    for $i = 1, \ldots, k$ (with $r_0 = 0$ and $U_{0:n} = 0$).

    (ii) Prove that the following r.v.'s are independent:

$$U_{r_1:n}/U_{r_2:n}, \ldots, U_{r_{k-1}:n}/U_{r_k:n}, U_{r_k:n}.$$

    Moreover,

$$U_{r_i:n}/U_{r_{i+1}:n} \stackrel{d}{=} U_{r_i:r_{i+1}-1}$$

    for $i = 1, \ldots, k$ (with $r_{k+1} = n + 1$ and $U_{n+1:n} = 1$).

    (iii) Prove that the following r.v.'s are independent:

$$U_{r_1:n}, (U_{r_2:n} - U_{r_1:n})/(1 - U_{r_1:n}), \ldots, (U_{r_k:n} - U_{r_{k-1}:n})/(1 - U_{r_{k-1}:n}).$$

    Moreover,

$$(U_{r_i:n} - U_{r_{i-1}:n})/(1 - U_{r_{i-1}:n}) \stackrel{d}{=} U_{r_i-r_{i-1}:n-r_{i-1}}$$

    for $i = 1, \ldots, k$ (with $r_0 = 0$ and $U_{0:n} = 0$).

    (iv) Prove that the following r.v.'s are independent:

$$(U_{r_2:n} - U_{r_1:n})/U_{r_2:n}, \ldots, (U_{r_k:n} - U_{r_{k-1}:n})/U_{r_k:n}, 1 - U_{r_k:n}.$$

    Moreover,

$$(U_{r_{i+1}:n} - U_{r_i:n})/U_{r_{i+1}:n} \stackrel{d}{=} U_{r_{i+1}-r_i:r_{i+1}-1}$$

    for $i = 1, \ldots, k$ (with $r_{k+1} = n + 1$ and $U_{n+1:n} = 0$).

17. Denote by ξ_i independent standard normal r.v.'s. It is well known that $(\xi_1^2 + \xi_2^2)/2$ is a standard exponential r.v. Prove that

$$(U_{1:n}, \ldots, U_{n:n}) \overset{d}{=} \left(\left(\sum_{i=1}^{2r} \xi_i^2 \right) \bigg/ \left(\sum_{i=1}^{2(n+1)} \xi_i^2 \right) \right)_{r=1}^{n}.$$

18. Let $\xi_1, \ldots, \xi_{k+1}$ be independent gamma r.v.'s with parameters $s_1, \ldots, s_{k+1}$.
    (i) Then, $(\xi_i / \sum_{j=1}^{k+1} \xi_j)_{i=1}^{k}$ has a k-variate Dirichlet distribution with parameter vector $(s_1, \ldots, s_{k+1})$.

(Wilks, 1962)

   (ii) Show that for $0 = r_0 < r_1 < \cdots < r_k < r_{k+1} = n + 1$,

$$(U_{r_i:n} - U_{r_{i-1}:n})_{i=1}^{k} \overset{d}{=} \left(\xi_i \bigg/ \sum_{j=1}^{k+1} \xi_j \right)_{i=1}^{k}$$

with $s_i = r_i - r_{i-1}$.

19. Let F_n denote the sample d.f. of n i.i.d. $(0, 1)$-uniformly distributed r.v.'s, and $\eta_1, \ldots, \eta_{n+1}$ independent standard exponential r.v.'s. Then,

$$F_n(t) \overset{d}{=} n^{-1} \sum_{i=1}^{n} 1_{(-\infty, t]} \left(\sum_{j=1}^{i} \eta_j \bigg/ \sum_{j=1}^{n+1} \eta_j \right).$$

20. (i) Let $X_{i:n}$ denote the ith order statistic of n i.i.d. random variables with common density f. As an extension of Theorem 1.6.1 one obtains that $(X_{r:n} - X_{r-1:n})_{r=1}^{n}$ has the density

$$\mathbf{x} \to n! \left(\prod_{i=1}^{n} f\left(\sum_{j=1}^{i} x_j \right) \right), \qquad x_j > 0, i = 1, \ldots, n,$$

and the density is zero, otherwise.
   (ii) The density of $(U_{r:n} - U_{r-1:n})_{r=1}^{n}$ is given by

$$\mathbf{x} \to n! \qquad \text{if } x_j > 0, i = 1, \ldots, n, \text{ and } \sum_{j=1}^{n} x_j < 1,$$

and the density is zero, otherwise.
   (iii) For $1 \le r < s \le n$ the density of $(U_{r:n} - U_{r-1:n}, U_{s:n} - U_{s-1:n})$ is given by

$$\mathbf{x} \to n(n - 1)(1 - x - y)^{n-2} \qquad \text{if } x, y > 0 \text{ and } x + y < 1,$$

and the density is zero, otherwise.

21. (Convolutions of gamma r.v.'s)
    (i) Give a direct proof of Lemma 1.6.6 by induction over n and by using the convolution formula $P\{\xi + \eta \le t\} = \int G(t - s) \, dF(s)$ where ξ and η are independent r.v.'s with d.f.'s G and F.
    (ii) It is clear that $\xi + \eta$ is a gamma r.v. with parameter $m + n$ if ξ and η are gamma r.v.'s with parameters m and n.

22. Let $\alpha > 0$ and $i = 1$ or $i = 2$. Prove that the sample minimum of n i.i.d. random variables with common generalized Pareto d.f. $W_{i,\alpha}$ has the d.f. $W_{i,n\alpha}$.

23. Prove that

$$EU_{r:n}^{-j} = \prod_{m=1}^{j} (n - m + 1)/(r - m) \qquad \text{if } 1 \le j < r.$$

[Hint: Use the method of the proof to Lemma 1.7.1.]

24. Put $\lambda_r = r/(n+1)$, $U_{n+1:n} = 1$ and $U_{0:n} = 0$. Prove that

(i)
$$E(U_{r:n} - \lambda_r)^2/(U_{s:n} - U_{r:n}) = \lambda_r(1 - \lambda_r)/(s - r - 1)$$

if $1 \leq r < s \leq n+1$, and

(ii)
$$E(U_{s:n} - \lambda_s)^2/(U_{s:n} - U_{r:n}) = \lambda_s(1 - \lambda_s)/(s - r - 1)$$

if $0 \leq r < s \leq n$.

25. For $0 = r_0 < r_1 < \cdots < r_k < r_{k+1} = n+1$ and reals a_i, $i = 1, \ldots, k$,

$$\sum_{i=1}^{k+1} (r_i - r_{i-1} - 1)E\frac{a_i(U_{r_i:n} - \lambda_{r_i})^2 - a_{i-1}(U_{r_{i-1}:n} - \lambda_{r_{i-1}})^2}{U_{r_i:n} - U_{r_{i-1}:n}} = 0$$

where $a_0 = a_{k+1} = 0$.

26. Let $X_{r:n}$ be the rth order statistic of n i.i.d. random variables with common d.f. $F(x) = 1 - 1/\log x$ for $x \geq e$. Then, for every positive integer k,

$$E|X_{r:n}|^k = \infty.$$

27. For the order statistics $X_{1:1}$ and $X_{1:2}$ from the Pareto d.f. $W_{1,1}$ we get

$$EX_{1:1} = \infty \quad \text{and} \quad EX_{1:2} = 2.$$

28. Let $M_{r,n}$ be the randomized sample median as defined in (1.7.19) and

$$N_{r,n} = X_{r:n}1_{(1/2,1)}(\eta) + X_{n-r+1:n}1_{(0,1/2]}(\eta)$$

where η is a $(0,1)$-uniformly distributed r.v. that is independent from $(\xi_1, \ldots, \xi_n)$. Show that the distributions of $M_{r,n}$ and $N_{r,n}$ are equal.

29. (Conditional distribution of $(\xi_1, \ldots, \xi_n)$ given $(X_{1:n}, \ldots, X_{n:n})$)
Let $X_{i:n}$ be the order statistics of n i.i.d. random variables $\xi_1, \ldots, \xi_n$. Let S_n denote the group of permutations on $\{1, \ldots, n\}$. Then, the conditional distribution of $(\xi_1, \ldots, \xi_n)$ given $(X_{1:n}, \ldots, X_{n:n})$ is defined by

$$P((\xi_1, \ldots, \xi_n) \in A|(X_{1:n}, \ldots, X_{n:n})) = (n!)^{-1} \sum_{\tau \in S_n} 1_A(X_{\tau(1):n}, \ldots, X_{\tau(n):n}).$$

Thus, the conditional expectation of $f(\xi_1, \ldots, \xi_n)$ given $(X_{1:n}, \ldots, X_{n:n})$ is defined by

$$E(f(\xi_1, \ldots, \xi_n)|(X_{1:n}, \ldots, X_{n:n})) = (n!)^{-1} \sum_{\tau \in S_n} f(X_{\tau(1):n}, \ldots, X_{\tau(n):n}).$$

30. (Rank statistic and order statistic)
The rank of ξ_i is defined by $R_{i,n} = nF_n(\xi_i)$ where F_n is the sample d.f. based on $\xi_1, \ldots, \xi_n$. Moreover, $R_n = (R_{1,n}, \ldots, R_{n,n})$ is the rank statistic. Suppose that $(\xi_1, \ldots, \xi_n)$ has the density g. Then:

(i)
$$\xi_i = X_{R_{i,n}:n} \quad \text{w.p. 1.}$$

(ii) The conditional distribution of R_n given $X_n = (X_{1:n}, \ldots, X_{n:n})$ is defined by

$$P(R_n = \kappa|X_n) = g(X_{\kappa(1):n}, \ldots, X_{\kappa(n):n})\Big/\sum_{\tau \in S} g(X_{\tau(1):n}, \ldots, X_{\tau(n):n})$$

for $\kappa = (\kappa(1), \ldots, \kappa(n)) \in S_n$.

(iii) If, in addition, $\xi_1, \ldots, \xi_n$ are i.i.d. random variables then R_n and X_n are independent and $P\{R_n = \kappa\} = 1/n!$ for every $\kappa \in S_n$.

(Hájek & Sidák, 1967, pages 36–38)

31. (Positive dependence of order statistics)

Let $X_{i:n}$ denote the ith order statistic of n i.i.d. random variables with common continuous d.f. F. Assume that $E|X_{i:n}| < \infty$, $E|X_{j:n}| < \infty$ and $E|X_{i:n}X_{j:n}| < \infty$. Then, $\text{Cov}(X_{i:n}, X_{j:n}) \geq 0$.

(Proved by P. Bickel (1967) under stronger conditions.)

32. (Conditional independence under Markov property)

Let $Y_1, \ldots, Y_n$ be real-valued r.v.'s which possess the Markov property. Let $1 \leq r_1 < \cdots < r_k \leq n$. Then, conditioned on $Y_{r_1}, \ldots, Y_{r_k}$, the random vectors $(Y_1, \ldots, Y_{r_1}), (Y_{r_1+1}, \ldots, Y_{r_2}), \ldots, (Y_{r_k+1}, \ldots, Y_n)$ are independent; that is, the product measure

$$P((Y_1, \ldots, Y_{r_1}) \in \cdot | Y_{r_1}) \times P((Y_{r_1+1}, \ldots, Y_{r_2}) \in \cdot | (Y_{r_1}, Y_{r_2})) \times \cdots$$
$$\cdots \times P((Y_{r_k+1}, \ldots, Y_n) \in \cdot | Y_{r_k})$$

is the conditional distribution of $(Y_1, \ldots, Y_n)$ given $(Y_{r_1}, \ldots, Y_{r_k})$.

33. Let F, r_i, x_{r_i} and $F_{i,x}$ be as in Theorem 1.8.1.

(i) For $i \in I := \{j : 1 \leq j \leq k+1, r_j - r_{j-1} > 1\}$ define the random vector $(\zeta_{r_{i-1}+1}, \ldots, \zeta_{r_i-1})$ by the original r.v.'s ξ_i lying strictly between $X_{r_{i-1}:n}$ and $X_{r_i:n}$ in the original order of the outcome.

Then, the conditional distribution of $(\zeta_{r_{i-1}+1}, \ldots, \zeta_{r_i-1})$, $i \in I$, given $X_{r_1:n} = x_{r_1}, \ldots, X_{r_k:n} = x_{r_k}$ is the joint distribution of the independent random vectors $(\eta_{r_{i-1}+1}, \ldots, \eta_{r_i-1})$, $i \in I$, where for every $i \in I$ the components of the vector are i.i.d. with common d.f. $F_{i,x}$.

(ii) Notice that

$$(\zeta_{r_{i-1}+1}, \ldots, \zeta_{r_i-1}) = (\zeta_{j(1)}, \ldots, \xi_{j(r_i-r_{i-1}-1)})$$

whenever $1 \leq j(1) < \cdots < j(r_i - r_{i-1} - 1) \leqslant n$, and

$$X_{r_{i-1}:n} < \min(\xi_{j(1)}, \ldots, \xi_{j(r_i-r_{i-1}-1)}) \leq \max(\xi_{j(1)}, \ldots, \xi_{j(r_i-r_{i-1}-1)}) < X_{r_i:n}.$$

34. (Conditional d.f. of exceedances)

Let F_n be the sample d.f. of r.v.'s with common uniform d.f. on $(0, 1)$. $nF_n(t)$, $0 \leq t \leq 1$, is a Markov process such that $nF_n(t)$, $x_0 \leq t \leq 1$, conditioned on $nF_n(x_0) = k$, is distributed as

$$(n - k)F_{n-k}(t - x_0) + k, \qquad x_0 \leq t \leq 1.$$

Bibliographical Notes

Ordering of observations according to their magnitude and identifying central or extreme events belongs to the most simple human activities. Thus, one can give early reference to the subject of order statistics by quotations from any number of ancient books. For example, J. Tiago de Oliveira gives reference to the age of Methuselah (Genesis, The Bible) in the preface of *Statistical Extremes and Applications* (1984). By the way, Methuselah is reported to have lived 969 years. This should not merely be regarded as a curiosity but also as a comment indicating the difficulties for the proper choice of a model; here in connection with the question (compare with E.J. Gumbel (1933), *Das Alter des Methusalem*): Does the distribution of mortality have a bounded support?

An exhaustive chronological bibliography on order statistics of pre-1950 and 1950–1959 publications with summaries, references and citations has been compiled by L. Harter. The first relevant result is that of Nicolas Bernoulli (1709) which may be interpreted as the expectation of the maximum of uniform random variables.

In the early period, the sample median was of some importance because of its property of minimizing the sum of absolute deviations. It is noteworthy that Laplace (1818) proved the asymptotic normality of the sample median. This result showed that the sample median, as an estimator of the center of the normal distribution, is asymptotically inefficient w.r.t. the sample mean.

From our point of view, the statistical theory in the 19th century may be characterized by (a) the widely accepted role of the normal distribution as a "universal" law and (b) the beginning of a critical phase which arose from the fact that extremes often do not fit that assumption. Extremes were regarded as doubtful, outlying observations (outliers) which had to be rejected. The attitude toward extremes at that time may be interpreted as an attempt to "immunize" the normality assumption against experience.

Modern statistical theory is connected with the name of R.A. Fisher who in 1921 discussed the problem of outliers: "..., the rejection of observations is too crude to be defended; an unless there are other reasons for rejection than mere divergences from the majority, it would be more philosophical to accept these extreme values, not as gross errors, but as indications that the distribution of errors is not normal."

A paper by L. von Bortkiewicz in 1922 aroused the interest of some of his contemporaries (E.L. Dodd (1923), R. von Mises (1923), L.H.C. Tippett (1925)). Von Bortkiewicz studied the sample range of normal random variables. An important step toward the asymtotic theory of extremes was made by E.L. Dodd and R. von Mises. Both authors studied the asymptotic behavior of the sample maximum of normal and non-normal random variables. The article of von Mises is written in a very attractive, modern style. Under weak regularity conditions, e.g. satisfied by the normal d.f., von Mises proved that the expectation of the sample maximum is asymptotically equal to $F^{-1}(1 - 1/n)$; moreover, he proved that

$$P\{|X_{n:n} - F^{-1}(1 - 1/n)| \le \varepsilon\} \to 1, \qquad n \to \infty, \text{ for every } \varepsilon > 0.$$

A similar result was also deduced by Dodd for various classes of distributions.

This development was culminated in the article of R.A. Fisher and L.H.C. Tippett (1928), who derived the three types of extreme value distributions and discussed the stability problem. The limiting d.f. $G_{1,\alpha}$ was independently discovered by M. Fréchet (1927). As mentioned by Wilks (1948), Fréchet's result and that of Fisher and Tippett actually appeared almost simultaneously in 1928.

We mention some of the early results obtained for central order statistics: In 1902, K. Pearson derived the expectation of a spacing under a continuous d.f. (Galton difference problem) and, in 1920, investigated the performance of "systematic statistics" as estimators of the median by computing asymptotic expectations and covariances of sample quantiles. Craig (1932) established densities of sample quantiles in special cases. Thompson (1936) treated confidence intervals for the q-quantile. Compared to the development in extreme value theory the results concerning central order statistics were obtained more sporadically than systematically.

It is clear that the considerations in this book concerning exact distributions of order statistics are not exhaustive. For example, it is worthwhile studying distributions of order statistics in the discrete case as it was done by Nagaraja (1982, 1986), Arnold et al. (1984), and Rüschendorf (1985a). B.C. Arnold and his co-authors showed that order statistics of a sample of size $n \geq 3$ possess the Markov property if, and only if, there does not exist an atom x of the underlying d.f. F such that $0 < F(x^-)$ and $F(x) < 1$. In that paper one may also find expressions for the density of order statistics in the discrete case. We also note that densities of order statistics in case of a random sample size are given in an explicit form by Consul (1984); see also Smith (1984, pages 631, 632). Further results concerning exact distributions of order statistics may be found in the books mentioned below.

Apart from the books of E.J. Gumbel (1958), L. de Haan (1970), H.A. David (1981), J. Galambos (1987), M.R. Leadbetter et al. (1983), and S.I. Resnick (1987), mentioned in the various sections, we refer to the books of Johnson and Kotz (1970, 1972) (order statistics for special distributions), Barnett and Lewis (1978) (outliers), and R.R. Kinnison (1985) (applied aspects of extreme value theory). The reading of survey articles about order statistics written by S.S. Wilks (1948), A. Rényi (1953), and J. Galambos (1984) can be highly recommended. For an elementary, enjoyable introduction to classical results of extreme value theory we refer to de Haan (1976).

CHAPTER 2

Multivariate Order Statistics

This chapter is primarily concerned with the marginal ordering of the observations. Thus, the restriction to one component again leads to the order statistics dealt with in Chapter 1. Our treatment of multivariate order statistics will not be as exhaustive as that in the univariate case because of the technical difficulties and the complicated formulae for d.f.'s and densities.

There is one exception, namely, the case of multivariate maxima of i.i.d. random vectors with d.f. F. This case is comparatively easy to deal with since the d.f. of the multivariate maximum is again given by F^n, and the density is consequently of a simple form.

2.1. Introduction

Multivariate order statistics (including extremes) will be defined by taking order statistics componentwise (in other words, we consider marginal ordering). It is by no means self-evident to define order statistics and extremes in this particular way and we do not deny that other definitions of multivariate order statistics are perhaps of equal importance. Some other possibilities will be indicated at the end of this section. One reason why our emphasis is laid on this particular definition is that it favorably fits to our present program and purposes.

In this sequel, the relations and arithmetic operations are always taken componentwise. Given $\mathbf{x} = (x_1, \ldots, x_d)$ and $\mathbf{y} = (y_1, \ldots, y_d)$ we write

$$\mathbf{x} \leq \mathbf{y} \quad \text{if} \quad x_i \leq y_i, \qquad i = 1, \ldots, d, \qquad (2.1.1)$$

and

$$\mathbf{x} + \mathbf{y} = (x_1 + y_1, x_2 + y_2, \ldots, x_d + y_d). \qquad (2.1.2)$$

The Definition of Multivariate Order Statistics

Let $\xi_1, \ldots, \xi_n$ be n random vectors of dimension d where $\xi_i = (\xi_{i,1}, \xi_{i,2}, \ldots, \xi_{i,d})$. The ordered values of the jth components $\xi_{1,j}, \xi_{2,j}, \ldots, \xi_{n,j}$ are denoted by

$$X_{1:n}^{(j)} \leq X_{2:n}^{(j)} \leq \cdots \leq X_{n:n}^{(j)}. \tag{2.1.3}$$

Using the map $Z_{r:n}$ as defined in (1.1.4) we have

$$X_{r:n}^{(j)} = Z_{r:n}(\xi_{1,j}, \xi_{2,j}, \ldots, \xi_{n,j}). \tag{2.1.4}$$

We also write

$$\mathbf{X}_{r:n} = (X_{r:n}^{(1)}, X_{r:n}^{(2)}, \ldots, X_{r:n}^{(d)}). \tag{2.1.5}$$

Using the order relation as defined in (2.1.1) we obtain

$$\mathbf{X}_{1:n} \leq \mathbf{X}_{2:n} \leq \cdots \leq \mathbf{X}_{n:n}. \tag{2.1.6}$$

Notice that

$$\mathbf{X}_{1:n} = (X_{1:n}^{(1)}, X_{1:n}^{(2)}, \ldots, X_{1:n}^{(d)}) \tag{2.1.7}$$

is the d-variate sample minimum, and

$$\mathbf{X}_{n:n} = (X_{n:n}^{(1)}, X_{n:n}^{(2)}, \ldots, X_{n:n}^{(d)}) \tag{2.1.8}$$

is the d-variate sample maximum.

Observe that realizations of $\mathbf{X}_{j:n}$ are not realizations of $\xi_1, \ldots, \xi_n$ in general.

The Relation to Frequencies

For certain problems the results of the previous sections can easily be extended to the multivariate set-up. As an example we mention that (1.1.7) implies that

$$P\{\mathbf{X}_{r:n} \leq \mathbf{t}\} = P\left\{ \sum_{i=1}^{n} (1_{(-\infty, t_1]}(\xi_{i,1}), \ldots, 1_{(-\infty, t_d]}(\xi_{i,d})) \geq \mathbf{r} \right\} \tag{2.1.9}$$

where $\mathbf{t} = (t_1, t_2, \ldots, t_d)$ and $\mathbf{r} = (r, r, \ldots, r)$. Notice that in (2.1.9) we obtain a sum of independent random vectors if the random vectors $\xi_1, \xi_2, \ldots, \xi_n$ are independent. It makes no effort to extend (2.1.9) to any subclass of the r.v.'s $X_{r:n}^{(j)}$. For $I \subset \{(j, r): j = 1, \ldots, d \text{ and } r = 1, \ldots, n\}$ we have

$$P\{X_{r:n}^{(j)} \leq t_{j,r}, (j, r) \in I\} = P\left\{ \sum_{i=1}^{n} 1_{(-\infty, t_{j,r}]}(\xi_{i,j}) \geq r, (j, r) \in I \right\}. \tag{2.1.10}$$

Thus, again the joint distribution of the r.v.'s $X_{r:n}^{(j)}$, $(j, r) \in I$, can be represented by means of the distribution of a sum of independent random vectors if the random vectors $\xi_1, \ldots, \xi_n$ are independent. Note that a similar result holds if maxima

$$X_{n(1):n(1)}^{(1)}, \ldots, X_{n(d):n(d)}^{(d)}$$

are treated with different sample sizes for each component.

Further Concepts of Multivariate Ordering

A particular characteristic of univariate order statistics was that the ordered values no longer contain any information about the order of their outcome. Recall that this information is presented by the rank statistic R_n (see P.1.30). The corresponding general formulation of this aspect in the Euclidean d-space is given by the definition of the order statistic via sets of observations. Thus, given r.v.'s or random vectors $\xi_1, \ldots, \xi_n$ we also may call the set $\{\xi_1, \ldots, \xi_n\}$ the order statistic. It is well known that for i.i.d. random vectors these random sets form a minimal sufficient statistic.

Other concepts are more related to the ordering according to the magnitude of the observations like in the univariate case. Our enthusiasm for this topic is rather limited because no successful theory exists (besides the particular case of sample maxima and sample minima as defined in (2.1.7) and (2.1.8)). However, this topic meets an increasing interest since Barnett's brilliant paper in 1976 which is full of ideas, suggestions and applications. Some brief comments about the different concepts of multivariate ordering:

(a) The convex hull of the data points and the subsequent "peeling" of the multi-dimensional sample entails one possibility of a multivariate ordering. This concept is nice from a geometric point of view. The convex hull can e.g. be used as an estimator of the distribution's support.
(b) The concomitants are obtained (in the bivariate case) by arranging the data in the second component according to the ordering in the first component.
(c) The multivariate sample median is a solution of the equation

$$\sum_{i=1}^{n} \|\mathbf{x}_i - \mathbf{x}\|_2 = \min_{\mathbf{x}}! \tag{2.1.11}$$

where $\| \ \|_2$ denotes the Euclidean norm. The median of a multivariate probability measure Q is defined by

$$\int \|\mathbf{y} - \mathbf{x}\|_2 \, dQ(\mathbf{y}) = \min_{\mathbf{x}}!. \tag{2.1.12}$$

Total ψ-Ordering

Last but not least, we mention the ordering of multivariate data according to the ranking method everyone is familiar with in his daily life. The importance of this concept is apparent.

Following Plackett (1976) we introduce a total order of the points $\mathbf{x}_1, \ldots, \mathbf{x}_n$ by means of a real-valued function ψ. Define

$$\mathbf{x} \leq_\psi \mathbf{y} \tag{2.1.13}$$

if

$$\psi(\mathbf{x}) \le \psi(\mathbf{y}). \tag{2.1.14}$$

Usually one is not only interested in the ranking of the data $\mathbf{x}_1, \ldots, \mathbf{x}_n$ expressed in numbers $1, \ldots, n$ but also in the total information contained in $\mathbf{x}_1, \ldots, \mathbf{x}_n$, thus getting the representation of the original data by

$$\tilde{x}_{1:n} \le_\psi \cdots \le_\psi \tilde{x}_{n:n}. \tag{2.1.15}$$

One advantage of this type of ordering compared to the marginal ordering is that $\tilde{x}_{i:n}$ is a point of the original sample. It is clear that the ordering (2.1.15) heavily depends on the selection procedure represented by the function ψ.

As an example, consider the function $\psi(\mathbf{x}) = \|\mathbf{x} - \mathbf{x}_0\|_2$. Other reasonable functions ψ may be found in Barnett (1976) and Plackett (1976). Given the random vectors $\xi_1, \ldots, \xi_n$ let

$$\tilde{X}_{1:n} \le_\psi \cdots \le_\psi \tilde{X}_{n:n} \tag{2.1.16}$$

denote the ψ-order statistics defined according to (2.1.15) with $\psi(\mathbf{x}) = \|\mathbf{x} - \mathbf{x}_0\|_2$. Define

$$R_{k:n} = \|\tilde{X}_{k:n} - \mathbf{x}_0\|_2 \tag{2.1.17}$$

which is the distance of the kth largest ψ-order statistic from the center $\mathbf{x}_0$. Obviously,

$$R_{k:n} = Z_{k:n}(\|\xi_1 - \mathbf{x}_0\|_2, \ldots, \|\xi_n - \mathbf{x}_0\|_2) \tag{2.1.18}$$

is the kth largest order statistic of the n i.i.d. univariate r.v.'s $\|\xi_1 - \mathbf{x}_0\|_2, \ldots, \|\xi_n - \mathbf{x}_0\|_2$ with common d.f.

$$F(\mathbf{x}_0, \cdot) = P\{\xi_1 \in B(\mathbf{x}_0, \cdot)\}. \tag{2.1.19}$$

Here

$$B(\mathbf{x}_0, r) = \{\mathbf{x}: \|\mathbf{x} - \mathbf{x}_0\|_2 \le r\}$$

is the ball with center $\mathbf{x}_0$ and radius r.

Notice that the probability

$$P\{\tilde{X}_{k:n} \in B(\mathbf{x}_0, r)\} \tag{2.1.20}$$

may easily be computed since this quantity is equal to $P\{R_{k:n} \le r\}$.

We also mention a result related to that of Corollary 1.8.4 in the univariate case.

By rearranging $\tilde{X}_{n-k+1:n}, \ldots, \tilde{X}_{n:n}$ in the original order of their outcome we obtain the k exceedances, say, $\zeta_1, \ldots, \zeta_k$ of the random vectors $\xi_1, \ldots, \xi_n$. It is well known that the conditional distribution of the exceedances $\zeta_1, \ldots, \zeta_k$ given $R_{n-k:n} = r$ is the joint distribution of k i.i.d. random vectors $\eta_1, \ldots, \eta_k$ with common distribution equal to the original distribution of ξ_1 truncated outside of

$$C(\mathbf{x}_0, r) = \{\mathbf{x}: \|\mathbf{x} - \mathbf{x}_0\|_2 > r\}. \tag{2.1.21}$$

The author is grateful to Peter Hall for communicating a 3-line sketch of the proof of this result. An extension can be found in P.2.1.

If $F(\mathbf{x}_0, \cdot)$ is continuous then we deduce from Theorem 1.5.1 that for the ψ-maximum $\tilde{X}_{n:n}$ the following identities hold:

$$
\begin{aligned}
P\{\tilde{X}_{n:n} \in B\} &= \int P(\tilde{X}_{n:n} \in B \mid R_{n-1:n}) \, dP \\
&= n(n-1) \int P\{\xi_1 \in B \cap C(\mathbf{x}_0, \cdot)\} F(\mathbf{x}_0, \cdot)^{n-2} \, dF(\mathbf{x}_0, \cdot).
\end{aligned}
\tag{2.1.22}
$$

The construction in (2.1.16) can be generalized to the case where $\mathbf{x}_0$ is replaced by a random vector ξ_0 leading to the kth ordered distance r.v. $R_{k:n}$ as studied in Dziubdziela (1976) and Reiss (1985b). Now the ranking is carried out according to the random function $\psi(\mathbf{x}) = \|\mathbf{x} - \xi_0\|_2$. A possible application of such a concept is the definition of an α-trimmed mean

$$
(n\alpha)^{-1} \sum_{i=1}^{[n\alpha]} \tilde{X}_{i:n}
\tag{2.1.23}
$$

centered at the random vector ξ_0.

2.2. Distribution Functions and Densities

From (2.1.9) and (2.1.10) it is obvious that the joint d.f. of order statistics $X_{r:n}^{(j)}$ can be established by means of multinomial probabilities of appropriate "cell frequency vectors" $N_1, \ldots, N_k$ where $N_j = \sum_{i=1}^{n} 1_{R_j}(\xi_i)$ and the $R_1, \ldots, R_k$ form a partition of the Euclidean d-space. Note that

$$
P\{N_1 = i_1, \ldots, N_k = i_k\} = \frac{n! \, p_1^{i_1} \cdot \ldots \cdot p_k^{i_k}}{i_1! \cdot \ldots \cdot i_k!}
$$

where $i_j \geq 0$, $\sum_{j=1}^{k} i_j = n$ and $p_j = P\{\xi_1 \in R_j\}$.

The D.F. of Multivariate Extremes

Let $\xi, \xi_1, \xi_2, \ldots, \xi_n$ be i.i.d. random vectors. We start with a simple result concerning the d.f. of multivariate order statistics. For the sample maximum $\mathbf{X}_{n:n}$ based on $\xi_1, \xi_2, \ldots, \xi_n$ we obtain as an extension of (1.3.2) that

$$
P\{\mathbf{X}_{n:n} \leq \mathbf{t}\} = F^n(\mathbf{t}).
\tag{2.2.1}
$$

This becomes obvious by writing

$$
\begin{aligned}
P\{\mathbf{X}_{n:n} \leq \mathbf{t}\} &= P\{X_{n:n}^{(1)} \leq t_1, \ldots, X_{n:n}^{(d)} \leq t_d\} \\
&= P\{\max\{\xi_{1,1}, \ldots, \xi_{n,1}\} \leq t_1, \ldots, \max\{\xi_{1,d}, \ldots, \xi_{n,d}\} \leq t_d\} \\
&= P\{\xi_1 \leq \mathbf{t}, \ldots, \xi_n \leq \mathbf{t}\} = F^n(\mathbf{t}).
\end{aligned}
$$

The extension of (2.2.1) to the case of i.n.n.i.d. r.v.'s is straightforward. Moreover, in analogy to (2.2.1) one gets for the sample minimum $\mathbf{X}_{1:n}$ the formula

$$P\{\mathbf{X}_{1:n} > \mathbf{t}\} = L(\mathbf{t})^n \tag{2.2.2}$$

where $L(\mathbf{t}) = P\{\xi > \mathbf{t}\}$ is the survivor function.

For $d = 2$, the following representation for the bivariate survivor function holds:

$$L(x, y) = P\{\xi > (x, y)\} = 1 - F_1(x) - F_2(y) + F(x, y)$$

with F_i denoting the marginal d.f.'s of F. Hence,

$$F(x, y) = 1 - (1 - F_1(x)) - (1 - F_2(y)) + L(x, y).$$

An extension of this representation to the d-variate d.f. may be found in P.2.5.

Formula (2.2.2) in conjunction with (1.3.3) yields

$$P\{\mathbf{X}_{1:n} \le (x, y)\} = 1 - (1 - F_1(x))^n - (1 - F_2(y))^n + L(x, y)^n. \tag{2.2.3}$$

If a d.f. on the Euclidean d-space has d continuous partial derivatives then we know (see e.g. Bhattacharya and Rao (1976), Theorem A.2.2) that the dth partial derivative $\partial^d F/(\partial t_1 \ldots \partial t_d)$ is a density of F. Thus, if f is a density of F then, if $d = 2$,

$$f_{(n,n):n} = nF^{n-1}f + n(n-1)F^{n-2}\frac{\partial F}{\partial x}\frac{\partial F}{\partial y} \tag{2.2.4}$$

is the density of the sample maximum $\mathbf{X}_{n:n} = (X_{n:n}^{(1)}, X_{n:n}^{(2)})$ for $n \ge 2$.

The density of the sample minimum $\mathbf{X}_{1:n} = (X_{1:n}^{(1)}, X_{1:n}^{(2)})$ is given by

$$nL^{n-1}f + n(n-1)L^{n-2}\frac{\partial L}{\partial x}\frac{\partial L}{\partial y}. \tag{2.2.5}$$

For an extension and a reformulation of (2.2.4) we refer to (2.2.7) and (2.2.8).

The D.F. of Bivariate Order Statistics

The exact joint d.f. and joint density of order statistics $X_{r:n}^{(j)}$ can be established via multinomial random vectors. The joint distribution of $X_{r:n}^{(1)}$ and $X_{s:n}^{(2)}$ will be examined in detail.

Let again $\xi_i = (\xi_{i,1}, \xi_{i,2})$, $i = 1, \ldots, n$, be independent copies of the random vector $\xi = (\xi_1, \xi_2)$ with common d.f. F and marginals F_i. Thus, $F(x, y) = P\{\xi \le (x, y)\}$, $F_1(x) = P\{\xi_1 \le x\}$ and $F_2(y) = P\{\xi_2 \le y\}$.

A partition of the plane into the four quadrants

$$R_1 = (-\infty, x] \times (-\infty, y], \qquad R_2 = (-\infty, x] \times (y, \infty),$$

$$R_3 = (x, \infty) \times (-\infty, y], \qquad R_4 = (x, \infty) \times (y, \infty)$$

(where the dependence of R_i on (x, y) will be suppressed) leads to the configuration

$$
\begin{array}{c|c}
R_2 & R_4 \\
\hline
R_1 & R_3
\end{array}\Bigg|_{(x,y)} .
$$

Put

$$L_i = P\{\xi \in R_i\}.$$

Notice that L_4 is the bivariate survivor function as mentioned above. We have

$$F(x, y) = L_1(x, y), \quad F_1(x) = L_1(x, y) + L_2(x, y), \quad F_2(y) = L_1(x, y) + L_3(x, y)$$

and hence

$$L_1(x, y) = F(x, y), \quad L_2(x, y) = F_1(x) - F(x, y), \quad L_3(x, y) = F_2(y) - F(x, y),$$

and as noted above

$$L_4(x, y) = 1 - F_1(x) - F_2(y) + F(x, y).$$

Denote by N_j the frequency of the ξ_i in R_j; thus,

$$N_j = \sum_{i=1}^{n} 1_{R_j}(\xi_i).$$

From (1.1.7) it is immediate that

$$P\{X_{r:n}^{(1)} \le x, X_{s:n}^{(2)} \le y, N_1 = m\}$$

$$= P\left\{\sum_{i=1}^{n} 1_{(-\infty, x]}(\xi_{i,1}) \ge r, \sum_{i=1}^{n} 1_{(-\infty, y]}(\xi_{i,2}) \ge s, N_1 = m\right\}$$

$$= P\{N_1 + N_2 \ge r, N_1 + N_3 \ge s, N_1 = m\}$$

$$= \sum_{k=r}^{n} \sum_{l=s}^{n} P\{N_1 = m, N_2 = k - m, N_3 = l - m\}.$$

Inserting the probabilities of the multinomial random vector (N_1, N_2, N_3, N_4) we get

Lemma 2.2.1. *The d.f. $F_{(r, s):n}$ of $(X_{r:n}^{(1)}, X_{s:n}^{(2)})$ is given by*

$$F_{(r,s):n} = \sum_{k=r}^{n} \sum_{l=s}^{n} \sum_{m=\max(k+l-n, 0)}^{\min(k,l)} \frac{n! L_1^m L_2^{k-m} L_3^{l-m} L_4^{n-k-l+m}}{m!(k-m)!(l-m)!(n-k-l+m)!}.$$

The Density of Bivariate Order Statistics

If $F_{(r, s):n}$ possesses two partial derivatives, one may use the representation $(\partial^2/\partial x \partial y)F_{(r, s):n}$ of the density of $F_{(r, s):n}$, however, it is difficult to arrange the terms in an appropriate way.

A different method will allow us to compute the density of $(X_{r:n}^{(1)}, X_{s:n}^{(2)})$ under the condition that F has a density, say, f. To make the proof rigorous one has to use the Radon–Nikodym theorem and Lebesgue's differentiation theorem for integrals.

In a first step we shall prove that a density of $F_{(r,s):n}$ exists if F has a density. Notice that for every Borel set B we have

$$P\{(X_{r:n}^{(1)}, X_{s:n}^{(2)}) \in B\} \le \sum_{i,j=1}^{n} P\{(\xi_{i,1}, \xi_{j,2}) \in B\}$$

$$= \sum_{\substack{i,j=1 \\ i \ne j}}^{n} \int_B f_1(x) f_2(y)\, dx\, dy + \sum_{i=1}^{n} \int_B f(x, y)\, dx\, dy$$

where $f_1 = \int f(\cdot, v)\, dv$ and $f_2 = \int f(u, \cdot)\, du$ are the densities of F_1 and F_2. Thus, if B has Lebesgue measure zero then $P\{(X_{r:n}^{(1)}, X_{s:n}^{(2)}) \in B\} = 0$, and hence the Radon–Nikodym theorem implies that $F_{(r,s):n}$ has a (Lebesgue) density.

The proof of Lemma 2.2.2 below will be based on the fact that for every integrable function g on the Euclidean k-space almost all $\mathbf{x} = (x_1, \ldots, x_k)$ are Lebesgue points of g, that is,

$$\lim_{h \to 0} (2h)^{-k} \int_{x_1-h}^{x_1+h} \cdots \int_{x_k-h}^{x_k+h} g(\mathbf{z})\, d\mathbf{z} = g(\mathbf{x}) \qquad (2.2.6)$$

for (Lebesgue) almost all $\mathbf{x}$ (see e.g. Floret (1981), page 276).

The following lemma was established in cooperation with W. Kohne.

Lemma 2.2.2. *If the bivariate i.i.d. random vectors $\xi_1, \xi_2, \ldots, \xi_n$ have the common density f then the random vector $(X_{r:n}^{(1)}, X_{s:n}^{(2)})$ has the density*

$$f_{(r,s):n} = n! \sum_{m=0}^{n} \frac{L_1^m}{m!} [L_2^{r-1-m} L_3^{s-1-m} L_4^{n-r-s+m+1} f/$$

$$(r-1-m)!(s-1-m)!(n-r-s+m+1)!$$

$$+ L_2^{r-2-m} L_3^{s-1-m} L_4^{n-r-s+m+1} L_5 L_6/$$

$$(r-2-m)!(s-1-m)!(n-r-s+m+1)!$$

$$+ L_2^{r-2-m} L_3^{s-2-m} L_4^{n-r-s+m+2} L_5 L_8/$$

$$(r-2-m)!(s-2-m)!(n-r-s+m+2)!$$

$$+ L_2^{r-1-m} L_3^{s-1-m} L_4^{n-r-s+m} L_6 L_7/$$

$$(r-1-m)!(s-1-m)!(n-r-s+m)!$$

$$+ L_2^{r-1-m} L_3^{s-2-m} L_4^{n-r-s+m+1} L_7 L_8/$$

$$(r-1-m)!(s-2-m)!(n-r-s+m+1)!]$$

with the convention that the terms involving negative factorials are replaced by zeros. The functions $L_1, \ldots, L_4$ are defined as above. Moreover,

$$L_5(x, y) = \int_{-\infty}^{x} f(u, y) \, du, \qquad L_6(x, y) = \int_{y}^{\infty} f(x, v) \, dv,$$

$$L_7(x, y) = \int_{x}^{\infty} f(u, y) \, du, \qquad L_8(x, y) = \int_{-\infty}^{y} f(x, v) \, dv.$$

Notice that $\sum_{m=0}^{n}$ *can be replaced by* $\sum_{m=0}^{\min(r,s)-1}$. *Moreover,*

$$L_6(x, y) = f_1(x) - L_8(x, y) \quad and \quad L_7(x, y) = f_2(y) - L_5(x, y).$$

PROOF. Put $S_{0,h}(x, y) = (x - h, x + h] \times (y - h, y + h]$ where the indices h, x, y will be suppressed as far as no confusion can arise. According to (2.2.6) it suffices to show that

$$(2h)^{-2} P\{(X_{r:n}^{(1)}, X_{s:n}^{(2)}) \in S_{0,h}(x, y)\} \to f_{(r,s):n}(x, y) \tag{1}$$

as $h \downarrow 0$ for almost all (x, y). To compute $P\{(X_{r:n}^{(1)}, X_{s:n}^{(2)}) \in S_0\}$ we shall make use of the following configuration

S_2	$\vdots$ S_6	S_4
$S_5 \cdots$	$\vdots$ $\cdots \vdots$ $\vdots$ (x, y)	$\cdots S_7$ $\Big\}\, 2h.$
S_1	$\vdots$ S_8	S_3

Put $N_j = \sum_{i=1}^{n} 1_{S_j}(\xi_i)$ and $q_j = P\{\xi \in S_j\} = \int_{S_j} f(u, v) \, du \, dv$ for $0 \le j \le 8$. Obviously, $q_j \to L_j$ as $h \to 0$ for $j = 1, \ldots, 4$. Moreover, by applying (2.2.6) it is straightforward to prove that almost everywhere:

$$(2h)^{-2} q_0 \to f \quad and \quad (2h)^{-1} q_j \to L_j \quad as \ h \to 0 \tag{2}$$

for $j = 5, \ldots, 8$. First, observe that for all (x, y) such that (2) holds we have

$$h^{-2} P\{N_0 \ge 2\} \to 0, \qquad h^{-2} P\Big\{N_0 = 1, \sum_{j=5}^{8} N_j \ge 1\Big\} \to 0,$$

and

$$h^{-2} P\Big\{N_0 = 0, \sum_{j=5}^{8} N_j \ge 2\Big\} \to 0$$

as $h \to 0$ and hence it remains to prove that

$$(2h)^{-2} \Bigg[P\{(X_{r:n}^{(1)}, X_{s:n}^{(2)}) \in S_0, N_0 = 1, N_5 = N_6 = N_7 = N_8 = 0\}$$

$$+ P\Big\{(X_{r:n}^{(1)}, X_{s:n}^{(2)}) \in S_0, N_0 = 0, \sum_{j=5}^{8} N_j < 2\Big\} \Bigg] \to f_{(r,s):n} \tag{3}$$

as $h \to 0$ almost everywhere.

Applying (1.1.7) we conclude that

$$\{(X_{r:n}^{(1)}, X_{s:n}^{(2)}) \in S_0\}$$

$$= \{x - h < X_{r:n}^{(1)} \le x + h, \, y - h < X_{s:n}^{(2)} \le y + h\}$$

$$= \left\{ \sum_{i=1}^{n} 1_{(-\infty, x-h]}(\xi_{i,1}) < r \le \sum_{i=1}^{n} 1_{(-\infty, x+h]}(\xi_{i,1}), \right.$$

$$\left. \sum_{i=1}^{n} 1_{(-\infty, y-h]}(\xi_{i,2}) < s \le \sum_{i=1}^{n} 1_{(-\infty, y+h]}(\xi_{i,2}) \right\} \tag{4}$$

$$= \{N_1 + N_2 + N_5 < r \le N_1 + N_2 + N_5 + N_0 + N_6 + N_8,$$

$$N_1 + N_3 + N_8 < s \le N_1 + N_3 + N_8 + N_0 + N_5 + N_7\}.$$

Thus, for $m = 0, \dots, n$,

$$\{(X_{r:n}^{(1)}, X_{s:n}^{(2)}) \in S_0, N_0 = 1, N_5 = N_6 = N_7 = N_8 = 0, N_1 = m\}$$

$$= \{N_1 + N_2 < r \le N_1 + N_2 + 1, \, N_1 + N_3 < s \le N_1 + N_3 + 1,$$

$$N_0 = 1, N_5 = N_6 = N_7 = N_8 = 0, N_1 = m\} \tag{5}$$

$$= \{N_0 = 1, N_1 = m, N_2 = r - 1 - m, N_3 = s - 1 - m,$$

$$N_5 = N_6 = N_7 = N_8 = 0\}.$$

By (4) we also get for $m = 0, \dots, n$,

$$\left\{(X_{r:n}^{(1)}, X_{s:n}^{(2)}) \in S_0, N_0 = 0, \sum_{j=5}^{8} N_j \le 2, N_1 = m\right\}$$

$$= \{N_0 = 0, N_1 + N_2 + N_5 = r - 1, N_1 + N_3 + N_8 = s - 1,$$

$$N_6 + N_8 = 1, N_5 + N_7 = 1, N_1 = m\}$$

$$= \{N_0 = 0, N_1 = m, N_2 = r - 2 - m, N_3 = s - 1 - m,$$

$$N_7 = 0, N_8 = 0, N_5 = 1, N_6 = 1\}$$

$$+ \{N_0 = 0, N_1 = m, N_2 = r - 2 - m, N_3 = s - 2 - m, \tag{6}$$

$$N_6 = 0, N_7 = 0, N_5 = 1, N_8 = 1\}$$

$$+ \{N_0 = 0, N_1 = m, N_2 = r - 1 - m, N_3 = s - 1 - m,$$

$$N_5 = 0, N_8 = 0, N_6 = 1, N_7 = 1\}$$

$$+ \{N_0 = 0, N_1 = m, N_2 = r - 1 - m, N_3 = s - 2 - m,$$

$$N_5 = 0, N_6 = 0, N_7 = 1, N_8 = 1\}.$$

Now (3) is immediate from (2), (5), and (6). The proof is complete. □

In the special case of the sample maximum (that is, $r = n$ and $s = n$) we have

$$f_{(n,n):n} = nF^{n-1}f + n(n-1)F^{n-2}L_5 L_8 \tag{2.2.7}$$

which is a generalization of (2.2.4) in the bivariate case. If the partial derivatives exist then $f = \partial^2 F/\partial x \partial y$,

$$L_5(x, y) = \int_{-\infty}^{x} f(u, y)\, du = (\partial F/\partial y)(x, y),$$

and

$$L_8(x, y) = \int_{-\infty}^{y} f(x, v)\, dv = (\partial F/\partial x)(x, y).$$

Let $\xi = (\xi_1, \xi_2)$ again be a random vector with d.f. F and density f. Let $f_1(x) = \int f(x, v)\, dv$ and $f_2(y) = \int f(u, y)\, du$ be the marginal densities, and let

$$F_1(x|y) = P(\xi_1 \le x|\xi_2 = y) = L_5(x, y)/f_2(y)$$

and

$$F_2(y|x) = P\{\xi_2 \le y|\xi_1 = x\} = L_8(x, y)/f_1(x)$$

be the conditional d.f.'s. Now, (2.2.7) may be written

$$f_{(n,n):n}(x, y)$$

$$= nF^{n-1}(x, y)f(x, y) + n(n - 1)F^{n-2}(x, y)F_1(x|y)F_2(y|x)f_1(x)f_2(y). \tag{2.2.8}$$

The Partial Maxima Process

A vector of extremes with different sample sizes in the different components has to be treated in connection with the partial maxima process X_n defined by

$$X_n(t) = \begin{cases} a_n^{-1}(X_{[nt]:[nt]} - b_n) & t \ge \dfrac{1}{n} \\ \qquad\qquad\text{if} \\ a_n^{-1}(X_{1:1} - b_n) & 0 < t < \dfrac{1}{n} \end{cases} \tag{2.2.9}$$

for $t > 0$ where the reals b_n and $a_n > 0$ are appropriate normalizing constants. In order to calculate the finite dimensional marginal d.f.'s of X_n one needs the following.

Lemma 2.2.3. *Let* $1 \le s_1 < s_2 < \cdots < s_k$ *be integers and* $\xi_1, \ldots, \xi_{s_k}$ *i.i.d. random variables with common d.f.* F. *Then,*

$$P\{X_{s_1:s_1} \le x_1, \ldots, X_{s_k:s_k} \le x_k\} = F^{s_1}(y_1)F^{s_2-s_1}(y_2)\ldots F^{s_k-s_{k-1}}(y_k) \tag{2.2.10}$$

where $y_j = \min(x_j, x_{j+1}, \ldots, x_k)$.

PROOF. Obvious by noting that

$$\{X_{s_1:s_1} \le x_1, \ldots, X_{s_k:s_k} \le x_k\}$$

$$= \{X_{s_1:s_1} \le y_1, \ldots, X_{s_k:s_k} \le y_k\}$$

$$= \{\max(\xi_1, \ldots, \xi_{s_1}) \le y_1, \ldots, \max(\xi_{s_{k-1}+1}, \ldots, \xi_{s_k}) \le y_k\}. \qquad \square$$

We remark that a corresponding formula for sample minima can be established via the equality

$$\{X_{1:s_1} > x_1, \ldots, X_{1:s_k} > x_k\}$$
$$= \{\min(\xi_1, \ldots, \xi_{s_1}) > y_1, \ldots, \min(\xi_{s_{k-1}+1}, \ldots, \xi_{s_k}) > y_k\} \quad (2.2.11)$$

where $y_j = \max(x_j, x_{j+1}, \ldots, x_k)$.

Multivariate Extreme Value Distributions

In Section 1.3 we mentioned that the limiting (thus, also stable) d.f.'s of the univariate maximum $X_{n:n}$ are the Fréchet, Weibull, and Gumbel d.f.'s $G_{i,\alpha}$. The situation in the multivariate case is much more complex. First, we mention two trivial examples of limiting multivariate d.f.'s.

EXAMPLES 2.2.4. Let $\mathbf{X}_{n:n} = (X_{n:n}^{(1)}, \ldots, X_{n:n}^{(d)})$ be the sample maximum based on i.i.d. random vectors $\xi_1, \ldots, \xi_n$ which are distributed like $\xi = (\eta_1, \ldots, \eta_d)$.

(i) (Complete dependence)

Our first example concerns the case that the components $\eta_1, \ldots, \eta_d$ of ξ are identical; i.e. we have

$$\eta_1 = \eta_2 = \cdots = \eta_d.$$

Let F_1 denote the d.f. of η_1. Then, the d.f. F of ξ is given by

$$F(\mathbf{t}) = F_1(\min(t_1, \ldots, t_d))$$

and hence

$$P\{\mathbf{X}_{n:n} \le \mathbf{t}\} = F^n(\mathbf{t}) = F_1^n(\min(t_1, \ldots, t_d)). \quad (2.2.12)$$

If $F_1 = G_{i,\alpha}$ then with c_n and d_n as in (1.3.13):

$$F^n(c_n t_1 + d_n, \ldots, c_n t_d + d_n) = G_{i,\alpha}^n(c_n \min(t_1, \ldots, t_d) + d_n)$$
$$= G_{i,\alpha}(\min(t_1, \ldots, t_d)) = F(\mathbf{t}).$$

(ii) (Independence)

Secondly, assume that the components $\eta_1, \ldots, \eta_d$ of ξ are independent. Then it is clear that $X_{n:n}^{(1)}, \ldots, X_{n:n}^{(d)}$ are independent. If $G_{i(j),\alpha(j)}$ is the d.f. of η_j then with $c_{n,j}$ and $d_{n,j}$ as in (1.3.13):

$$F^n(c_{n,1} t_1 + d_{n,1}, \ldots, c_{n,d} t_d + d_{n,d}) = F(\mathbf{t}) = \prod_{j=1}^{d} G_{i(j),\alpha(j)}(t_j). \quad (2.2.13)$$

(iii) (Asymptotic independence)

Given $\xi = (-\xi, \xi)$, we have

$$\mathbf{X}_{n:n} = (X_{n:n}^{(1)}, X_{n:n}^{(2)}) = (-X_{1:n}, X_{n:n})$$

where $X_{1:n}$ and $X_{n:n}$ are the sample minimum and sample maximum based on the independent copies $\xi_1, \ldots, \xi_n$ of ξ. In Section 4.2 we shall see that $X_{1:n}$ and $X_{n:n}$ (and, thus, $X_{n:n}^{(1)}$ and $X_{n:n}^{(2)}$) are asymptotically independent. Thus, again we are getting independent r.v.'s in the limit.

Contrary to the univariate case the multivariate extreme value d.f.'s form a nonparametric family of distributions. There is a simple device which enables us to check whether a given d.f. is a multivariate extreme value d.f.

We say that a d-variate d.f. G is nondegenerate if the univariate marginals are nondegenerate. A nondegenerate d-variate d.f. G is a limiting d.f. of sample maxima if, and only if, G is max-stable, that is,

$$G^n(b_{n,1} + a_{n,1}x_1, \ldots, b_{n,d} + a_{n,d}x_d) = G(x_1, \ldots, x_d) \qquad (2.2.14)$$

for some normalizing constants $a_{n,j} > 0$ and $b_{n,j}$ (compare e.g. with Galambos (1987), page 295, or Resnick (1987), Proposition 5.9).

If a d-variate d.f. is max-stable then it is easy to show that the univariate marginals are max-stable and, hence, these d.f.'s have to be of the type $G_{1,\alpha}$, $G_{2,\alpha}$ or G_3 with $\alpha > 0$.

On the other hand, if the jth univariate marginal d.f. is $G_{i(j),\alpha(j)}$ for $j = 1, \ldots, d$, one can take the normalizing constants as given in (1.3.13) to verify the max-stability.

Again the transformation technique works: Let G be a stable d.f. with univariate marginals $G_{i(j),\alpha(j)}$ for $j = 1, \ldots, d$. Writing again $T_{i,\alpha} = G_{i,\alpha}^{-1} \circ G_{2,1}$ we obtain that

$$G(T_{i(1),\alpha(1)}(x_1), \ldots, T_{i(d),\alpha(d)}(x_d)), \qquad x_1 < 0, \ldots, x_d < 0, \qquad (2.2.15)$$

defines a stable d.f. with univariate marginal d.f.'s $G_{2,1}$ (the standard exponential d.f. on the negative half-line).

EXAMPLE 2.2.5. Check that G defined by

(i) $$G(x,y) = G_{2,1}(x)G_{2,1}(y)\exp\left(-\frac{x \cdot y}{x + y}\right), \qquad x, y < 0,$$

is an extreme value d.f. with "negative" exponential marginals $G_{2,1}$, and

(ii) $$G(x,y) = G_3(x)G_3(y)\exp[(e^x + e^y)^{-1}]$$

is the corresponding extreme value d.f. with Gumbel marginals.

A bivariate d.f. with marginals $G_{2,1}$ is max-stable if and only if the Pickands (1981) representation holds; that is

$$G(x,y) = \exp\left(\int_{[0,1]} \min(ux, (1-u)y)\, dv(u)\right), \qquad x, y < 0, \quad (2.2.16)$$

where v is any finite measure having the property

$$\int_{[0,1]} u\, dv(u) = \int_{[0,1]} (1-u)\, dv(u) = 1. \qquad (2.2.17)$$

Recall that the marginals are given by $G_1(x) = \lim_{y \to \infty} G(x, y)$ and $G_2(y) = \lim_{x \to \infty} G(x, y)$ and hence (2.2.17) immediately implies that, in fact, the marginals in (2.2.16) are equal to $G_{2,1}$.

If v is the Dirac measure putting mass 2 on the point $\frac{1}{2}$ then $G(x, y) = \exp(\min(x, y))$. If v is concentrated on $\{0, 1\}$ and puts masses 1 on the points 0 and 1 then $G(x, y) = G_{2,1}(x)G_{2,1}(y)$.

The transformation technique immediately leads to the corresponding representations for marginals different from $G_{2,1}$. Check that e.g.

$$G(x, y) = \exp\left(-\int_{[0,1]} \max(ue^{-x}, (1 - u)e^{-y}) \, dv(u)\right) \qquad (2.2.18)$$

is the representation in case of standard Gumbel marginals if again (2.2.17) holds.

For the extension of (2.2.16) to higher dimensions we refer to P.2.10.

Multivariate D.F.'s

This section will be concluded with some general remarks about multivariate d.f.'s.

First recall that multivariate d.f.'s are characterized by the following three properties:

(a) F is right continuous;
   that is, if $\mathbf{x}_n \downarrow \mathbf{x}_0$ then $F(\mathbf{x}_n) \downarrow F(\mathbf{x}_0)$.
(b) F is normed;
   that is, if $\mathbf{x}_n = (x_{n,1}, \ldots, x_{n,d})$ are such that $x_{n,i} \uparrow \infty$ for every $i = 1, \ldots, d$ then $F(\mathbf{x}_n) \uparrow 1$; moreover, if $\mathbf{x}_n \geq \mathbf{x}_{n+1}$ and $x_{n,i} \downarrow -\infty$ for some $i \in \{1, \ldots, d\}$ then $F(\mathbf{x}_n) \to 0$, $n \to \infty$.
(c) F is Δ-monotone;
   that is, for all $\mathbf{a} = (a_1, \ldots, a_d)$ and $\mathbf{b} = (b_1, \ldots, b_d)$,

$$\Delta_{\mathbf{a}}^{\mathbf{b}} F := \sum_{\mathbf{m} \in \{0,1\}^d} (-1)^{d - \sum_{i=1}^d m_i} F(b_1^{m_1} a_1^{1-m_1}, \ldots, b_d^{m_d} a_d^{1-m_d}) \geq 0. \qquad (2.2.19)$$

Recall that if Q is the probability measure corresponding to F then

$$Q(\mathbf{a}, \mathbf{b}] = \Delta_{\mathbf{a}}^{\mathbf{b}} F.$$

From the representations (2.2.16) and (2.2.17) we already know that multivariate extreme value d.f.'s are continuous. However, notice that the continuity is a simple consequence of the fact that the univariate marginal d.f.'s are continuous. This is immediate from inequality (2.2.20).

Lemma 2.2.6. Let F be a d-variate d.f. with univariate marginal d.f.'s F_i, $i = 1, \ldots, d$. Then, for every $\mathbf{x}, \mathbf{y}$,

$$|F(\mathbf{x}) - F(\mathbf{y})| \leq \sum_{i=1}^d |F_i(x_i) - F_i(y_i)|. \qquad (2.2.20)$$

PROOF. Let Q be the probability measure pertaining to F. Given $\mathbf{x}, \mathbf{y}$ we write

$$B_i = \begin{cases} (x_i, y_i] \\ (y_i, x_i] \end{cases} \quad \text{if} \quad \begin{matrix} x_i \le y_i \\ x_i > y_i. \end{matrix}$$

We get

$$|F(\mathbf{x}) - F(\mathbf{y})| = \left| \sum_{i=1}^{d} [F(y_1, \ldots, y_{i-1}, x_i, \ldots, x_d) - F(y_1, \ldots, y_i, x_{i+1}, \ldots, x_d)] \right|$$

$$\le \sum_{i=1}^{d} Q\left(\left(\bigtimes_{j=1}^{i-1} (-\infty, y_j] \right) \times B_i \times \left(\bigtimes_{j=i+1}^{d} (-\infty, x_j] \right) \right)$$

$$\le \sum_{i=1}^{d} |F_i(x_i) - F_i(y_i)|. \qquad \square$$

P.2. Problems and Supplements

1. Let $\xi_1, \ldots, \xi_n$ be i.i.d. random vectors with common continuous d.f. F. For $i \in I := \{j: 1 \le j \le k + 1, r_j - r_{j-1} > 1\}$ define the random vectors $\zeta_{r_{i-1}+1}, \ldots, \zeta_{r_i-1}$ by the original random vectors ξ_i (in the order of their outcome) which have the property

$$R_{r_{i-1}:n} < \|\xi_i - \mathbf{x}_0\|_2 < R_{r_i:n}$$

(with the convention that $R_{r_0:n} = 0$ and $R_{r_{k+1}:n} = \infty$). Then the conditional distribution of $(\zeta_{r_{i-1}+1}, \ldots, \zeta_{r_i-1})$, $i \in I$, given $R_{r_1:n} = z_1, \ldots, R_{r_k:n} = z_k$ is the joint distribution of the independent random vectors $(\eta_{r_{i-1}+1}, \ldots, \eta_{r_i-1})$, $i \in I$, where for every $i \in I$ the components of the vector are i.i.d. random vectors with common distribution equal to the distribution of ξ_1 truncated to $\{\mathbf{x}: z_{i-1} < \|\mathbf{x} - \mathbf{x}_0\|_2 < z_i\}$ with $z_0 = 0$ and $z_{k+1} = \infty$.

2. (Distribution of ψ-order statistics)
   (i) Prove the analogue of (2.1.21) for the kth ψ-order statistic $\tilde{X}_{k:n}$.
   (ii) (Problem) Derive the asymptotic distributions of central and extreme ψ-order statistics $\tilde{X}_{k:n}$.
   (iii) (Problem) Derive the asymptotic distribution of the trimmed mean in (2.1.22) for different centering random vectors ξ_0.

3. Let $A_i \in \mathscr{A}$, $i = 1, \ldots, n$ and $m \in \{0, \ldots, n\}$. With $S_0 = 1$ and

$$S_j = \sum_{1 \le i_1 < \cdots < i_j \le n} P(A_{i_1} \cap \cdots \cap A_{i_j}), \qquad j = 1, \ldots, n$$

   one gets

   (i)

   $$P\left\{ \sum_{i=1}^{n} 1_{A_i} = m \right\} = \sum_{j=m}^{n} \binom{j}{m} (-1)^{j-m} S_j$$

   and

   (ii)

   $$P\left\{ \sum_{i=1}^{n} 1_{A_i} \ge m \right\} = \sum_{j=m}^{n} \binom{j-1}{m-1} (-1)^{j-m} S_j.$$

(iii)
$$P\left\{\sum_{i=1}^{n} 1_{A_i} = m\right\} \begin{array}{l} \le \\ \ge \end{array} \sum_{j=m}^{k} \binom{j}{m}(-1)^{j-m}S_j \quad \text{if} \quad \begin{array}{ll} k-m & \text{even} \\ k-m & \text{odd.} \end{array}$$

4. (i)
$$P\bigcup_{i=1}^{n} A_i = \sum_{j=1}^{n} (-1)^{j-1}S_j.$$

(ii) (Bonferroni inequality)
$$P\bigcup_{i=1}^{n} A_i \begin{array}{l} \le \\ \ge \end{array} \sum_{j=1}^{k} (-1)^{j-1}S_j \quad \text{if} \quad \begin{array}{ll} k & \text{odd} \\ k & \text{even.} \end{array}$$

5. Let $\xi = (\xi_1, \ldots, \xi_d)$ be a random vector with d.f. F.
   (i) Prove that
   $$1 - F(\mathbf{t}) = \sum_{j=1}^{d} (-1)^{j+1} h_j(\mathbf{t})$$

   for $\mathbf{t} = (t_1, \ldots, t_d)$ where
   $$h_j(\mathbf{t}) = \sum_{1 \le i_1 < \cdots < i_j \le d} P\{\xi_{i_1} > t_{i_1}, \ldots, \xi_{i_j} > t_{i_j}\}, \quad j = 1, \ldots, d.$$

(ii) Moreover,
   $$1 - F(\mathbf{t}) \begin{array}{l} \le \\ \ge \end{array} \sum_{j=1}^{k} (-1)^{j+1} h_j(\mathbf{t}) \quad \text{if} \quad \begin{array}{ll} k & \text{odd} \\ k & \text{even.} \end{array}$$

(iii) Find $C > 0$ such that for every positive integer n and $x \in [0, 1]$,
   $$\exp(-nx) - Cn^{-1} \le (1 - x)^n \le \exp(-nx).$$

(iv) Check that
   $$F(\mathbf{t})^n \le \exp\left(\sum_{j=1}^{k} (-1)^j nh_j(\mathbf{t})\right) \quad \text{if } k \text{ even or } k = d.$$

   Moreover, for some universal constant $C > 0$,
   $$F(\mathbf{t})^n \ge \exp\left(\sum_{j=1}^{k} (-1)^j nh_j(\mathbf{t})\right) - Cn^{-1} \quad \text{if } k \text{ odd or } k = d.$$

6. (Uniform Distribution on $A = \{(x, y): x, y \ge 0, x + y \le 1\}$)
   The density $f_{(n,n):n}$ of $(X_{n:n}^{(1)}, X_{n:n}^{(2)})$ under the uniform distribution on A is given by
   $$f_{(n,n):n}(x, y) = 2^n n(xy)^{n-1} 1_A(x, y) + 4n(n - 1)F^{n-2}(x, y)\min(x, 1 - y)\min(1 - x, y)$$
   for $0 \le x, y \le 1$ where F is the underlying d.f. given by
   $$F(x, y) = \begin{cases} 2xy \\ 2xy - (x + y - 1)^2 \end{cases} \quad \text{if} \quad \begin{array}{l} x + y \le 1 \\ x + y \ge 1 \end{array}$$
   for $0 \le x, y \le 1$.

7. Let the underlying density be given by $f(x, y) = x + y$ for $0 \le x, y \le 1$ and $f(x, y) = 0$ otherwise. Then, the d.f. F is given by
   $$F(x, y) = (x^2 y + xy^2)/2, \quad 0 \le x, y \le 1.$$

The density $f_{(n,n):n}$ of $(X_{n:n}^{(1)}, X_{n:n}^{(2)})$ is given by

$$f_{(n,n):n}(x, y) = nF^{n-1}(x, y)f(x, y) + n(n-1)F^{n-1}(x, y)(xy + x^2/2)(xy + y^2/2)$$

for $0 \le x, y \le 1$.

8. (Problem) Let (ξ_1, ξ_2) be a random vector with continuous d.f. F. Denote by F_1 and F_2 the d.f.'s of ξ_1 and ξ_2. Extend (2.2.8) to

$$P\{(X_{n:n}^{(1)}, X_{n:n}^{(2)}) \in B\}$$

$$= \int_B nF^{n-1}(x, y)\, dF(x, y) + \int_B n(n-1)F^{n-2}(x, y)F_1(x|y)F_2(y|x)\, d(F_1 \times F_2)(x, y).$$

9. (i) Prove that a bivariate extreme value d.f. G with standard "negative" exponential marginals (see (2.2.16)) can be written

$$G(x, y) = \exp\left[(x + y)d\left(\frac{y}{x + y}\right)\right], \qquad x, y < 0.$$

where the "dependence" function d is given by

$$d(w) = \int_{[0, 1]} \max(u(1 - w), (1 - u)w)\, dv(u)$$

and v is a finite measure on $[0, 1]$ satisfying condition (2.2.17).
(ii) Check that $d(0) = d(1) = 1$. Moreover, $d \equiv 1$ under independence and $d(w) = \max(1 - w, w)$ under complete dependence.
(iii) Check that $d(w) = 1 - w + w^2$ in Example 2.2.5(ii).

10. A d-variate d.f. with marginals $G_{2,1}$ is max-stable if, and only if,

$$G(\mathbf{x}) = \exp\left(\int_S \min(u_1 x_1, \dots, u_d x_d)\, d\mu(\mathbf{u})\right)$$

where μ is a finite measure on the d-variate unit simplex

$$S := \left\{\mathbf{u}: \sum_{i=1}^d u_i = 1, u_i \ge 0\right\}$$

having the property

$$\int_S u_i\, d\mu(\mathbf{u}) = 1 \quad \text{for } i = 1, \dots, d.$$

(Pickands, 1981; for the proof see Galambos, 1987)

11. (Pickands estimator of dependence function)
(i) Let (η_1, η_2) have the d.f. G as given in P.2.9(i). Prove that for every $t < 0$ and $w \in (0, 1)$,

$$P\left\{\max\left(\frac{\eta_1}{1 - w}, \frac{\eta_2}{w}\right) \le t\right\} = \exp[td(w)].$$

(ii) Let $(\eta_{1,i}, \eta_{2,i}), i = 1, \dots, n$, be i.i.d. random vectors with common d.f. G as given in P.2.9(i). Define

$$\hat{d}_n(w) = \left[n^{-1} \sum_{i=1}^{n} \min\left(\frac{|\eta_{1,i}|}{1-w}, \frac{|\eta_{2,i}|}{w} \right) \right]^{-1}$$

as an estimator of the dependence function d. Prove that

$$E(1/\hat{d}_n(w)) = 1/d(w)$$

and

$$\text{Variance}(1/\hat{d}_n(w)) = 1/(nd(w)^2).$$

12. (Multivariate transformation technique)
Let $\xi = (\xi_1, \ldots, \xi_d)$ be a random vector with continuous d.f. F. We use the notation

$$F_i(\cdot | x_{i-1}, \ldots, x_1) = P(\xi_i \le \cdot | \xi_{i-1} = x_{i-1}, \ldots, \xi_1 = x_1)$$

for the conditional d.f. of ξ_i given $\xi_{i-1} = x_{i-1}, \ldots, \xi_1 = x_1$.
(i) Put

$$T(\mathbf{x}) = (T_1(\mathbf{x}), \ldots, T_d(\mathbf{x}))$$

$$= (F_1(x_1), F_2(x_2|x_1), \ldots, F_d(x_d|x_{d-1}, \ldots, x_1)).$$

Prove that $T_1(\xi), \ldots, T_d(\xi)$ are i.i.d. $(0, 1)$-uniformly distributed r.v.'s.
(ii) Define $T^{-1}(\mathbf{q}) = (S_1(\mathbf{q}), \ldots, S_d(\mathbf{q}))$ by

$$S_1(\mathbf{q}) = F_1^{-1}(q_1)$$

$$S_i(\mathbf{q}) = F_i^{-1}(q_i | S_{i-1}(\mathbf{q}), \ldots, S_1(\mathbf{q})) \quad \text{for } i = 2, \ldots, d.$$

Prove that $P\{T^{-1}(T(\xi)) = \xi\} = 1$. Moreover, if $\eta_1, \ldots, \eta_d$ are i.i.d. $(0, 1)$-uniformly distributed r.v.'s then

$$T^{-1}(\eta_1, \ldots, \eta_d) \quad \text{has the d.f. } F.$$

13. Compute the probability

$$P\{\mathbf{X}_{n:n} = \xi_j \quad \text{for some } j \in \{1, \ldots, n\}\}.$$

Bibliographical Notes

It is likely that Gini and Galvani (1929) were the first who considered the bivariate median defined by the property of minimizing the sum of the deviations w.r.t. the Euclidean norm (see (2.1.11)). This is the "spatial" median as dealt with by Oja and Niinimaa (1985). In that paper the asymptotic performance of a "generalized sample median" as an estimator of the symmetry center of a multivariate normal distribution is investigated. Another notable article related to this is Isogai (1985).

The result concerning the conditional distribution of exceedances (see (2.1.21)) and its extension in P.2.1 was e.g. applied by Moore and Yackel (1977) and Hall (1983) in connection with nearest neighbor density estimators; however, a detailed proof does not seem to exist.

A new insight in the asymptotic, stochastic behavior of the convex hull of data points is obtained by the recent work of Eddy and Gale (1981) and Brozius and de Haan (1987). This approach connects the asymptotic treatment of convex hulls with that of multivariate extremes (w.r.t. the marginal ordering).

For a different representation of the density of multivariate order statistics we refer to Galambos (1975).

In the multivariate set-up we only made use of the transformation technique to transform a multivariate extreme value d.f. to a d.f. with predetermined margins. P.2.12 describes the multivariate transformation technique as developed by Rosenblatt (1952), O'Reilly and Quesenberry (1973), Raoult et al. (1983), and Rüschendorf (1985b). It does not seem to be possible to make this technique applicable to multivariate order statistics (with the exception of concomitants).

Further references concerning multivariate order statistics will be given in Chapter 7.

CHAPTER 3

Inequalities and the Concept of Expansions

In order to obtain rough estimates of probabilities of certain events which involve order statistics, we shall apply exponential bound theorems. These bounds correspond to those for sums of independent r.v.'s. In Section 3.1 such bounds are established in the particular case of order statistics of i.i.d. random variables with common uniform d.f. on $(0, 1)$. This section also contains two applications to moments of order statistics.

Apart from the basic notion of expansions of finite length, Section 3.2 will provide some useful auxiliary results for the treatment of expansions.

In Parts II and III of this volume we shall make extensive use of inequalities for the distance between probability measures. As pointed out before, the variational distance will be central to our investigations. However, we shall also need the Hellinger distance, a weighted L_2-distance (in other words, χ^2-distance), and the Kullback–Leibler distance.

In Section 3.3 our main interest will be focused on bounds for the distance between product measure via the distance between single components. We shall start with some results connected to the Scheffé lemma.

3.1. Inequalities for Distributions of Order Statistics

In this section we deduce exponential bounds for the distributions of order statistics from the corresponding result for binomial r.v.'s. By applying this result we shall also obtain bounds for moments of order statistics.

Let us start with the following well-known exponential bound (see Loève (1963), page 255) for the distribution of sums of i.i.d. random variables $\xi_1, \ldots, \xi_n$ with $E\xi_i = 0$ and $|\xi_i| \leq 1$: We have

$$P\left\{\sum_{i=1}^{n} \xi_i \geq \varepsilon\tau_n\right\} \leq \exp[-t\varepsilon + \tfrac{3}{4}t^2] \qquad (3.1.1)$$

for every $\varepsilon \geq 0$ and $0 \leq t \leq \tau_n$ where $\tau_n^2 = \sum_{i=1}^{n} E\xi_i^2$. Because of relation (1.1.8) between distributions of order statistics and binomial probabilities one can expect that a result similar to (3.1.1) also holds for order statistics in place of sums.

Exponential Bounds for Order Statistics of Uniform R.V.'s

First, our result will be formulated for order statistics $U_{1:n} \leq \cdots \leq U_{n:n}$ of i.i.d. random variables η_i which are uniformly distributed on $(0, 1)$. The transformation technique leads to the general case of i.i.d. random variables ξ_i with common d.f. F.

Lemma 3.1.1. *For every* $\varepsilon \geq 0$ *and* $r \in \{1, \ldots, n\}$ *we have*

$$P\left\{\frac{n^{1/2}}{\sigma}(U_{r:n} - \mu) \begin{array}{l} \leq -\varepsilon \\ \geq \varepsilon \end{array}\right\} \leq \exp\left(-\frac{\varepsilon^2}{3(1 + \varepsilon/(\sigma n^{1/2}))}\right) \qquad (3.1.2)$$

where $\mu = r/(n + 1)$ *and* $\sigma^2 = \mu(1 - \mu)$.

PROOF. (I) First, we prove the upper bound of $P\{(n^{1/2}/\sigma)(U_{r:n} - \mu) \leq -\varepsilon\}$. W.l.g. assume that $\alpha = \mu - \varepsilon\sigma/n^{1/2} > 0$. Otherwise, the upper bound in (3.1.2) is trivial. In particular, $\alpha \in (0, 1)$. By (1.1.8), putting $\varepsilon_0 = (r - n\alpha)/(n\alpha(1 - \alpha))^{1/2}$ and $\xi_i = 1_{(-\infty, \alpha]}(\eta_i) - \alpha$, we get

$$P\{(n^{1/2}/\sigma)(U_{r:n} - \mu) \leq -\varepsilon\} = P\left\{\sum_{i=1}^{n} 1_{(-\infty, \alpha]}(\eta_i) \geq r\right\}$$

$$= P\left\{\sum_{i=1}^{n} \xi_i \geq r - n\alpha\right\} \leq \exp(-\varepsilon_0 t + \tfrac{3}{4}t^2)$$

if $0 \leq t \leq (n\alpha(1 - \alpha))^{1/2}$ where the last step is an application of (3.1.1) to ξ_i and $\varepsilon = \varepsilon_0$. It is easy to see that $t = 2\varepsilon(\alpha(1 - \alpha))^{1/2}/(3\sigma(1 + \varepsilon/(\sigma n^{1/2})))$ fulfills the condition $0 \leq t \leq (n\alpha(1 - \alpha))^{1/2}$. Moreover, $-\varepsilon_0 t + (3/4)t^2 \leq -\varepsilon^2/(3(1 + \varepsilon/ (\sigma n^{1/2})))$ since $\varepsilon_0 \geq \varepsilon\sigma/(\alpha(1 - \alpha))^{1/2}$ and $\alpha(1 - \alpha)/\sigma^2 \leq 1 + \varepsilon/(\sigma n^{1/2})$. This proves the first inequality.

(II) Secondly, recall that $U_{r:n} \stackrel{d}{=} 1 - U_{n-r+1:n}$ (see Example 1.2.2), hence we obtain from part (I) that

$$P\{(n^{1/2}/\sigma)(U_{r:n} - \mu) \geq \varepsilon\}$$

$$= P\{(n^{1/2}/\sigma)(1 - U_{n-r+1:n} - \mu) \geq \varepsilon\}$$

$$= P\{(n^{1/2}/\sigma)(U_{n-r+1:n} - (n - r + 1)/(n + 1)) \leq -\varepsilon\}$$

$$\leq \exp(-\varepsilon^2/3(1 + \varepsilon/(\sigma n^{1/2}))). \qquad \square$$

The right-hand side of (3.1.2) can be written in a simpler form for a special choice of ε. We have

$$P\{[n^{1/2}/\max\{\sigma, (6s(\log n)/n)^{1/2}\}]|U_{r:n} - \mu| \ge (6s \log n)^{1/2}\} \le 2n^{-s}. \quad (3.1.3)$$

Moreover, a crude estimate is obtained by

$$P\{(n^{1/2}/\sigma)|U_{r:n} - \mu| \ge \varepsilon\} \le 2 \exp(-\varepsilon/5), \qquad \varepsilon \ge 0. \quad (3.1.4)$$

Notice that $2\exp(-\varepsilon/5) \ge 1$ whenever $\varepsilon \le 1$. It is apparent that (3.1.4) is weaker than (3.1.3) for small and moderate ε.

As a supplement to Lemma 3.1.1 we shall prove another bound of $P\{U_{r:n} \le \delta\}$ that is sharp for small $\delta > 0$. Note that $P\{U_{r:n} \le \delta\} \downarrow 0$ as $\delta \downarrow 0$, however, this cannot be deduced from Lemma 3.1.1.

Lemma 3.1.2. *If $U_{r:n}$ and μ are as above then for every $\varepsilon \ge 0$:*

$$P\{U_{r:n} \le \mu\varepsilon\} \le e^{1/r}(e\varepsilon)^r/(2\pi r)^{1/2}.$$

PROOF. From Theorem 1.3.2 and Sterling's formula we get

$$P\{U_{r:n} \le \mu\varepsilon\} = [n!/(r-1)!(n-r)!] \int_0^{\mu\varepsilon} x^{r-1}(1-x)^{n-r}\,dx$$

$$\le [n^r/(r-1)!] \int_0^{\mu\varepsilon} x^{r-1}\,dx \le (r^r/r!)\varepsilon^r$$

$$= (\exp(r + \theta(r)/r)/(2\pi r)^{1/2})\varepsilon^r$$

where $|\theta(r)| < 1$. Now the proof can easily be completed. $\qquad\square$

Extension to the General Case

The investigation of exponential bounds for distributions of order statistics will be continued in Section 4.7 where local limit results are established. To prove these results we need, however, the inequalities above. The extension of inequality (3.1.2) to arbitrary d.f.'s is accomplished by means of Corollary 1.2.7. For order statistics $X_{1:n}, \ldots, X_{n:n}$ of n i.i.d. random variables with common d.f. F we have

$$P\left\{[n^{1/2}g(\mu)/\sigma](X_{r:n} - F^{-1}(\mu)) \begin{array}{l} \le -\varepsilon \\ \ge \varepsilon \end{array}\right\} \le P\left\{(n^{1/2}/\sigma)(U_{r:n} - \mu) \begin{array}{l} \le h(-\varepsilon) \\ \ge h(\varepsilon) \end{array}\right\} \quad (3.1.5)$$

where $g(\mu)$ is a nonnegative constant and $h(x) = (n^{1/2}/\sigma)[F(F^{-1}(\mu) + x\sigma/(g(\mu)n^{1/2})) - \mu]$. Thus, upper bounds for the left-hand side of (3.1.5) can be deduced from (3.1.2) by using bounds for $h(-\varepsilon)$ and $h(\varepsilon)$. Notice that if F has a bounded second derivative on a neighborhood of $F^{-1}(\mu)$ then, by taking

$g(\mu) = F'(F^{-1}(\mu))$, we get

$$h(x) = x + O(x^2 \sigma / g^2(\mu) n^{1/2}). \tag{3.1.6}$$

If one needs an upper bound of the left-hand side of (3.1.5) for a fixed sample size n then one has to formulate the smoothness condition for F in a more explicit way so that the capital O in (3.1.6) can be replaced by a constant. This should always be done for the given specific problem.

Inequalities for Moments of Order Statistics

Let $U_{r:n}$, μ and σ be given as in Lemma 3.1.1. From (1.7.5) we know that $E((U_{r:n} - \mu)^2) = \sigma^2/(n + 2)$. The following lemma due to Wellner (1977) gives upper bounds for absolute central moments of $U_{r:n}$.

Lemma 3.1.3. *For every positive integer j and $r \in \{1, \ldots, n\}$:*

$$E|U_{r:n} - \mu|^j \le 2j! 5^j \sigma^j n^{-j/2}.$$

PROOF. By partial integration (or Fubini's theorem) we obtain for every d.f. G with bounded support that

$$\int_0^\infty x^j \, dG(x) = j \int_0^\infty x^{j-1} (1 - G(x)) \, dx$$

so that, by writing $G(x) = P\{(n^{1/2}/\sigma)|U_{r:n} - \mu| \le x\}$, the exponential bound in (3.1.4) applied to $1 - G(x)$ yields

$$E|(n^{1/2}/\sigma)(U_{r:n} - \mu)|^j = \int_0^\infty x^j \, dG(x)$$

$$= j \int_0^\infty x^{j-1} (1 - G(x)) \, dx$$

$$\le 2j \int_0^\infty x^{j-1} \exp(-x/5) \, dx = 2j! 5^j. \qquad \square$$

To prove an expansion of the kth absolute moment $E|X_{r:n}|^k$ (see Section 6.1) we shall use an expansion of $E(|X_{r:n}|^k 1_{\{|X_{r:n}| \le u\}})$ and, furthermore, an upper bound of $E(|X_{r:n}|^k 1_{\{|X_{r:n}| > u\}})$ for appropriately chosen numbers u. Such a bound can again be derived from the exponential bound (3.1.2).

Lemma 3.1.4. *Let $X_{i:n}$ be the ith order statistic of n i.i.d. random variables with common d.f. F. Assume that $E|X_{s:j}| < \infty$ for some positive integers j and $s \in \{1, \ldots, j\}$.*

*Then there exists a constant $C > 0$ such that for every real u and integers n,
k and $r \in \{1, \ldots, n\}$ with $1 \leq i := r - ks \leq m := n - (j + 1)k$ the following two
inequalities hold:*

$$E(|X_{r:n}|^k 1_{\{X_{r:n} \geq u\}}) \leq C^k \frac{b(i, m - i + 1)}{b(r, n - r + 1)} P\{X_{i:m} \gtrless u\}.$$

PROOF. We shall only verify the upper bound of $E(|X_{r:n}|^k 1_{\{X_{r:n} > u\}})$. The other
inequality may be established in a similar way.

Since $X_{r:n} \stackrel{d}{=} F^{-1}(U_{r:n})$ and $F^{-1}(q) > u$ iff $q > F(u)$ we get

$$E(|X_{r:n}|^k 1_{\{X_{r:n} > u\}})$$

$$= E(|F^{-1}(U_{r:n})|^k 1_{\{F^{-1}(U_{r:n}) > u\}})$$

$$= \frac{1}{b(r, n - r + 1)} \int_{F(u)}^1 |F^{-1}(x)|^k x^{r-1}(1 - x)^{n-r} dx$$

$$= \frac{1}{b(r, n - r + 1)} \int_{F(u)}^1 (|F^{-1}(x)| x^s (1 - x)^{j-s+1})^k x^{i-1}(1 - x)^{m-i} dx$$

$$\leq \frac{b(i, m - i + 1)}{b(r, n - r + 1)} C^k P\{U_{i:m} > F(u)\}$$

where C is the constant of (1.7.11). Since $P\{U_{i:m} > F(u)\} = P\{X_{i:m} > u\}$ the
proof is complete. $\qquad\square$

Bounds for the Maximum Deviation of Sample Q.F.'s

This section will be concluded with some simple applications of inequality
(3.1.3) to the sample q.f. Let G_n^{-1} be the sample q.f. based on n i.i.d. $(0, 1)$-
uniformly distributed r.v.'s. The first result concerns the maximum deviation
of G_n^{-1} from the underlying q.f. $G^{-1}(q) = q$.

Lemma 3.1.5. *For every $s > 0$ there exists a constant $B(s) > 0$ such that*

$$P\{|G_n^{-1}(q) - q| > (\log n / n)^{1/2} \kappa(q, s, n) \text{ for some } q \in (0, 1)\} \leq B(s) n^{-s}$$

where $\kappa(q, s, n) = (7(s + 1) \max\{q(1 - q), 7(s + 1)(\log n)/n\})^{1/2}$.

PROOF. By (3.1.3)

$$P\{|G_n^{-1}(q) - q| > (\log n / n)^{1/2} \tilde{\kappa}(q, s, n) \text{ for some } q \in (0, 1)\} \leq 2n^{-s}$$

where $\tilde{\kappa}(q, s, n) = 6s \max\{\sigma(q), (6s(\log n)/n)^{1/2}\} + 1/n$, with $\sigma^2(q) = (r(q)/(n + 1))(1 - r(q)/(n + 1))$ and $r(q) = nq$ if nq is an integer and $r(q) = [nq] + 1$,
otherwise. Now check that $\kappa(q, s, n) \geq \tilde{\kappa}(q, s, n)$ for sufficiently large n. $\qquad\square$

From Lemma 3.1.5 it is immediate that

$$P\left\{\frac{n^{1/2}|G_n^{-1}(q) - q|}{\max\{(q(1-q))^{1/2}, ((\log n)/n)^{1/2}\}}\right.$$

$$\left. > C(s)(\log n)^{1/2} \text{ for some } q \in (0,1)\right\} \le B(s)n^{-s} \qquad (3.1.7)$$

for some constant $C(s) > 0$.

Oscillation of Sample Q.F.

From Theorem 1.6.7 we know that the spacing $U_{s:n} - U_{r:n}$ of n i.i.d. $(0,1)$-uniformly distributed r.v.'s has the same distribution as $U_{s-r:n}$. This relation makes (3.1.3) applicable to spacings, too. The details of the proof can be left to the reader.

Lemma 3.1.6. *For every $s > 0$ there exist constants $B(s) > 0$ and $C(s) > 0$ such that*

$$P\left\{\sup_{0 < p_1 < p_2 < 1} \frac{n^{1/2}|G_n^{-1}(p_2) - G_n^{-1}(p_1) - (p_2 - p_1)|}{\max\{(p_2 - p_1)^{1/2}, ((\log n)/n)^{1/2}\}}\right.$$

$$\left. > C(s)(\log n)^{1/2}\right\} < B(s)n^{-s}.$$

Extensions of (3.1.7) and Lemma 3.1.6 will be proved under appropriate smoothness conditions on the underlying q.f. F^{-1}.

Lemma 3.1.7. *Assume that the q.f. F^{-1} has a derivative on the interval $(q_1 - \varepsilon, q_2 + \varepsilon)$ for some $\varepsilon > 0$. Put*

$$D_1 := \sup_{q_1 - \varepsilon < p < q_2 + \varepsilon} |(F^{-1})'(p)|.$$

Then, for every $s > 0$ there exist constants $B(s, \varepsilon) > 0$ and $C(s, \varepsilon) > 0$ (only depending on s and ε) such that

(i) $$P\left\{\sup_{q_1 \le p \le q_2} \frac{n^{1/2}|F_n^{-1}(p) - F^{-1}(p)|}{\max\{(p(1-p))^{1/2}, ((\log n)/n)^{1/2}\}}\right.$$

$$\left. > C(s, \varepsilon)D_1(\log n)^{1/2}\right\} < B(s, \varepsilon)n^{-s},$$

and if, in addition, the derivative $(F^{-1})'$ satisfies a Lipschitz condition of order $\beta \in [1/2, 1]$, that is,

$$|(F^{-1})'(p_2) - (F^{-1})'(p_1)| \le D_2|p_2 - p_1|^{\beta} \quad \text{for } q_1 - \varepsilon < p_1, p_2 \le q_2 + \varepsilon$$

for some $D_2 > 0$, then

(ii) $P\left\{\displaystyle\sup_{q_1 \le p_1 \le p_2 \le q_2} \frac{n^{1/2}|F_n^{-1}(p_2) - F_n^{-1}(p_1) - (F^{-1}(p_2) - F^{-1}(p_1))|}{\max\{(p_2 - p_1)^{1/2}, ((\log n)/n)^{1/2}\}}\right.$

$$\left. > C(s, \varepsilon)(D_1 + D_2)(\log n)^{1/2}\right\} < B(s, \varepsilon)n^{-s}.$$

PROOF. In view of the quantile transformation we may take the version $F^{-1}(G_n^{-1})$ of the sample q.f. F_n^{-1} where G_n^{-1} is defined as in Lemma 3.1.5. Now, applying (3.1.7) and the inequality

$$|F^{-1}(G_n^{-1}(p)) - F^{-1}(p)| \le D_1|G_n^{-1}(p) - p|$$

we obtain (i).

Using the auxiliary function

$$\psi(y) = F^{-1}(p_2 + y(G_n^{-1}(p_2) - p_2)) - F^{-1}(p_1 + y(G_n^{-1}(p_1) - p_1))$$

we obtain the representation

$$F^{-1}(G_n^{-1}(p_2)) - F^{-1}(G_n^{-1}(p_1)) - [F^{-1}(p_2) - F^{-1}(p_1)]$$

$$= \psi(1) - \psi(0)$$

$$= (F^{-1})'(p_2 + \theta(G_n^{-1}(p_2) - p_2))[G_n^{-1}(p_2) - p_2]$$

$$- (F^{-1})'(p_1 + \theta(G_n^{-1}(p_1) - p_1))[G_n^{-1}(p_1) - p_1]$$

with $0 < \theta < 1$. Now, standard calculations and Lemma 3.1.6 lead to (ii). $\square$

From the proof of Lemma 3.1.7 it is obvious that (i) still holds if F^{-1} satisfies a Lipschitz condition of order 1.

3.2. Expansions of Finite Length

When analyzing higher order approximations one realizes that in many cases these approximations have a similar structure. As an example, we mention the Edgeworth expansions which occur in connection with the central limit theorem. In this case, a normal distribution is necessarily the leading term of the expansion. The concept of Edgeworth expansions is not general enough to cover the higher order approximation as studied in the present context. Apart from the fact that our attention is not restricted to sequences of distributions one also has to consider non-normal limiting distributions in the field of extreme order statistics. Thus, an extension of the notion of Edgeworth expansions to the more general notion of expansions of finite length is necessary.

It is not the purpose of this section to develop a theory for expansions of finite length, and it is by no means necessary to have this notion in mind to understand our results concerning order statistics. However, at least in this section, we want to make clear what is meant by speaking of expansions. Moreover, this notion can serve as a guide for finding higher order approximations.

A Definition of Expansions of Finite Length

Let g_γ and $g_{0,\gamma}, \ldots, g_{m-1,\gamma}$ be real-valued functions with domain A for every index $\gamma \in \Gamma$ so that $\sum_{i=0}^{m-1} g_{i,\gamma}$ can be regarded as an approximation to g_γ.

We say that $g_\gamma, \gamma \in \Gamma$, admits the expansion $\sum_{i=0}^{m-1} g_{i,\gamma}$ of length m arranged in powers of $h(\gamma) > 0$ if for every $x \in A$ there exists a constant $C(x) > 0$ such that

$$\left| g_\gamma(x) - \sum_{i=0}^{j} g_{i,\gamma}(x) \right| \le C(x) h(\gamma)^{j+1}, \qquad \gamma \in \Gamma, \tag{3.2.1}$$

for every $j = 0, \ldots, m - 1$.

The expansion is said to hold uniformly over $A_0 \subset A$ if $\sup\{C(x): x \in A_0\} < \infty$.

If $\sup\{h(\gamma): \gamma \in \Gamma\} < \infty$, we may assume w.l.g. that

$$|g_{i,\gamma}| \le C h(\gamma)^i$$

by putting $C \sup\{1 + h(\gamma): \gamma \in \Gamma\}$ in place of C.

In our context the functions g_γ etc. will mainly be d.f.'s or probability measures.

A significant feature of an expansion of length $m + 1$ is that the first m terms coincide with the expansion of length m. Thus, one always has the choice between the simplicity and the accuracy of an approximation. The first term of an expansion (giving the simplest approximation) is usually known from a limit theorem; an error bound for the limit theorem leads to an expansion of length one. One purpose of asymptotic expansions is to give a better insight into the remainder term of the limit theorem.

EXAMPLE 3.2.1. If a real-valued function f defined on the real line has m bounded derivatives then

$$\left| f(\gamma) - \sum_{i=0}^{m-1} \frac{f^{(i)}(\gamma_0)}{i!} (\gamma - \gamma_0)^i \right| \le C|\gamma - \gamma_0|^m$$

and hence the Taylor expansion $\sum_{i=0}^{m-1} f^{(i)}(\gamma_0)(\gamma - \gamma_0)^i/i!$ of f about γ_0 is an expansion arranged in powers of $h(\gamma) = |\gamma - \gamma_0|$.

EXAMPLE 3.2.2. Let Φ denote the standard normal d.f. and $\varphi = \Phi'$. By noting that $\gamma\varphi(\gamma) = -\varphi'(\gamma)$ one easily obtains by partial integration and by using the induction scheme that

$$1 - \Phi(\gamma) = \frac{\varphi(\gamma)}{\gamma}\left(1 + \sum_{i=1}^{m-1}(-1)^i \frac{1\cdot 3\cdot 5\cdots(2i-1)}{\gamma^{2i}}\right)$$

$$+ (-1)^m 3\cdot 5\cdots(2m-1)\int_\gamma^\infty \frac{\varphi(y)}{y^{2m}}dy \qquad (3.2.2)$$

for every positive integer m and $\gamma > 0$ (where $\sum_{i=1}^0$ equals zero by convention).
An application of (3.2.2) in the cases $m = 1$ and $m = 2$ leads to

$$\varphi(\gamma)(1/\gamma - 1/\gamma^3) \le 1 - \Phi(\gamma) \le \varphi(\gamma)/\gamma \qquad (3.2.3)$$

for $\gamma > 0$.

By means of (3.2.2) we get an expansion of $(1 - \Phi(\gamma))\gamma/\varphi(\gamma)$ in powers of $h(\gamma) = \gamma^{-2}$. We have

$$\left|\frac{(1 - \Phi(\gamma))\gamma}{\varphi(\gamma)} - \left(1 + \sum_{i=1}^{m-1}(-1)^i \frac{1\cdot 3\cdot 5\cdot\ldots\cdot(2i-1)}{\gamma^{2i}}\right)\right| \le C_m\gamma^{-2m}. \quad (3.2.4)$$

Notice that $(1 - \Phi(\gamma))\gamma/\varphi(\gamma)$ cannot be represented by means of the formal series $1 + \sum_{i=1}^\infty(-1)^i 3\cdot 5\cdots(2i-1)/\gamma^{2i}$ since $3\cdot 5\cdots(2i-1)\to\infty$ as $i\to\infty$. However, (3.2.4) provides a useful inequality if γ is large. Moreover, the approximation for $m + 1$ is more accurate than that for m if γ is sufficiently large.

EXAMPLE 3.2.3. A sequence of d.f.'s H_n admits an Edgeworth expansion of length m if

$$\sup_t \left|H_n(t) - \left(\Phi(t) + \varphi(t)\sum_{i=1}^{m-1} n^{-i/2}L_i(t)\right)\right| \le Cn^{-m/2} \qquad (3.2.5)$$

where L_i are polynomials. This is a special case of (3.2.1) with $g_n = H_n$, $g_{0,n} = \Phi$, $g_{i,n} = n^{-i/2}\varphi L_i$ for $i = 1, \ldots, m - 1$ and $h(n) = n^{-1/2}$.

Expansions of Probability Measures

We will primarily be interested in expansions

$$P_{0,\gamma} + \sum_{i=1}^{m-1} v_{i,\gamma}$$

of probability measure P_γ which hold uniformly over all measurable sets. If the probability measure $P_{0,\gamma}$ is the first order approximation to P_γ then the approximation can be improved by adding to $P_{0,\gamma}$ an approximation, say $v_{1,\gamma}$ to $P_\gamma - P_{0,\gamma}$. Since $P_\gamma - P_{0,\gamma}$ is a signed measure with total mass equal to zero it is clear that the set function $v_{1,\gamma}$ will typically also have this property.

Lemma 3.2.4. *Let P_γ and $P_{0,\gamma}$ be probability measures and let $v_{i,\gamma}$ be finite signed measures on a measurable space $(S, \mathscr{B})$.*
If $P_{0,\gamma} + \sum_{i=1}^{m-1} v_{i,\gamma}$ is an expansion of P_γ, $\gamma \in \Gamma$, uniformly over $\mathscr{B}$ arranged

in powers of $h(\gamma)$, $\gamma \in \Gamma$, *then there exists an expansion* $P_{0,\gamma} + \sum_{i=1}^{m-1} \mu_{i,\gamma}$ *such that* $\mu_{i,\gamma}$ *are finite signed measures with* $\mu_{i,\gamma}(S) = 0$. *Moreover, one may take*

$$\mu_{i,\gamma} = \nu_{i,\gamma} - \nu_{i,\gamma}(S), \qquad i = 1, \ldots, m-1 \tag{3.2.6}$$

or

$$\mu_{i,\gamma} = \nu_{i,\gamma} - \nu_{i,\gamma}(S)P_{0,\gamma}, \qquad i = 1, \ldots, m-1. \tag{3.2.7}$$

PROOF. Straightforward by using the fact that $|\sum_{i=1}^{j} \nu_{i,\gamma}(S)| \le Ch(\gamma)^{j+1}$ for some constant $C > 0$. $\qquad\qquad\qquad\qquad\qquad\qquad\qquad\qquad\qquad\qquad\qquad\qquad\square$

According to Lemma 3.2.4. we can assume w.l.g. that the term $\nu_{i,\gamma}$ of an expansion has the property $\nu_{i,\gamma}(S) = 0$. Another useful tool in this context is the following.

Lemma 3.2.5. *Let* P_γ, $P_{0,\gamma}$ *and* $\nu_{i,\gamma}$ *be as in Lemma* 3.2.4. *Suppose that there exists* $C > 0$ *such that*

$$\sup_{B \in \mathscr{B}} \left| P_\gamma(B) - \frac{P_{0,\gamma}(B) + \sum_{i=1}^{j} \nu_{i,\gamma}(B)}{1 + \sum_{i=1}^{j} \nu_{i,\gamma}(S)} \right| \le Ch(\gamma)^{j+1} \tag{3.2.8}$$

and $|\nu_{i,\gamma}(S)| \le Ch(\gamma)^{i}$ *for every* $j = 0, \ldots, m-1$ *and* $\gamma \in \Gamma$ *[where* (3.2.8) *has to hold whenever* $1 + \sum_{i=1}^{j} \nu_{i,\gamma}(S) > 0$]*.*

Then, $P_{0,\gamma} + \sum_{i=1}^{m-1} \mu_{i,\gamma}$, $\gamma \in \Gamma$, *is an expansion of* P_γ, $\gamma \in \Gamma$, *uniformly over* $\mathscr{B}$ *arranged in powers of* $h(\gamma)$ *where* $\mu_{i,\gamma}$ *is inductively defined by*

$$\mu_{i,\gamma} = \nu_{i,\gamma} - \nu_{i,\gamma}(S)P_{0,\gamma} - \sum_{k=1}^{i-1} \nu_{k,\gamma}(S)\mu_{i-k,\gamma}, \qquad i = 1, \ldots, m-1. \tag{3.2.9}$$

PROOF. First notice that from the inequality $|\nu_{i,\gamma}(S)| \le Ch(\gamma)^{i}$ it is immediate by induction over $i = 1, \ldots, m-1$ that

$$|\nu_{i,\gamma}|(S) \le Ch(\gamma)^{i} \tag{3.2.10}$$

where C will be used as a generic constant which only depends on m.

The triangle inequality and (3.2.10) yield

$$\left| P_\gamma - \left(P_{0,\gamma} + \sum_{i=1}^{j} \mu_{i,\gamma} \right) \right| \le Ch(\gamma)^{j+1} + \left| \frac{P_{0,\gamma} + \sum_{i=1}^{j} \nu_{i,\gamma}}{1 + \sum_{i=1}^{j} \nu_{i,\gamma}(S)} - \left(P_{0,\gamma} + \sum_{i=1}^{j} \mu_{i,\gamma} \right) \right|$$

$$\le Ch(\gamma)^{j+1} \left(1 + \left(1 + \sum_{i=1}^{j} \nu_{i,\gamma}(S) \right)^{-1} \right)$$

since

$$P_{0,\gamma} + \sum_{i=1}^{j} \nu_{i,\gamma} - \left(1 + \sum_{i=1}^{j} \nu_{i,\gamma}(S) \right) \left(P_{0,\gamma} + \sum_{i=1}^{j} \mu_{i,\gamma} \right)$$

$$= \sum_{i=1}^{j} \sum_{k=1}^{i-1} \nu_{k,\gamma}(S)\mu_{i-k,\gamma} - \left(\sum_{k=1}^{j} \nu_{k,\gamma}(S) \right) \left(\sum_{i=1}^{j} \mu_{i,\gamma} \right)$$

$$= \sum_{i=j+1}^{2j} \sum_{k=1}^{i-1} \nu_{k,\gamma}(S)\mu_{i-k,\gamma}.$$

Thus, the assertion is proved for those γ for which $h(\gamma)$ is sufficiently small. By (3.2.10) again it can easily be seen that, otherwise, the assertion trivially holds by choosing the constant C sufficiently large. $\square$

By induction over $i = 1, \ldots, m - 1$ it is easy to see that the signed measures $\mu_{i,\gamma}$ in Lemma 3.2.5 already fulfill the condition $\mu_{i,\gamma}(S) = 0$.

Expansions of D.F.'s

An expansion of probability measures which holds uniformly over all measurable sets on the real line yields an expansion

$$P_{0,\gamma}(-\infty, t] + \sum_{i=1}^{m-1} v_{i,\gamma}(-\infty, t]$$

of d.f.'s.

Assume that $P_{0,\gamma} = N_{(0,1)}$, $v_{i,\gamma}$ has a density $\varphi R_{i,\gamma}$ where $R_{i,\gamma}$ is a polynomial and the mass of $v_{i,\gamma}$ is equal to zero. Then, the expansion of the d.f.'s can always be written in the form

$$\Phi(t) + \varphi(t) \sum_{i=1}^{m-1} L_{\gamma,i}(t)$$

where $L_{\gamma,i}$ are polynomials. This is immediate from the following lemma which yields that one can find polynomials $L_{\gamma,i}$ such that $(\varphi L_{\gamma,i})' = \varphi R_{\gamma,i}$.

Lemma 3.2.6. *For every positive integer* k,

$$\varphi(x)\left(x^{2k} - \int x^{2k} \varphi(x) \, dx \right) = \left[-\varphi(x) \sum_{i=1}^{k} a_i x^{2i-1} \right]' \qquad (3.2.11)$$

where $a_k = 1$ *and* $a_i = (2i + 1)a_{i+1}$, $i = 1, \ldots, k - 1$. *Secondly,*

$$\varphi(x) x^{2k-1} = \left[-\varphi(x) \sum_{i=1}^{k} a_i x^{2(i-1)} \right]' \qquad (3.2.12)$$

where $a_k = 1$ *and* $a_i = 2i a_{i+1}$, $i = 1, \ldots, k - 1$.

PROOF. (3.2.11) and (3.2.12) can be proved in a straightforward way. Observe that a_1 in (3.2.11) is given by

$$a_1 = 1 \cdot 3 \cdot 5 \cdot \ldots \cdot (2k - 1) = \int x^{2k} \varphi(x) \, dx. \qquad (3.2.13) \quad \square$$

Two further technical lemmas that provide the basic tools for proving expansions for extreme and central order statistics will be given in Appendix 2.

3.3. Distances of Measures: Convergence and Inequalities

Given the r.v.'s ξ and η with values in a measurable space $(S, \mathcal{B})$ [in our context, S will be the real line or the Euclidean k-space] the variational distance is defined by

$$\sup_{B \in \mathcal{B}} |P\{\xi \in B\} - P\{\eta \in B\}|. \tag{3.3.1}$$

In this sequel, we shall write $\sup_B$ in place of $\sup_{B \in \mathcal{B}}$. Let Q_0 and Q_1 denote the distributions of ξ and η. Then, we write

$$\|Q_0 - Q_1\| = \sup_B |Q_0(B) - Q_1(B)|. \tag{3.3.2}$$

Since the variational distance is difficult to deal with, we shall also introduce related distances as the L_1-distance, the Hellinger distance, a weighted L_2-distance and the Kullback–Leibler distance. These distances will enable us to establish important estimates of the variational distance.

The Variational Distance and the L_1-Distance

Representing the probability measures by their μ-densities f_i [in our context, the f_i are usually Lebesgue-densities] one obtains the following well-known relation between the variational and the L_1-distance.

Lemma 3.3.1.

$$\|Q_0 - Q_1\| = 2^{-1} \int |f_0 - f_1| \, d\mu. \tag{3.3.3}$$

PROOF. Check that

$$\int |f_0 - f_1| \, d\mu = \int (f_0 - f_1) \, d\mu + 2 \int (f_1 - f_0)^+ \, d\mu = 2 \int (f_1 - f_0)^+ \, d\mu$$

where f^+ denotes the positive part of a function f. This implies for $B \in \mathcal{B}$,

$$\int |f_0 - f_1| \, d\mu = 2 \int_{\{f_0 > f_1\}} (f_0 - f_1) \, d\mu \geq 2(Q_0(B) - Q_1(B))$$

with "$=$" for $B = \{f_0 > f_1\}$. Hence

$$\sup_B (Q_0(B) - Q_1(B)) = 2^{-1} \int |f_0 - f_1| \, d\mu.$$

This yields the assertion.   □

The Scheffé Lemma and Related Results

We continue our calculations with some simple results concerning the pointwise convergence of densities and the convergence w.r.t. the L_1-distance.

Lemma 3.3.2. *For every nonnegative integer n, let f_n be the μ-density of the probability measure Q_n. Then,*

$$f_n \underset{n}{\to} f_0 \quad \mu - a.e. \text{ implies } \int |f_n - f_0| \, d\mu \underset{n}{\to} 0.$$

PROOF. We know (compare with the proof above) that

$$\int |f_n - f_0| \, d\mu = 2 \int (f_0 - f_n)^+ \, d\mu. \tag{1}$$

Moreover, $f_0 \geq (f_0 - f_n)^+ \geq 0$ and $(f_0 - f_n)^+ \underset{n}{\to} 0 \ \mu -$ a.e. Therefore, the dominated convergence theorem implies that

$$\int (f_0 - f_n)^+ \, d\mu \underset{n}{\to} 0.$$

This together with (1) yields the assertion. □

A short look at the proof above reveals also that the following extension holds.

Lemma 3.3.3. *Let f_n be a nonnegative, μ-integrable function. If*

$$\limsup_n \int f_n \, d\mu \leq \int f_0 \, d\mu \quad and \quad \lim_n f_n = f_0 \quad \mu - a.e.$$

then

$$\int |f_n - f_0| \, d\mu \underset{n}{\to} 0.$$

It is well known (and easy to show by examples) that the conditions of Lemma 3.3.3 are not necessary for the L_1-convergence of f_n to f_0. We also prove the following stronger version of the Scheffé lemma.

Lemma 3.3.4. *With f_n as in Lemma 3.3.3 the following conditions (i)–(iii) are equivalent:*

(i)
$$\lim_n \int |f_n - f_0| \, d\mu = 0.$$

(ii)
$$\lim_n \int f_n \, d\mu = \int f_0 \, d\mu,$$

and for every subsequence $i(n)$ there exists a subsequence $k(n) = i(j(n))$ such that

$$\lim_n f_{k(n)} = f_0 \quad \mu - a.e.$$

(iii) *For every subsequence $i(n)$ there exists a subsequence $k(n) = i(j(n))$ such that*

$$\limsup_n \int f_{k(n)} \, d\mu \le \int f_0 \, d\mu,$$

and

$$\liminf_n f_{k(n)} \ge f_0 \quad \mu - a.e.$$

PROOF. We prove (i) $\Rightarrow$ (ii) $\Rightarrow$ (iii) $\Rightarrow$ (i).

(i) $\Rightarrow$ (ii): It is immediate that $\lim_n \int f_n \, d\mu = \int f_0 \, d\mu$. Moreover, for every subsequence $i(n)$ there exists a subsequence $k(n) = i(j(n))$ such that

$$\sum_{n=1}^{\infty} \int |f_{k(n)} - f_0| \, d\mu = \int \sum_{n=1}^{\infty} |f_{k(n)} - f_0| \, d\mu < \infty.$$

This implies $\sum_{n=1}^{\infty} |f_{k(n)} - f_0| < \infty$ μ − a.e. and hence $\lim_n f_{k(n)} = f_0$ μ − a.e.

(ii) $\Rightarrow$ (iii): Obvious.

(iii) $\Rightarrow$ (i): It suffices to prove that for every subsequence $i(n)$ there exists a subsequence $k(n) = i(j(n))$ such that

$$\lim_n \int |f_{k(n)} - f_0| \, d\mu = 0.$$

Condition (iii) implies that there exists $k(n) = i(j(n))$ such that

$$\lim_n (f_0 - f_{k(n)})^+ = 0 \quad \mu - a.e.$$

Thus, by repeating the arguments of the proof of Lemma 3.3.2 we obtain the desired conclusion. $\qquad\square$

The following version of the Scheffé lemma will be particularly useful in cases where the measurable space varies with n.

Lemma 3.3.5. *Let g_n and f_n be nonnegative, measurable functions. Assume that $\int g_n \, d\mu_n$, $n = 1, 2, 3, \ldots$ is a bounded sequence, and that $\lim_n \int (g_n - f_n) \, d\mu_n = 0$. Then the following three conditions are equivalent:*

(i)
$$\lim_n \int |g_n - f_n| \, d\mu_n = 0,$$

(ii)
$$\lim_n \int |f_n/g_n - 1| g_n \, d\mu_n = 0,$$

(iii)
$$\lim_n \int_{\{|f_n/g_n - 1| \ge \varepsilon\}} g_n \, d\mu_n = 0 \quad \textit{for every } \varepsilon > 0.$$

PROOF. (i) $\Rightarrow$ (ii) $\Rightarrow$ (iii): Obvious from

$$\int_{\{|f_n/g_n - 1| \geq \varepsilon\}} g_n \, d\mu_n \leq \varepsilon^{-1} \int |f_n/g_n - 1| g_n \, d\mu_n \leq \varepsilon^{-1} \int |g_n - f_n| \, d\mu_n.$$

(iii) $\Rightarrow$ (i): For $\varepsilon > 0$ put $B = B(n, \varepsilon) = \{g_n > 0, |f_n/g_n - 1| < \varepsilon\}$. If (iii) holds then

$$\int_B g_n \, d\mu_n = \int g_n \, d\mu_n - \int_{\{|f_n/g_n - 1| \geq \varepsilon\}} g_n \, d\mu_n \geq \int g_n \, d\mu_n - \varepsilon \qquad (1)$$

for sufficiently large n. Moreover,

$$\left| \int_B f_n \, d\mu_n - \int_B g_n \, d\mu_n \right| \leq \int_B |f_n - g_n| \, d\mu_n = \int_B |f_n/g_n - 1| g_n \, d\mu_n$$

$$\leq \varepsilon \int_B g_n \, d\mu_n. \qquad (2)$$

Combining (1) and (2),

$$\int_B f_n \, d\mu_n \geq \int_B g_n \, d\mu_n - \varepsilon - \varepsilon \int_B g_n \, d\mu_n \geq \int f_n \, d\mu_n - 2\varepsilon - \varepsilon \int_B g_n \, d\mu_n \qquad (3)$$

if n is sufficiently large. By (1)–(3),

$$\int |f_n - g_n| \, d\mu_n \leq \int_B |f_n - g_n| \, d\mu_n + \int f_n \, d\mu_n - \int_B f_n \, d\mu_n$$

$$+ \int g_n \, d\mu_n - \int_B g_n \, d\mu_n \leq 2\varepsilon \int g_n \, d\mu_n + 3\varepsilon$$

if n is sufficiently large. Since ε is arbitrary this implies (i). $\qquad \square$

Finally, Lemma 3.3.5 will be formulated for the particular case of probability measures.

Corollary 3.3.6. *For probability measures Q_n and P_n with μ_n-densities f_n and g_n the following two assertions are equivalent:*

(i) $$\lim_n \int |g_n - f_n| \, d\mu_n = 0,$$

(ii) $$\lim_n P_n\{|f_n/g_n - 1| \geq \varepsilon\} = 0 \quad \text{for every } \varepsilon > 0.$$

The Variational Distance between Product Measures

The aim of the following is to prove estimates of the variational distance between products of probability measures in terms of distances between the single components. Our starting point is an upper bound in terms of the variational distances of the components. The technical details and a generalization of the present result to signed measures can be found in Appendix 3.

Lemma 3.3.7. *For probability measures Q_i and P_i, $i = 1, \ldots, k$,*

$$\left\| \bigtimes_{i=1}^{k} Q_i - \bigtimes_{i=1}^{k} P_i \right\| \leq \sum_{i=1}^{k} \|Q_i - P_i\|. \tag{3.3.4}$$

The following example shows that the inequality is sharp as far as the order of the upper bound is concerned. However, we will realize later that this is not the typical situation.

EXAMPLE 3.3.8. Let Q_t be the uniform distribution on the interval $[0, t]$. We show that for $0 \leq s \leq k$:

$$s + O(s^2) = 1 - \exp(-s) \leq \|Q_1^k - Q_{1/(1-s/k)}^k\| \leq k\|Q_1 - Q_{1/(1-s/k)}\| = s.$$

The two upper bounds are immediate from (3.3.4) and the identity

$$\|Q_1^k - Q_t^k\| = 1 - t^{-k}.$$

This also implies

$$\|Q_1^k - Q_{1/(1-s/k)}^k\| = 1 - (1 - s/k)^k \geq 1 - \exp(-s).$$

The Hellinger Distance and Other Distances

To obtain sharp estimates of the variational distance of product measures we introduce further distances and show their relation to the variational distance. Let again Q_i be a probability measure with μ-density f_i. Put

$$H(Q_0, Q_1) = \left[\int (f_0^{1/2} - f_1^{1/2})^2 \, d\mu \right]^{1/2} \qquad \text{"Hellinger distance"}$$

$$D(Q_0, Q_1) = \left[\int (f_1/f_0 - 1)^2 \, dQ_0 \right]^{1/2} \qquad \text{"}\chi^2 - \text{distance"}$$

$$K(Q_0, Q_1) = \int (-\log f_1/f_0) \, dQ_0. \qquad \text{"Kullback–Leibler distance"}$$

It can be shown that these distances are independent of the particular choice of the dominating measure μ and of the densities f_0 and f_1. Keep in mind that the distances $\|\cdot\|$ and H are symmetrical whereas, this does not hold for the distances D and K.

Notice that $\|Q_0 - Q_1\| \leq 1$ and $H(Q_0, Q_1) \leq 2^{1/2}$. Moreover, $\|Q_0 - Q_1\| = 1$ and $H(Q_0, Q_1) = 2^{1/2}$ if the densities f_0 and f_1 have disjoint supports. We remark that, in literature, $2^{-1/2} H$ is also used as the definition of the Hellinger distance.

The definition of the χ^2-distance will be extended to finite signed measures in Appendix 3.

Check that $H(Q_0, Q_1) \leq (2\|Q_0 - Q_1\|)^{1/2}$ and

$$H(Q_0, Q_1) = \left[2 \left(1 - \int (f_0 f_1)^{1/2} \, d\mu \right) \right]^{1/2}. \qquad (3.3.5)$$

Lemma 3.3.9. (i) $\qquad\qquad \|Q_0 - Q_1\| \leq H(Q_0, Q_1). \qquad (3.3.6)$

(ii) *If Q_1 is dominated by Q_0 then*

$$H(Q_0, Q_1) \leq D(Q_0, Q_1). \qquad (3.3.7)$$

PROOF. Ad (i): (3.3.3) and the Schwarz inequality yield

$$\|Q_0 - Q_1\| = 2^{-1} \int |f_0 - f_1| \, d\mu = 2^{-1} \int |f_0^{1/2} - f_1^{1/2}| |f_0^{1/2} + f_1^{1/2}| \, d\mu$$

$$\leq 2^{-1} \left[\int (f_0^{1/2} - f_1^{1/2})^2 \, d\mu \right]^{1/2} \left[\int (f_0^{1/2} + f_1^{1/2})^2 \, d\mu \right]^{1/2}$$

$$= H(Q_0, Q_1) \left[2 \left(1 + \int (f_0 f_1)^{1/2} \, d\mu \right) \right]^{1/2} \bigg/ 2 \leq H(Q_0, Q_1).$$

Ad (ii): Let f_1 be a Q_0-density of Q_1. We have

$$H(Q_0, Q_1)^2 = \int (1 - f_1^{1/2})^2 \, dQ_0 \leq \int [(1 - f_1^{1/2})(1 + f_1^{1/2})]^2 \, dQ_0$$

$$= D(Q_0, Q_1)^2. \qquad \square$$

Note that (3.3.7) does not hold if the condition that Q_1 is dominated by Q_0 is omitted. Without this condition one can easily prove (use (3.3.5)) that $H(Q_0, Q_1) \leq [2D(Q_0, Q_1)]^{1/2}$.

Under the condition of Lemma 3.3.9 it is clear that $\|Q_0 - Q_1\| \leq D(Q_0, Q_1)$. This inequality can slightly be improved by applying the Schwarz inequality to $\int |1 - f_1| \, dQ_0$. We have

$$\|Q_0 - Q_1\| \leq 2^{-1} D(Q_0, Q_1). \qquad (3.3.8)$$

Another bound for the Hellinger distance (and thus for the variational distance) can be constructed by using the Kullback–Leibler distance. This bound is nontrivial if Q_0 is dominated by Q_1. We have

$$H(Q_0, Q_1) \leq K(Q_0, Q_1)^{1/2}. \qquad (3.3.9)$$

A modification and the proof of this result can be found in Appendix 3. The use of the Kullback–Leibler distance has the following advantages: If f_1/f_0 is the product of several terms, say, g_i then we get an upper bound of $\log(f_1/f_0)$ by summing up estimates of $\log(g_i)$. Moreover, it will be extremely useful in applications that only integrals of bounds of $\log(g_i)$ have to be treated.

Further Inequalities for Distances of Product Measures

In this sequel, it is understood that for every $i = 1, \ldots, k$ the probability measures Q_i and P_i are defined on the same measurable space.

Lemma 3.3.10. (i) $\qquad\qquad H\left(\underset{i=1}{\overset{k}{\times}} Q_i, \underset{i=1}{\overset{k}{\times}} P_i\right) \leq \left(\sum_{i=1}^{k} H(Q_i, P_i)^2\right)^{1/2}.$

(ii) $\qquad\qquad\qquad K\left(\underset{i=1}{\overset{k}{\times}} Q_i, \underset{i=1}{\overset{k}{\times}} P_i\right) = \sum_{i=1}^{k} K(Q_i, P_i).$

(iii) *If, in addition, P_i is dominated by Q_i for $i = 1, \ldots, k$, then*

$$D\left(\underset{i=1}{\overset{k}{\times}} Q_i, \underset{i=1}{\overset{k}{\times}} P_i\right) \leq \exp\left[2^{-1} \sum_{i=1}^{k} D(Q_i, P_i)^2\right]\left(\sum_{i=1}^{k} D(Q_i, P_i)^2\right)^{1/2}.$$

PROOF. Ad (i): Suppose that Q_i and P_i have the μ_i-densities f_i and g_i. By (3.3.5),

$$H\left(\underset{i=1}{\overset{k}{\times}} Q_i, \underset{i=1}{\overset{k}{\times}} P_i\right)^2 = 2\left[1 - \int\left[\prod_{i=1}^{k} (f_i g_i)^{1/2}(x_i)\right]\left(d \underset{i=1}{\overset{k}{\times}} \mu_i\right)(x_1, \ldots, x_k)\right]$$

$$= 2\left[1 - \prod_{i=1}^{k} \int (f_i g_i)^{1/2} d\mu_i\right]$$

$$= 2\left[1 - \prod_{i=1}^{k} (1 - 2^{-1} H(Q_i, P_i)^2)\right] \leq \sum_{i=1}^{k} H(Q_i, P_i)^2$$

where the final inequality is immediate from

$$\prod_{i=1}^{k} (1 - u_i) \geq 1 - \sum_{i=1}^{k} u_i$$

for $0 \leq u_i \leq 1$.

Ad (ii): Obvious.

Ad (iii): Since $D(Q_i, P_i)^2 = \int f_i^2 dQ_i - 1$ where f_i is the Q_i-density of P_i we obtain by straightforward calculations that

$$D\left(\underset{i=1}{\overset{k}{\times}} Q_i, \underset{i=1}{\overset{k}{\times}} P_i\right)^2 = \prod_{i=1}^{k} [1 + D(Q_i, P_i)^2] - 1$$

$$\leq \exp\left[\sum_{i=1}^{k} D(Q_i, P_i)^2\right] - 1$$

$$\leq \exp\left[\sum_{i=1}^{k} D(Q_i, P_i)^2\right]\left(\sum_{i=1}^{k} D(Q_i, P_i)^2\right). \qquad \square$$

Combining the results above we get

Corollary 3.3.11.

$$\left\| \overset{k}{\underset{i=1}{\text{\Large X}}} Q_i - \overset{k}{\underset{i=1}{\text{\Large X}}} P_i \right\| \le \left(\sum_{i=1}^{k} H(Q_i, P_i)^2 \right)^{1/2} \le \left(\sum_{i=1}^{k} D(Q_i, P_i)^2 \right)^{1/2}. \quad (3.3.10)$$

Recall that the second inequality in (3.3.10) only holds if P_i is dominated by Q_i.

If $Q_i = Q$ and $P_i = P$ for $i = 1, \ldots, k$ then by (3.3.4),

$$\|Q^k - P^k\| \le k\|Q - P\|, \quad (3.3.11)$$

and by (3.3.10),

$$\|Q^k - P^k\| \le k^{1/2} H(Q, P). \quad (3.3.12)$$

Thus, if $\|Q - P\|$ and $H(Q, P)$ are of the same order (Example 3.3.8 treats an exceptional case where this is not true) then (3.3.12) provides a more accurate inequality than (3.3.11). From (3.3.10) it is obvious that also $\|Q^k - P^k\| \le k^{1/2} D(Q, P)$. A refinement of this inequality will be studied in Appendix 3.

Distances of Induced Probability Measures

Let Q and P be probability measures on the same measurable space and T a measurable map into another measurable space. Denote by TQ the probability measure induced by Q and T; we have

$$TQ(B) = Q\{T \in B\}.$$

Thus, in this context, the symbol T also denotes a map from one family of probability measures into another family.

The following result is obvious.

Lemma 3.3.12. $\qquad \|TQ - TP\| \le \|Q - P\|.$

To highlight the relevance of this inequality let us consider the statistic $T(X_{r:n}, \ldots, X_{s:n})$ based on the order statistics $X_{r:n}, \ldots, X_{s:n}$. If Q is an approximation to the distribution P of $(X_{r:n}, \ldots, X_{s:n})$ then TQ is an approximation to the distribution TP of $T(X_{r:n}, \ldots, X_{s:n})$. An upper bound for the error $\|TQ - TP\|$ of this approximation is given by $\|Q - P\|$.

In view of the results above it is also desirable to obtain corresponding results for the distances H and D.

Lemma 3.3.13. $\qquad H(TQ, TP) \le H(Q, P).$

PROOF. We repeat in short the arguments in Pitman [1979, (2.2)]. Let g_0 and f_0 be μ-densities of Q and P where w.l.g. μ is a probability measure. If $g_1 \circ T$ and $f_1 \circ T$ are conditional expectations of g_0 and f_0 given T (relative to μ) then g_1 and f_1 are densities of TQ and TP w.r.t. $T\mu$.

Thus, by applying the Schwarz inequality for conditional expectations [see e.g. Chow and Teicher (1978), page 215] to the conditional expectation of $(g_0 f_0)^{1/2}$ given T we obtain in a straightforward way that

$$\int (g_0 f_0)^{1/2} \, d\mu \le \int (g_1 f_1)^{1/2} d(T\mu)$$

which implies the assertion according to (3.3.5).   □

Lemma 3.3.14. *Under the condition that P is dominated by Q,*

$$D(TQ, TP) \le D(Q, P).$$

PROOF. Check that $\int (f_1)^2 \, dTQ \le \int (f_0)^2 \, dQ$ where f_0 is a Q-density of P and f_1 is a TQ-density of TP. Moreover, use arguments similar to those in the proof to Lemma 3.3.13.   □

P.3. Problems and Supplements

1. (i) For every $x > k/n$,

$$P\{U_{k:n} > x\} \le \exp[-n(x - k/n)^2/3].$$

   (ii) Let $x > 0$ be fixed. Then, for every positive integer m we find a constant $C(m, x)$ such that for every n and $k \le n$,

$$P\{U_{k:n} > x\} \le C(m, x)(k/n)^m.$$

2. Let $X_{n:n}$ be the maximum of the r.v.'s $\xi_1, \ldots, \xi_n$. For $k = 1, \ldots, n$:

$$P\{X_{n:n} \le x\} \begin{array}{c} \le \\ \ge \end{array} 1 + \sum_{j=1}^{k} (-1)^j S_j(x) \quad \text{if} \quad \begin{array}{c} k \quad \text{odd} \\ k \quad \text{even} \end{array}$$

with

$$S_j(x) = \sum_{1 \le i_1 < \cdots < i_j \le n} P\{\xi_{i_1} > x, \ldots, \xi_{i_j} > x\}, \qquad j = 1, \ldots, n.$$

3. Prove that

$$N_{(\mu_1, \sigma_1^2)}(B) - N_{(\mu_0, \sigma_0^2)}(B) = (\sigma_0/\sigma_1 - 1) \int_B (1 - ((x - \mu_1)/\sigma_0)^2 \, dN_{(\mu_1, \sigma_0^2)}$$

$$+ ((\mu_1 - \mu_0)/\sigma_0^2) \int_B (x - \mu_0) \, dN_{(\mu_0, \sigma_0^2)}(x)$$

$$+ O[((\mu_1 - \mu_0)/\sigma_0)^2 + (\sigma_0/\sigma_1 - 1)^2].$$

                                                                (see Falk and Reiss, 1988)

4. For $n = 0, 1, 2, \ldots$ let P_n be unimodal probability measures which are dominated by the Lebesgue measure. Then,

$$\|P_n - P_0\| \to 0, n \to \infty \quad \text{iff} \quad P_n \to P_0 \text{ weakly.}$$

(see Ibragimov, 1956, and Reiss, 1973)

5. Let v_1 and v_2 be finite signed measures on a measurable space $(S, \mathscr{B})$. Let $\mathscr{M}$ be a system of $[0, 1]$-valued, $\mathscr{B}$-measurable functions defined on S.
   (i) Define

$$\mathscr{T} = \{\psi^{-1}(t, 1]: t \in [0, 1], \psi \in \mathscr{M}\}.$$

   Then,

$$\sup_{\psi \in \mathscr{M}} \left| \int \psi \, dv_1 - \int \psi \, dv_2 \right| = \sup_{B \in \mathscr{T}} |v_1(B) - v_2(B)|.$$

   (ii) As a special case we obtain for the system $\mathscr{M}$ of all $\mathscr{B}$-measurable, $[0, 1]$-valued functions that

$$\sup_{\psi \in \mathscr{M}} \left| \int \psi \, dv_1 - \int \psi \, dv_2 \right| = \sup_{B \in \mathscr{B}} |v_1(B) - v_2(B)|.$$

   (iii) If $\mathscr{M}$ is the system of all $[0, 1]$-valued, unimodal functions on the real line then

$$\sup_{\psi \in \mathscr{M}} \left| \int \psi \, dv_1 - \int \psi \, dv_2 \right| = \sup_{I \in \mathscr{I}} |v_1(I) - v_2(I)|$$

   where $\mathscr{I}$ is the system of all intervals on the real line.

6. Let F_0 be a d.f. Then for every positive integer m there exists a finite set A_m such that for every d.f. F_1 the following inequality holds:

$$\sup_t |F_0(t) - F_1(t)| \le m^{-1} + \max_{t \in A_m} (0, F_0(t) - F_1(t), F_1(t^-) - F_0(t^-)).$$

7. Prove that

$$\sup_B |P\{\xi_1 \in B\} - P\{\xi_2 \in B\}| \le P\{\xi_1 \ne \xi_2\}.$$

8. Let $Q_{0,n}$ and $Q_{1,n}$ be probability measures such that $Q_{1,n}$ is dominated by $Q_{0,n}$. Find conditions under which
   (i)

$$\|P_{0,n}^n - P_{1,n}^n\| = 2\Phi(2^{-1} n^{1/2} D(P_{0,n}, P_{1,n})) - 1 + O(n^{-1/2}),$$

(Reiss, 1980)

   (ii) the most powerful test of level α for testing $P_{0,n}^n$ against $P_{1,n}^n$ has the power

$$\Phi(\Phi^{-1}(\alpha) + n^{1/2} D(P_{0,n}, P_{1,n})) + O(n^{-1/2}).$$

(Weiss, 1974; Reiss, 1980)

9. (Jensen inequality)
   Let h be a convex function on an open interval I and ξ a r.v. with range I such that ξ and $h(\xi)$ are finitely integrable. Then,

$$h(E\xi) \le Eh(\xi).$$

(see e.g. Ferguson, 1967, Lemma 1, page 76)

10. (Dvoretzky, Kiefer, Wolfowitz inequality)

   Let G_n^{-1} be the sample q.f. in Lemma 3.1.5. Then for every $\varepsilon > 0$,

$$P\left\{ \sup_{q \in (0,1)} n^{1/2} |G_n^{-1} - q| > \varepsilon \right\} = P\left\{ \sup_{q \in (0,1)} n^{1/2} |G_n - q| > \varepsilon \right\} \leq C \exp[-2\varepsilon^2]$$

   for some $C > 0$.

                                                                 (see e.g. Serfling, 1980, page 59)

Bibliographical Notes

This chapter is not central to our considerations and so it suffices to only make some short remarks.

Exponential bounds for order statistics related to (3.1.2) have been discovered and successfully applied by different authors (e.g. Reiss (1974a, 1975a), Wellner (1977)).

The upper bound for the variational distance using the Kullback–Leibler distance was established by Hoeffding and Wolfowitz (1958). In this context we also refer to Ikeda (1963, 1975) and Csiszár (1975). The upper bound for the variational distance between products of probability measures by using the variational distance between the single components was frequently proved in various articles, nevertheless, this inequality does not seem to be well known. It was established by Hoeffding and Wolfowitz (1958) and generalized by Blum and Pathak (1972) and Sendler (1975). The extension to signed measures (see Lemma A.3.3) was given in Reiss (1981b). Investigations along these lines allowing a deviation from the independence condition are carried out by Hillion (1983).

PART II

ASYMPTOTIC THEORY

CHAPTER 4

Approximations to Distributions of Central Order Statistics

Under weak conditions on the underlying d.f. it can be proved that central (as well as intermediate) order statistics are asymptotically normally distributed. This result easily extends to the case of the joint distribution of a fixed number of central order statistics. In Section 4.1 we shall discuss some conditions which yield the weak and strong asymptotic normality of central order statistics.

Expansions of distributions of single central order statistics will be established in Section 4.2. The leading term in such an expansion is the normal distribution, whereas, the higher order terms are given by integrals of polynomials w.r.t. the normal distribution. These expansions differ from the well-known Edgeworth expansions for distributions of sums of independent r.v.'s in the way that the higher order terms do not only depend on the sample size n but also on the index r of the order statistic. In the particular case of sample quantiles the accuracy of the normal approximation is shown to be of order $O(n^{-1/2})$.

In Section 4.3 it is proved that the usual normalization of joint distributions of order statistics makes these distributions asymptotically independent of the underlying d.f. This result still holds under conditions where the asymptotic normality is not valid.

In Section 4.4 we give a detailed description of the multivariate normal distribution which will serve as an approximation to the joint distribution of central order statistics.

Combining the results of the Sections 4.3 and 4.4, the asymptotic normality and expansions of the joint distribution of order statistics $X_{r_1:n}, \ldots, X_{r_k:n}$ (with $0 = r_0 < r_1 < \cdots < r_k < r_{k+1} = n + 1$) are proven in Section 4.5. It is shown that the accuracy of this approximation is of order

$$O\left(\sum_{i=1}^{k+1} (r_i - r_{i-1})^{-1}\right)^{1/2}$$

under weak regularity conditions. These approximations again hold w.r.t. the variational distance.

Some supplementary results concerning the d.f.'s of order statistics and moderate deviations are collected in the Sections 4.6 and 4.7.

4.1. Asymptotic Normality of Central Sequences

Convergence in Distribution of a Single Order Statistic

To begin with, let us consider the special case of order statistics $U_{1:n} \leq U_{2:n} \leq \cdots \leq U_{n:n}$ of n i.i.d. $(0, 1)$-uniformly distributed r.v.'s $\eta_1, \ldots, \eta_n$. If $r(n) \to \infty$ and $n - r(n) \to \infty$ as $n \to \infty$ then one can easily show that the order statistics $U_{r(n):n}$ (if appropriately normalized) converge in distribution to a standard normal r.v. as $n \to \infty$. Thus, with Φ denoting the standard normal d.f., we have

$$P\{a_{r(n),n}^{-1}(U_{r(n):n} - b_{r(n),n}) \leq t\} \to \Phi(t), \qquad n \to \infty, \tag{4.1.1}$$

for every t where $a_{r,n} = (r(n - r + 1))^{1/2}/(n + 1)^{3/2}$ and $b_{r,n} = r/(n + 1)$.

Since Φ is continuous we also know that the convergence in (4.1.1) holds uniformly in t. In this sequel, we prefer to write $a(n)$ and $b(n)$ instead of $a_{r(n),n}$ and $b_{r(n),n}$, thus suppressing the dependence on $r(n)$.

If $(r(n)/n - q) = o(n^{-1/2})$ for some $q \in (0, 1)$—a condition which is e.g. satisfied in the case of sample q-quantiles—another natural choice of the constants $a(n)$ and $b(n)$ is $a(n) = (q(1 - q))^{1/2}/n^{1/2}$ and $b(n) = q$.

Applying (1.1.8) we obtain

$$P\{a(n)^{-1}(U_{r(n):n} - b(n)) \leq t\}$$

$$= P\left\{-\sum_{i=1}^{n} [1_{(-\infty, \rho(n,t)]}(\eta_i) - \rho(n,t)] \leq -r(n) + n\rho(n,t)\right\} \tag{4.1.2}$$

where $\rho(n, t) = b(n) + ta(n)$. Since $(-r(n) + n\rho(n,t))/[n\rho(n,t)(1 - \rho(n,t))]^{1/2} \to t$ as $n \to \infty$, the convergence to $\Phi(t)$ is immediate from the central limit theorem for a triangular array of i.i.d. random variables (or some other appropriate limit theorem for binomial r.v.'s). It is easy to see that this method also applies to other r.v.'s. However, to extend (4.1.1) to other cases we shall follow another standard device, namely, to use the transformation technique.

If $X_{1:n} \leq X_{2:n} \leq \cdots \leq X_{n:n}$ are the order statistics of n i.i.d random variables with d.f. F then, according to Corollary 1.2.7, $P\{X_{r(n):n} \leq t\} = P\{U_{r(n):n} \leq F(t)\}$ and hence by (4.1.1),

$$P\{a'(n)^{-1}(X_{r(n):n} - b'(n)) \le t\} = P\{U_{r(n):n} \le F(b'(n) + ta'(n))\}$$
$$= \Phi[a(n)^{-1}[F(b'(n) + ta'(n)) - b(n)]] + o(1) = \Phi(t) + o(1)$$

(4.1.3)

if $a'(n)$ and $b'(n)$ are chosen so that

$$a(n)^{-1}[F(b'(n) + ta'(n)) - b(n)] \to t, \qquad n \to \infty.$$

Our first example concerns central order statistics.

EXAMPLE 4.1.1. Let $q \in (0, 1)$ be fixed. Assume that F is differentiable at $F^{-1}(q)$ and $F'(F^{-1}(q)) > 0$. If $n^{1/2}(r(n)/n - q) \to 0$, $n \to \infty$, then

$$P\left\{\frac{n^{1/2} F'(F^{-1}(q))}{(q(1 - q))^{1/2}}(X_{r(n):n} - F^{-1}(q)) \le t\right\} \to \Phi(t), \qquad n \to \infty, \quad (4.1.4)$$

for every t. This is immediate from (4.1.3) by taking $a(n) = (q(1 - q))^{1/2}/n^{1/2}$, $b(n) = q$, $a'(n) = a(n)/F'(F^{-1}(q))$, and $b'(n) = F^{-1}(q)$.

As a special case we have

$$P\left\{\frac{n^{1/2} F'(F^{-1}(q))}{(q(1 - q))^{1/2}}(F_n^{-1}(q) - F^{-1}(q)) \le t\right\} \to \Phi(t), \qquad n \to \infty. \quad (4.1.5)$$

The next example deals with upper intermediate order statistics.

EXAMPLE 4.1.2. Assume that $n - r(n) \to \infty$ and $r(n)/n \to 1$ as $n \to \infty$. Moreover, assume that $\omega(F) < \infty$ and that F has a derivative, say, f on the interval $(\omega(F) - \varepsilon, \omega(F))$ for some $\varepsilon > 0$ where f is uniformly continuous and bounded away from zero. These conditions are e.g. fulfilled for uniform r.v.'s. Then,

$$P\left\{\frac{(n + 1)^{3/2} f\left(F^{-1}\left(\dfrac{r(n)}{n + 1}\right)\right)}{(r(n)(n - r(n) + 1))^{1/2}}\left(X_{r(n):n} - F^{-1}\left(\frac{r(n)}{n + 1}\right)\right) \le t\right\}$$

$$\to \Phi(t), \qquad n \to \infty, \quad (4.1.6)$$

for every t. The proof is straightforward and can be left to the reader.

When treating intermediate order statistics the underlying d.f. F has to satisfy certain regularity conditions on a neighborhood of $\alpha(F)$ or $\omega(F)$. From this point of view intermediate order statistics are connected with extreme order statistics. The extreme value theory will provide conditions better tailored to this situation than those stated in Example 4.1.2 (see Theorem 5.1.7).

The Joint Asymptotic Normality

In a second step, consider the joint distribution of k order statistics where $k \ge 1$ is fixed. Our arguments above can easily be extended to the case of joint distributions. Here we shall restrict our attention to an extension of Example 4.1.1.

Theorem 4.1.3. *Let $0 < q_1 < q_2 < \cdots < q_k < 1$ be fixed. Assume that F is differentiable at $F^{-1}(q_i)$ and that $f(F^{-1}(q_i)) > 0$ for $i = 1, \ldots, k$ where $f = F'$. Then, if $(r(n, i)/n - q_i) = o(n^{-1/2})$ for every $i = 1, \ldots, k$ then*

$$P\{(n^{1/2}f(F^{-1}(q_i))(X_{r(n,i):n} - F^{-1}(q_i)))_{i=1}^k \leq \mathbf{t}\} \to \Phi_\Sigma(\mathbf{t}), \qquad n \to \infty, \quad (4.1.7)$$

for every $\mathbf{t} = (t_1, \ldots, t_k)$ where Φ_Σ is the d.f. of the k-variate normal distribution with mean vector zero and covariances $q_i(1 - q_j)$ for $1 \leq i \leq j \leq k$. As a special case we have

$$P\{(n^{1/2}f(F^{-1}(q_i))(F_n^{-1}(q_i) - F^{-1}(q_i)))_{i=1}^k \leq \mathbf{t}\} \to \Phi_\Sigma(\mathbf{t}), \qquad n \to \infty. \quad (4.1.8)$$

Convergence w.r.t. the Variational Distance

One of the advantages of the representation (4.1.2) is that one can treat the asymptotic behavior of the distribution of order statistics whenever a limit theorem for the r.v.'s $\sum_{i=1}^n 1_{(-\infty, \rho(n,t)]}(\eta_i)$ is at hand. The disadvantage of this approach is that the convergence cannot be proved in a stronger sense since we have to deal with discrete r.v.'s although the order statistics have a continuous d.f.

Another well-known method tackles this problem in a successful way. Let us return to the distribution of a single order statistic $U_{r(n):n}$. In the i.i.d. case we know the explicit form of the density. By showing that the density of $a(n)^{-1}(U_{r(n):n} - b(n))$ converges pointwise to the standard normal density (compare with (1.3.9)) we know from the Scheffé lemma that the convergence of the distributions holds w.r.t. the variational distance; that is

$$\sup_B |P\{a(n)^{-1}(U_{r(n):n} - b(n)) \in B\} - N_{(0,1)}(B)| \to 0, \qquad n \to \infty, \quad (4.1.9)$$

where $N_{(0,1)}$ denotes the standard normal distribution.

Notice that (4.1.9) is in fact stronger than (4.1.1) since (4.1.1) can be written

$$\sup_t |P\{a(n)^{-1}(U_{r(n):n} - b(n)) \in (-\infty, t]\} - N_{(0,1)}(-\infty, t]| \to 0, \qquad n \to \infty.$$

Next, the problem arises to extend (4.1.9) to a certain class of d.f.'s F. This is again possible by using the transformation technique.

Theorem 4.1.4. (i) *Let $q \in (0, 1)$ be fixed. Assume that F has a derivative, say, f on the interval $(F^{-1}(q) - \varepsilon, F^{-1}(q) + \varepsilon)$ for some $\varepsilon > 0$. Moreover, assume that f is continuous at $F^{-1}(q)$ and that $f(F^{-1}(q)) > 0$. Then, if $r(n)/n \to q$ as $n \to \infty$,*

$$\sup_B \left| P\left\{ \frac{(n+1)^{3/2} f\!\left(F^{-1}\!\left(\dfrac{r(n)}{n+1} \right) \right)}{(r(n)(n-r(n)+1))^{1/2}} \left[X_{r(n):n} - F^{-1}\!\left(\frac{r(n)}{n+1} \right) \right] \in B \right\} - N_{(0,1)}(B) \right| \to 0,$$

$$n \to \infty. \quad (4.1.10)$$

(ii) *Moreover, if* $(r(n)/n - q) = o(n^{-1/2})$ *then*

$$\sup_B \left| P\left\{ \frac{n^{1/2} f(F^{-1}(q))}{(q(1-q))^{1/2}} (X_{r(n):n} - F^{-1}(q)) \in B \right\} - N_{(0,1)}(B) \right| \to 0, \qquad n \to \infty.$$

$$(4.1.11)$$

(iii) *(4.1.10) also holds under the conditions of Example 4.1.2.*

Before sketching the proof of Theorem 4.1.4 let us examine an example which shows that we have to impose stronger regularity conditions on the underlying d.f. F than those in Example 4.1.1 to guarantee the convergence w.r.t. the variational distance.

EXAMPLE 4.1.5. Let F have the density

$$f = 1_{[-1/2, 0]} + \sum_i \frac{2i+1}{i+1} 1_{[1/(2i+1), 1/2i)}$$

where the summation runs over all positive integers i. By verifying the conditions of Example 4.1.1 we shall obtain that the d.f.'s of the standardized sample medians weakly converge to the standard normal d.f. Φ. Since

$$\sum_{i=1}^n \frac{2i+1}{i+1} \left(\frac{1}{2i} - \frac{1}{2i+1} \right) = \frac{n}{2(n+1)} \qquad (4.1.12)$$

it is easily seen that $\int f(x)\, dx = 1$. By (4.1.12),

$$F\left(\frac{1}{2n+1} \right) = F\left(\frac{1}{2(n+1)} \right) = \frac{1}{2} + \frac{1}{2(n+1)},$$

and hence, for every positive integer n

$$F(x) - \frac{1}{2} = \begin{cases} \dfrac{1}{2(n+1)} & x \in \left[\dfrac{1}{2(n+1)}, \dfrac{1}{2n+1} \right] \\[2ex] \dfrac{1}{2(n+1)} + \dfrac{2n+1}{n+1}\left(x - \dfrac{1}{2n+1} \right) & x \in \left[\dfrac{1}{2n+1}, \dfrac{1}{2n} \right]. \end{cases}$$

This implies that $x - x^2 \le F(x) - 1/2 \le x$ for $|x| \le 1/2$ showing that F is differentiable at $F^{-1}(1/2) = 0$ and $F^{(1)}(0) = 1$. Thus, by Example 4.1.1,

$$P\{2n^{1/2} X_{[n/2]:n} \le t\} \to \Phi(t), \qquad n \to \infty, \text{ for every } t,$$

which proves the weak convergence. On the other hand,

$$P\{2n^{1/2} X_{[n/2]:n} \in B_n\} = 0 < \liminf_k N_{(0,1)}(B_k) \qquad (4.1.13)$$

for every n where $B_n = \bigcup_i ((2n^{1/2}/2(i+1)), (2n^{1/2}/(2i+1)))$ with i taken over all positive integers.

To prove (4.1.13) verify that the Lebesgue measure of $B_n \cap (0,1)$ is $\geq \frac{1}{4}$ and that $f(x/2n^{1/2}) = 0$ for $x \in B_n$.

The proof of Theorem 4.1.4 starts with the representation

$$a'(n)^{-1}(X_{r(n):n} - b'(n)) \overset{d}{=} T_n[a(n)^{-1}(U_{r(n):n} - b(n))]$$

where $T_n(x) = a'(n)^{-1}[F^{-1}(b(n) + xa(n)) - b'(n)]$. According to (4.1.9)

$$\sup_B |P\{a'(n)^{-1}(X_{r(n):n} - b'(n)) \in B\} - P\{T_n(\eta) \in B\}| \to 0 \qquad (4.1.14)$$

as $n \to \infty$ where η is a standard normal r.v.

To complete the proof of Theorem 4.1.4 it suffices to examine functions of standard normal r.v.'s. Denote by S_n the inverse of T_n. Under appropriate regularity conditions, $S'_n(\varphi \circ S_n)$ is the density of $T_n(\eta)$. If $S_n(x) \to x$ and $S'_n(x) \to 1$ as $n \to \infty$ for every x then $S'_n(\varphi \circ S_n) \to \varphi$, $n \to \infty$. Therefore, the Scheffé lemma implies the convergence to the standard normal distribution w.r.t. the variational distance.

This idea will be made rigorous within some general framework. The following lemma should be regarded as a useful technicality.

Lemma 4.1.6. *Let $Y_{i:n}$ be the order statistics of n i.i.d random variables with common continuous d.f. F_0 and $X_{i:n}$ be the order statistics of n i.i.d. random variables with d.f. F_1. Let h and $g(h \circ G)$ be probability densities where h is assumed to be continuous at x for almost all x. Then, if*

$$\sup_B \left| P\{a(n)^{-1}(Y_{r(n):n} - b(n)) \in B\} - \int_B h(x)\,dx \right| \to 0, \qquad n \to \infty, \quad (4.1.15)$$

we have

$$\sup_B \left| P\{a'(n)^{-1}(X_{r(n):n} - b'(n)) \in B\} - \int_B g(x)h(G(x))\,dx \right| \to 0, \qquad n \to \infty,$$

$$(4.1.16)$$

provided the functions S_n defined by

$$S_n(x) = a(n)^{-1}[F_0^{-1}(F_1(b'(n) + xa'(n))) - b(n)]$$

are

(a) *strictly increasing and absolutely continuous on intervals $(\alpha(n), \beta(n))$ where $\alpha(n) \to -\infty$ and $\beta(n) \to \infty$, and*

(b) *$S_n(x) \to G(x)$ and $S'_n(x) \to g(x)$ as $n \to \infty$ for almost all x.*

PROOF. Write $T_n(x) = a'(n)^{-1}[F_1^{-1}(F_0(b(n) + xa(n))) - b'(n)]$. Since F_0 is continuous we obtain from Corollary 1.2.6 that

$$P\{a'(n)^{-1}(X_{r(n):n} - b'(n)) \in B\} = P\{T_n[a(n)^{-1}(Y_{r(n):n} - b(n))] \in B\}$$

and hence condition (4.1.15) yields

$$\sup_B \left| P\{a'(n)^{-1}(X_{r(n):n} - b'(n)) \in B\} - \int_B g(x)h(G(x))\,dx \right| \tag{4.1.17}$$

$$\leq \sup_B \left| \int_{\{T_n \in B\}} h(x)\,dx - \int_B g(x)h(G(x))\,dx \right| + o(n^0).$$

The image of $(\alpha(n), \beta(n))$ under S_n, say, J_n is an open interval, and $T_n|J_n$ is the inverse of $S_n|(\alpha(n), \beta(n))$. By P.1.11,

$$\int_{\{T_n \in B\}} h(x)\,dx = \int_B h_n(x)\,dx \tag{4.1.18}$$

for every Borel set $B \subset (\alpha(n), \beta(n))$ where $h_n = S_n'(h \circ S_n)1_{(\alpha(n), \beta(n))}$. Notice that w.l.g. S_n' can be assumed to be measurable. Since $\int h_n(x)\,dx \leq 1$ and $h_n \to g(h \circ G)$ almost everywhere the Scheffé lemma 3.3.2 yields

$$\sup_B \left| \int_B h_n(x)\,dx - \int_B g(x)h(G(x))\,dx \right| \to 0, \qquad n \to \infty.$$

This together with (4.1.18) yields

$$\sup_B \left| \int_{\{T_n \in B\}} h(x)\,dx - \int_B g(x)h(G(x))\,dx \right| \to 0, \qquad n \to \infty. \tag{4.1.19}$$

Combining (4.1.17) and (4.1.19) the proof is completed.   □

Whereas the constants $a(n)$ and $b(n)$ are usually predetermined the constants $a'(n)$ and $b'(n)$ should be chosen in a way such that S_n fulfills the required conditions. If $G(x) = x$ and $g(x) = 1$ (that is, the limiting expressions in (4.1.15) and (4.1.16) are equal) then a natural choice of the constants $a'(n)$ and $b'(n)$ is

$$b'(n) = F_1^{-1}(F_0(b(n))) \quad \text{and} \quad a'(n) = a(n)/(F_0^{-1} \circ F_1)'(b'(n)). \tag{4.1.20}$$

Then $S_n(0) = 0$ and $S_n'(0) = 1$ so that a Taylor expansion of S_n about 0 yields that $S_n(x)$ is approximately equal to x in a neighborhood of zero.

Now the proof of Theorem 4.1.4 will be a triviality.

PROOF OF THEOREM 4.1.4. We shall only prove (4.1.10) since (4.1.11) and (iii) follow in an analogous way.

Lemma 4.1.6 will be applied to F_0 being the uniform d.f. on $(0, 1)$, $F_1 = F$, $a(n) = (r(n)(n - r(n) + 1))^{1/2}/(n + 1)^{3/2}$, $b(n) = r(n)/(n + 1)$, $h = \varphi$, $g = 1$ and $G(x) = x$. (4.1.15) holds according to (4.1.9). Moreover, choose $b'(n) = F^{-1}(b(n))$ and $a'(n) = a(n)/f(b'(n))$. Since f is continuous at $F^{-1}(q)$ and $f(F^{-1}(q)) > 0$ we know that f is strictly positive on an interval $(F^{-1}(q) - \kappa,$

$F^{-1}(q) + \kappa)$ for some $\kappa > 0$. This implies that $S_n = a(n)^{-1}[F(b'(n) + xa'(n)) - b(n)]$ is strictly increasing and absolutely continuous on the interval $(-\kappa/2a'(n), \kappa/2a'(n))$, eventually, and hence condition (a) in Lemma 4.1.6 is satisfied. It is straightforward to verify condition (b). The proof is complete.   □

4.2. Expansions: A Single Central Order Statistic

The starting point for our study of expansions of distributions of central order statistics will be an expansion of the distribution of an order statistic $U_{r:n}$ of i.i.d. $(0, 1)$-uniformly distributed r.v.'s. The leading term in the expansion will be the standard normal distribution $N_{(0, 1)}$. The expansion will be ordered in powers of $(n/r(n - r))^{1/2}$. This shows that the accuracy of the approximation by $N_{(0, 1)}$ is bad if r or $n - r$ is small. The quantile transformation will lead to expansions in the case of order statistics of other r.v.'s.

Order Statistics of Uniform R.V.'s

For positive integers n and $r \in \{1, \ldots, n\}$ put $a_{r,n}^2 = r(n - r + 1)/(n + 1)^3$ and $b_{r,n} = r/(n + 1)$. Recall from Section 1.7 that $b_{r,n}$ and $a_{r,n}$ are the expectation and, approximately, the standard deviation of $U_{r:n}$.

Theorem 4.2.1. *For every positive integer m there exists a constant $C_m > 0$ such that for every n and $r \in \{1, \ldots, n\}$,*

$$\sup_B \left| P\{a_{r,n}^{-1}(U_{r:n} - b_{r,n}) \in B\} - \int_B \left(1 + \sum_{i=1}^{m-1} L_{i,r,n}\right) dN_{(0, 1)} \right|$$

$$\leq C_m(n/r(n - r))^{m/2} \quad (4.2.1)$$

where $L_{i,r,n}$ is a polynomial of degree $\leq 3i$.

PROOF. Throughout' this proof, the indices r and n will be suppressed. Moreover, C will be used as a generic constant which only depends on m. Put $\alpha = r$ and $\beta = n - r + 1$. From Theorem 1.3.2 it is immediate that the density of

$$a_{r,n}^{-1}(U_{r:n} - b_{r,n}) = ((\alpha + \beta)^{3/2}/(\alpha\beta)^{1/2})(U_{r:n} - \alpha/(\alpha + \beta))$$

is of the form ρg where ρ is a normalizing constant and

$$g(x) = [1 + (\beta/(\alpha + \beta)\alpha)^{1/2}x]^{\alpha-1}[1 - (\alpha/(\alpha + \beta)\beta)^{1/2}x]^{\beta-1}$$

if $-((\alpha + \beta)\alpha/\beta)^{1/2} < x < ((\alpha + \beta)\beta/\alpha)^{1/2}$. Notice that $\min[(\alpha + \beta)\alpha/\beta, (\alpha + \beta)\beta/\alpha] \geq \alpha\beta/(\alpha + \beta)$. Corollary A.2.3 yields

$$\left|\exp(x^2/2)g(x) - \left(1 + \sum_{i=1}^{m-1} h_i\right)\right| \le C[(\alpha + \beta)/\beta\alpha]^{m/2}(|x|^m + |x|^{3m}) \quad (1)$$

for $|x| \le [\alpha\beta/(\alpha + \beta)]^{1/6}$ where h_i are the polynomials as described in Corollary A.2.3. Define the signed measure v by

$$v(B) = \int_B \left(1 + \sum_{i=1}^{m-1} h_i\right) dN_{(0,1)} \Big/ \left[\int \left(1 + \sum_{i=1}^{m-1} h_i\right) dN_{(0,1)}\right].$$

W.l.g., by choosing the constant C sufficiently large, we may assume that the term $\int(1 + \sum_{i=1}^{m-1} h_i) dN_{(0,1)}$ is bounded away from zero. By (1), the exponential bound (3.1.2) and Lemma A.3.2 applied to the functions g and $f = \exp(-x^2/2)(1 + \sum_{i=1}^{m-1} h_i)$ and to the set $B = \{x: |x| \le [\alpha\beta/(\alpha + \beta)]^{1/6}\}$ we obtain

$$\sup_A |P\{((\alpha + \beta)^{3/2}/(\alpha\beta)^{1/2})(U_{r:n} - \alpha/(\alpha + \beta)) \in A\} - v(A)|$$

$$\le C((\alpha + \beta)/(\alpha\beta))^{m/2} \int (|x|^m + |x|^{3m}) dN_{(0,1)} \Big/ \int \left(1 + \sum_{i=1}^{m-1} h_i\right) dN_{(0,1)}$$

$$+ P\{((\alpha + \beta)^{3/2}/(\alpha\beta)^{1/2})(U_{r:n} - \alpha/(\alpha + \beta)) \notin B\} + |v|(B^c)$$

$$\le C((\alpha + \beta)/(\alpha\beta))^{m/2}.$$

Now the assertion is immediate from Lemma 3.2.5.   □

Addendum 4.2.2. *The application of Lemma 3.2.5 in the proof of Theorem 4.2.1 gives a more precise information about the polynomials $L_{i,r,n}$.*
(i) *The polynomials $L_{i,r,n}$ are recursively defined by*

$$L_{i,r,n} = h_{i,r,n} - \int h_{i,r,n} dN_{(0,1)} - \sum_{k=1}^{i-1} \left(\int h_{k,r,n} dN_{(0,1)}\right) L_{i-k,r,n}$$

where $h_{i,r,n} \equiv h_i$.
(ii) $\int L_{i,r,n} dN_{(0,1)} = 0, i = 1, \ldots, m - 1$.
(iii) *The coefficients of $L_{i,r,n}$ are of order $O((n/r(n - r))^{i/2})$.*
(iv) *For $i = 1, 2$ we have*

$$L_{1,r,n}(x) = \frac{n - 2r + 1}{(r(n - r + 1)(n + 1))^{1/2}}\left[\frac{x^3}{3} - x\right]$$

and   (4.2.2)

$$L_{2,r,n}(x) = \frac{1}{r(n - r + 1)(n + 1)}[(n - 2r + 1)^2(x^6 - 15)/18 -$$

$$[7(n - 2r + 1)^2 + 3r(n - r + 1)](x^4 - 3)/12 - (n - r + 1)^2(x^2 - 1)].$$

Before turning to the extension of Theorem 4.2.1 to a certain class of d.f.'s we make some comments:

(a) Perhaps the most important consequence of Theorem 4.2.1 is that we get a normal approximation with an error term of order $O((n/r(n - r))^{1/2})$. Thus, if $r = r(n) = [nq]$ where $0 < q < 1$ then the error bound is of order $O(n^{-1/2})$. In the intermediate case the approximation is less accurate and, moreover, if r or $n - r$ is fixed (that is, the case of extreme order statistics) we have no approximation at all.

(b) When taking the expansion of length 2—that is, we include the polynomial $L_{1,r,n}$ into our considerations—then the accuracy of the approximation improves considerably. We also get a better insight in the accuracy of the normal approximation.

For example, given the sample median $U_{n+1:2n+1}$ we see that the corresponding polynomial $L_{1,n+1,2n+1}$ is equal to zero and, thus, the accuracy of the normal approximation is of order $O(n^{-1})$. A similar conclusion can be made for order statistics which are close—as far as the indices are concerned—to the sample median. For sample quantiles different from the sample median the accuracy of the normal approximation cannot be better than $O(n^{-1/2})$.

Finally, we mention that for symmetric Borel sets B (that is, B has the property that $x \in B$ implies $-x \in B$) we have

$$\int_B L_{1,r,n} \, dN_{(0,1)} = 0,$$

so that for symmetric sets the normal approximation is of order $O(n/r(n - r))$.

(c) Numerical calculations show that for $n = 1, 2, \ldots, 250$ we can take $C_1 = .14$ and $C_2 = .12$ in Theorem 4.2.1.

The General Case

The extension of Theorem 4.2.1 to more general r.v.'s will be achieved by means of the transformation technique. If $X_{r:n}$ is the rth order statistic of n i.i.d. random variables with common d.f. F then $X_{r:n} \stackrel{d}{=} F^{-1}(U_{r:n})$. Notice that F^{-1} is monotone. Apart from this special case one is also interested in other monotone transformations of $U_{r:n}$.

As a refinement of the idea which led to Lemma 4.1.6 we get the following highly technical result.

Lemma 4.2.3. *Let m be a positive integer and $\varepsilon > 0$. Suppose that S is a function with the properties $S(0) = 0$, S is continuously differentiable on the interval $(-\varepsilon, \varepsilon)$, and*

$$\left| S'(x) - \left[1 + \sum_{i=1}^{m-1} \alpha_i x^i/i! \right] \right| \le \alpha_m |x^m|/m!, \qquad |x| < \varepsilon, \qquad (4.2.3)$$

with $|\alpha_i| \le \exp(-i\varepsilon)$ for $i = 1, \ldots, m$.

Moreover, let R_i be polynomials of degree $\le 3i$ so that the absolute values of the coefficients are $\le \exp(-i\varepsilon)$ for $i = 1, \ldots, m - 1$.

Then there exist constants $C > 0$ and $d \in (0, 1)$ [which only depend on m]
such that
(i) S *is strictly increasing on the interval* $I = (-d\varepsilon, d\varepsilon)$.
(ii) *For every monotone, real-valued function* T *such that the restriction of* T
to the set $S(I)$ *is the inverse of the restriction* $S|I$ *we have*

$$\sup_{B} \left| \int_{\{T \in B\}} \left(1 + \sum_{i=1}^{m-1} R_i\right) dN_{(0,1)} - \int_{B} \left(1 + \sum_{i=1}^{m-1} L_i\right) dN_{(0,1)} \right| \leq C \exp(-m\varepsilon)$$

where L_i is a polynomial of degree $\leq 3i$ and the absolute values of the coefficients
are $\leq C \exp(-i\varepsilon)$ for $i = 1, \ldots, m - 1$.
(iii) *We have*

$$L_1(x) = R_1(x) + \alpha_1(x - x^3/2),$$

and   (4.2.4)

$$L_2(x) = R_2(x) + \alpha_1[x^2 R_1'(x)/2 + (x - x^3/2)R_1(x)] + \alpha_1^2[x^6/8 - 5x^4/8]$$
$$+ \alpha_2[x^2/2 - x^4/6].$$

PROOF. Since $\varepsilon^\rho \exp(-\varepsilon)$ is uniformly bounded on $[0, \infty)$ for every $\rho \geq 1$ there
exists $d \in (0, 1)$ such that

$$S'(x) \geq 1 - \sum_{i=1}^{m} [d\varepsilon \exp(-\varepsilon)]^i/i! \geq 1/2, \qquad |x| \leq d\varepsilon. \tag{1}$$

The assertion (i) is immediate from (1).
    Moreover (1) implies that

$$S(0)(-d\varepsilon) \leq -d\varepsilon/2 \quad \text{and} \quad S(d\varepsilon) \geq d\varepsilon/2. \tag{2}$$

From the condition $S'(0) = 0$ and from (4.2.3) we deduce by integration
that

$$\left| S(x) - \left(x + \sum_{i=1}^{m-1} \frac{x^{i+1}}{(i+1)!} \alpha_i\right) \right| \leq \frac{|x|^{m+1}}{(m+1)!} \alpha_m, \qquad |x| < \varepsilon. \tag{3}$$

Using (3) we get in analogy to (1) that

$$(1 + |x|)|S(x) - x| \text{ is uniformly bounded over } |x| \leq d\varepsilon. \tag{4}$$

Applying the transformation theorem for densities (1.4.4) we obtain for
every Borel set $B \subset (-d\varepsilon, d\varepsilon)$ that

$$\int_{\{T \in B\}} \left(1 + \sum_{i=1}^{m-1} R_i\right) dN_{(0,1)} = \int_{B} h(x) \, dx \tag{5}$$

where

$$h(x) = S'(x)\varphi(S(x))\left(1 + \sum_{i=1}^{m-1} R_i(S'(x))\right). \tag{6}$$

Expanding φ about x we obtain from (4)

$$\left| \varphi(S(x)) - \varphi(x)\left(1 + \sum_{i=1}^{m-1} w_i(x)(S(x) - x)^i\right)\right|$$

$$\leq C\varphi(x)|w_m(x + \theta(S(x) - x))| |S(x) - x|^m \tag{7}$$

for $|x| \leq d\varepsilon$ and $\theta \in (0, 1)$. Moreover, $w_i = \varphi^{(i)}/(i!\varphi)$ is a polynomial of degree $\leq i$ and C denotes a generic constant which only depends on m. For $i = 1, 2$ we get

$$w_1(x) = -x \quad \text{and} \quad w_2(x) = (x^2 - 1)/2.$$

Writing

$$\psi(x) = \sum_{i=1}^{m-1} \frac{x^{i+1}}{(i+1)!}\alpha_i,$$

we obtain from (7) that

$$\left| h(x) - \varphi(x)[1 + \psi^{(1)}(x)]\left[1 + \sum_{i=1}^{m-1} w_i(x)\psi^{(i)}(x)\right]\left[1 + \sum_{i=1}^{m-1} R_i(x + \psi(x))\right]\right|$$

$$\leq C\varphi(x)\exp(-m\varepsilon)(1 + |x|^{6(m+1)^2}) \tag{8}$$

for $|x| < d\varepsilon$. From (8) we conclude that

$$\left| h(x) - \varphi(x)\left[1 + \sum_{i=1}^{m-1} L_i(x)\right]\right| \leq C\varphi(x)\exp(-m\varepsilon)(1 + |x|^{6(m+1)^2}) \tag{9}$$

for $|x| < d\varepsilon$ where L_i are polynomials which have the asserted property. From (5) and (9) we deduce by integration that

$$\left|\int_{\{T \in B\}} \left(1 + \sum_{i=1}^{m-1} R_i\right)dN_{(0,1)} - \int_B \left(1 + \sum_{i=1}^{m-1} L_i\right)dN_{(0,1)}\right|$$

$$\leq \int_{-\varepsilon}^{\varepsilon}\left|h(x) - \left(1 + \sum_{i=1}^{m-1} L_i(x)\right)\varphi(x)\right|dx \leq C\exp(-m\varepsilon) \tag{10}$$

for Borel sets $B \subset (-d\varepsilon, d\varepsilon)$. Moreover, for Borel sets $B \subset (-d\varepsilon, d\varepsilon)^c$ we get by (2)

$$\left|\int_{\{T \in B\}} \left(1 + \sum_{i=1}^{m-1} R_i\right)dN_{(0,1)} - \int_B \left(1 + \sum_{i=1}^{m-1} L_i\right)dN_{(0,1)}\right|$$

$$\leq \int_A \left|1 + \sum_{i=1}^{m-1} R_i\right|dN_{(0,1)} + \int_A \left|1 + \sum_{i=1}^{m-1} L_i\right|dN_{(0,1)} \leq C\exp(-m\varepsilon) \tag{11}$$

where A is the complement of $(-d\varepsilon/2, d\varepsilon/2)$.

Combining (10) and (11) the proof is complete. $\square$

Note that Lemma 4.2.3 still holds if the condition that S has a continuous derivative is replaced by the weaker condition that S is absolutely continuous.

Next, an expansion of length m will be established under the condition that the underlying d.f. F has $m + 1$ derivatives on some appropriate interval. Let again $a_{r,n}^2 = r(n - r + 1)/(n + 1)^3$ and $b_{r,n} = r/(n + 1)$. Based on Theorem 4.2.1 and Lemma 4.2.3 the proof of Theorem 4.2.4 will be a triviality.

Theorem 4.2.4. *For some $r \in \{1, \ldots, n\}$ let $X_{r:n}$ be the rth order statistic of n i.i.d. random variables with common d.f. F and density f. Assume that $f(F^{-1}(b_{r,n})) > 0$ and that the function $S_{r,n}$ defined by*

$$S_{r,n}(x) = a_{r,n}^{-1}(F[F^{-1}(b_{r,n}) + xa_{r,n}/f(F^{-1}(b_{r,n}))] - b_{r,n})$$

has $m + 1$ derivatives on the interval

$$I_{r,n} := \{x : |x| < 2^{-1} \log(r(n - r + 1)/(n + 1))\}.$$

Then there exists a constant $C_m > 0$ (only depending on m) such that

$$\sup_B \left| P\{a_{r,n}^{-1} f(F^{-1}(b_{r,n}))[X_{r:n} - F^{-1}(b_{r,n})] \in B\} - \int_B \left(1 + \sum_{i=1}^{m-1} L_{i,r,n}\right) dN_{(0,1)} \right|$$

$$\leq C_m \left[(n/r(n - r))^{m/2} + \max_{j=1}^{m} |\alpha_{j,r,n}|^{m/j} \right] \tag{4.2.5}$$

where $L_{i,r,n}$ is a polynomial of degree $\leq 3i$. Moreover, $\alpha_{j,r,n} = S_{r,n}^{(j+1)}(0)$ for $j = 1, \ldots, m - 1$ and $\alpha_{m,r,n} = \sup\{|S_{r,n}^{(m+1)}(x)| : x \in I_{r,n}\}$.

PROOF. Throughout the proof, the indices r and n will be suppressed. Writing

$$T(x) = a^{-1} f(F^{-1}(b))[F^{-1}(b + ax) - F^{-1}(b)] \tag{1}$$

and denoting by R_i the polynomials of Theorem 4.2.1 we obtain from Theorem 1.2.5 and Theorem 4.2.1 that for every Borel set B,

$$\left| P\{a^{-1} f(F^{-1}(b))[X_{r:n} - F^{-1}(b)] \in B\} - \int_{\{T \in B\}} \left(1 + \sum_{i=1}^{m-1} R_i\right) dN_{(0,1)} \right|$$

$$\leq C(n/r(n - r))^{m/2}.$$

It remains to prove that

$$\left| \int_{\{T \in B\}} \left(1 + \sum_{i=1}^{m-1} R_i\right) dN_{(0,1)} - \int_B \left(1 + \sum_{i=1}^{m-1} L_i\right) dN_{(0,1)} \right|$$

$$\leq C \left[(n/r(n - r))^{m/2} + \max_{j=1}^{m} |\alpha_{j,r,n}|^{m/j} \right]. \tag{2}$$

Put $\varepsilon = -\log[(n/r(n - r))^{1/2} + \max_{j=1}^{m} |\alpha_{j,r,n}|^{1/j}]$, and assume w.l.g. that $r(n - r)$ is sufficiently large so that $\varepsilon > 0$. A Taylor expansion of S' about zero yields that condition (4.2.3) is satisfied for ε and α_i. Moreover, $T|S(I)$ is the inverse of $S|I$. Thus, Lemma 4.2.3 implies (2). $\qquad \square$

Addendum 4.2.5. *From the proof to Theorem 4.2.4 we see that*
(i) $\int L_{i,r,n} \, dN_{(0,1)} = 0, \quad i = 1, \ldots, m - 1.$
(ii) *The coefficients of $L_{i,r,n}$ are of order*

$$O\left[(n/r(n - r))^{1/2} + \max_{j=1}^{i} |\alpha_{j,r,n}|^{i/j} \right].$$

(iii) *For $i = 1, 2$, we have (with $R_{i,r,n}$ denoting the polynomials of (4.2.2)),*

$$L_{1,r,n}(x) = R_{1,r,n}(x) + \alpha_{1,r,n}(x - x^3/2)$$

and

$$L_{2,r,n}(x) = R_{2,r,n}(x) + \alpha_{1,r,n}[x^2 R'_{1,r,n}(x)/2 + (x - x^3/2)R_{1,r,n}(x)]$$
$$+ \alpha_{1,r,n}^2 (x^6/8 - 5x^4/8) + \alpha_{2,r,n}(x^2/2 - x^4/6).$$

Notice that Theorem 4.2.1 is immediate from Theorem 4.2.4 applied to $S_{r,n}(x) = x$. In this case we have $\alpha_{j,r,n} = 0, j = 1, \ldots, m$.

EXAMPLE 4.2.6. In many cases one can omit the term $\max_{j=1}^{m} |\alpha_{j,r,n}|^{m/j}$ at the right-hand side of (4.2.5).

Let $0 < q_1 < q_2 < 1$ and suppose that the density is bounded away from zero on the interval $J = (F^{-1}(q_1) - \varepsilon, F^{-1}(q_2) + \varepsilon)$ for some $\varepsilon > 0$. If f has m bounded derivatives on J then $\max_{j=1}^{m} |\alpha_{j,r,n}|^{m/j} = O(n^{-m/2})$ uniformly over $r \in \{[nq_1], \ldots, [nq_2] + 1\}$.

Order Statistics of Exponential R.V.'s

Careful calculations will show that in the case of exponential r.v.'s the right-hand side of (4.2.5) is again of order $O((n/r(n - r))^{m/2})$.

Corollary 4.2.7. *Let $X_{i:n}$ be the ith order statistic of n i.i.d. standard exponential r.v.'s (having the d.f. $G(x) = 1 - e^{-x}$ and density $g(x) = e^{-x}, x \geq 0$). Let again $a_{r,n}^2 = r(n - r + 1)/(n + 1)^3$ and $b_{r,n} = r/(n + 1)$.*
Then there exists a constant $C_m > 0$ (only depending on m) such that

$$\sup_{B} \left| P\{a_{r,n}^{-1} g(G^{-1}(b_{r,n}))[X_{r:n} - G^{-1}(b_{r,n})] \in B\} - \int_{B} \left(1 + \sum_{i=1}^{m-1} L_{i,r,n}\right) dN_{(0,1)} \right|$$

$$\leq C_m (n/r(n - r))^{m/2} \tag{4.2.6}$$

where the polynomials $L_{i,r,n}$ are defined as in Theorem 4.2.4 with

$$\alpha_{i,r,n} = (-1)^i (r/(n + 1)(n - r + 1))^{i/2}.$$

In particular, for $i = 1, 2$,

$$L_{1,r,n}(x) = (r(n - r + 1)(n + 1))^{-1/2}[(2n - r + 2)x^3/6 - (n - r + 1)x],$$

and

$$L_{2,r,n}(x) = R_{2,r,n}(x) + ((n - r + 1)(n + 1))^{-1}[r(-5x^6/24 + 15x^4/8 - 5x^2/2)$$
$$- (n + 1)(-x^6/6 + 4x^4/3 - 3x^2/2)]$$

where $R_{2,r,n}$ is the corresponding polynomial in Theorem 4.2.1.

PROOF. Since $g^{(i)}(G^{-1}(q)) = (-1)^i(1 - q)$ it is immediate that $\alpha_{i,r,n}$ is of the desired form. Moreover, $|\alpha_{i,r,n}|^{1/i} \le (n/r(n - r + 1))^{1/2}$.

Let $S_{r,n}$ and $I_{r,n}$ be defined as in Theorem 4.2.4. Since $\log(1 + x) \le x$ for $x > -1$, and hence, also $\log x < x$, $x > 0$, we obtain

$$G^{-1}(b_{r,n}) - \log[(r(n - r + 1)/(n + 1))^{1/2}]a_{r,n}/g(G^{-1}(b_{r,n}))$$
$$\ge b_{r,n} - \log[(r(n - r + 1)/(n + 1))^{1/2}]a_{r,n}/(1 - b_{r,n}) > 0.$$

Using this inequality we see that $S_{r,n}$ has $m + 1$ derivatives on the interval $I_{r,n}$. Moreover, by straightforward calculations we obtain $\alpha_{m,r,n} \le C(n/r(n - r + 1))^{m/2}$ where C is a universal constant. Thus, Theorem 4.2.4 is applicable and yields the assertion. □

Numerical computations show that one can take $C_1 = .15$ and $C_2 = .12$ in Corollary 4.2.7 for $n = 1, \ldots, 250$. From the expansion of length 2 in Corollary 4.2.7 we obtain the following upper bound of the remainder term of the normal approximation:

$$\left[\int L_{1,r,n}^2 \, dN_{(0,1)}\right]^{1/2} + C_2 n/(r(n - r + 1)).$$

Moreover,

$$\int L_{1,r,n}^2 \, dN_{(0,1)} = \frac{8(n - r + 1)^2 + 8r(n - r + 1) + 5r^2}{12r(n - r + 1)(n + 1)} \le \frac{2(n + 1)}{3r(n - r + 1)}.$$

$$(4.2.7)$$

Stochastic Independence of Certain Groups of Order Statistics

This section will be concluded with an application of the expansion of length 2 of distributions of order statistics $U_{i:n}$. In the proof below we shall only indicate the decisive step which is based on the expansion of length 2.

Hereafter, let $1 \le s < n - m + 1$. Let $V_{s:n}$ and $V_{n-m+1:n}$ be independent r.v.'s such that $V_{s:n} \stackrel{d}{=} U_{s:n}$ and $V_{n-m+1:n} \stackrel{d}{=} U_{n-m+1:n}$. The basic inequality is given by

$$\sup_B |P\{(U_{s:n}, U_{n-m+1:n}) \in B\} - P\{(V_{s:n}, V_{n-m+1:n}) \in B\}|$$

$$(4.2.8)$$

$$\le C\left[\frac{sm}{n(n - s - m)}\right]^{1/2}$$

where $C > 0$ is a universal constant. Thus, if s and m are fixed then the upper bound is of order $O(n^{-1})$. If s is fixed and $(n - m)/n$ bounded away from 0 and 1 then the bound is of order $O(n^{-1/2})$. Finally, if s is fixed and $n - m = o(n)$ then the bound is of order $O((n - m)^{-1/2})$. This shows that extremes and intermediate order statistics are asymptotically independent.

The proof of (4.2.8) is based on Theorem 1.8.1 and Theorem 4.2.1. Conditioning on $U_{n-m+1:n}$ one obtains

$$P\{(U_{s:n}, U_{n-m+1:n}) \in B\} - P\{(V_{s:n}, V_{n-m+1:n}) \in B\} = ET(U_{n-m+1:n})$$

where   (4.2.9)

$$T(x) = P\{xU_{s:n-m} \in B_x\} - P\{U_{s:n} \in B_x\}$$

with B_x denoting the x-section of the set B.

The function T is of a rather complicated structure and has to be replaced by a simpler one. This can be achieved by expansions of length 2. The approximate representation of T as the difference of two expansions of length 2 simplifies further computations. We remark that a normal approximation instead of an expansion of length 2 leads to an inaccurate upper bound in (4.2.8). For details of the proof we refer to Falk and Reiss (1988) where the following two extensions of (4.2.8) can be also found.

Theorem 4.2.8. *Let $X_{i:n}$ be the ith order statistic of n i.i.d. random variables with common d.f. F. Given $1 \le s < n - m + 1 \le n$ we consider two vectors of order statistics, namely,*

$$X_l = (X_{1:n}, \ldots, X_{s:n}), \quad \text{and} \quad X_u = (X_{n-m+1:n}, \ldots, X_{n:n}).$$

Now let Y_l and Y_u be independent random vectors so that $Y_l \stackrel{d}{=} X_l$, and $Y_u \stackrel{d}{=} X_u$. Then,

$$\sup_B |P\{(X_l, X_u) \in B\} - P\{(Y_l, Y_u) \in B\}| \le C \left[\frac{sm}{n(n - s - m)} \right]^{1/2} \quad (4.2.10)$$

where C is the constant in (4.2.8).

A further extension is obtained when treating three groups of order statistics.

Theorem 4.2.9. *Let $X_{i:n}$ be as above. Given $1 \le k < r < s < n - m + 1 \le n$ we obtain three vectors of order statistics, namely,*

$$X_l = (X_{1:n}, \ldots, X_{k:n}), \quad X_c = (X_{r:n}, \ldots, X_{s:n}), \quad X_u = (X_{n-m+1:n}, \ldots, X_{n:n}).$$

Now let Y_l, Y_c and Y_u be independent random vectors so that $Y_l \stackrel{d}{=} X_l$, $Y_c \stackrel{d}{=} X_c$ and $Y_u \stackrel{d}{=} X_u$. Then there exists a universal constant $C > 0$ such that

$$\sup_{B} |P\{(X_l, X_c, X_u) \in B\} - P\{(Y_l, Y_c, Y_u) \in B\}|$$

$$\leq C\left[\frac{k(n-r)}{n(r-k)} + \frac{sm}{n(n-s-m)}\right]^{1/2}. \qquad (4.2.11)$$

Both theorems are deduced from (4.2.8) by means of the quantile transformation and by conditioning on order statistics.

4.3. Asymptotic Independence from the Underlying Distribution Function

From the preceding section we know that the normalized central order statistic $f(F^{-1}(b_{r,n}))(X_{r:n} - F^{-1}(b_{r,n}))$ is asymptotically normal—with expectation $\mu = 0$ and variance $a_{r,n}^2 = r(n - r + 1)/(n + 1)^3$—up to a remainder term of order $O(n^{-1/2})$ if, roughly speaking, the underlying density f is bounded away from zero. In the present section we shall primarily be interested in the property that the approximating normal distribution is independent from the underlying d.f. F. Consequently,

$$\sup_{B} |P\{f(F^{-1}(b_{r,n}))(X_{r:n} - F^{-1}(b_{r,n})) \in B\} - P\{(U_{r:n} - b_{r,n}) \in B\}|$$

$$= O(n^{-1/2}) \qquad (4.3.1)$$

where $U_{r:n}$ is the rth order statistic of n i.i.d. $(0, 1)$-uniformly distributed r.v.'s. Notice that the error bound above is sharp since the second term of the expansion of length two depends on the density f.

The Main Result

In analogy to (4.3.1) it will be shown in Theorem 4.3.1 that the variational distance between standardized joint distributions of k order statistics is of order $O((k/n)^{1/2})$. That means, after a linear transformation which depends on the underlying d.f. F the joint distribution of order statistics becomes independent from F within an error bound of order $O((k/n)^{1/2})$.

When treating the normal approximation, the situation is completely different. It is clear that the joint asymptotic normality of order statistics $X_{r:n}$ and $X_{s:n}$ implies that the spacings $X_{s:n} - X_{r:n}$ also have this property. However, if $s - r$ is fixed then spacings behave like extreme order statistics, and hence, the limiting distribution is different from the normal distribution.

Theorem 4.3.1. *Let $X_{i:n}$ be the ith order statistic of n i.i.d. random variables with common d.f. F and density f.*

Let $0 = r_0 < r_1 < \cdots < r_k < r_{k+1} = n + 1$ *with* $r_i - r_{i-1} \geq 4$ *for* $i = 1, 2, \ldots,$
$k + 1$. *Put* $b_i = r_i/(n + 1)$ *and* $\sigma_i^2 = b_i(1 - b_i)$ *for* $i = 1, \ldots, k$.

Assume that $f > 0$ *and* f *has three derivatives on the interval* I *where*
$I = (F^{-1}(b_1) - \varepsilon_1, F^{-1}(b_k) + \varepsilon_k)$ *with* $\varepsilon_i = 5n^{-1/2}(\log n)\sigma_i/f(F^{-1}(b_i))$ *for* $i = 1, k$.
Then, there exists a universal constant $C > 0$ *such that*

$$\sup_B |P\{[f(F^{-1}(b_i))(X_{r_i:n} - F^{-1}(b_i))]_{i=1}^k \in B\} - P\{[(U_{r_i:n} - b_i)]_{i=1}^k \in B\}|$$

$$\leq C(k/n)^{1/2}[c(f)^{1/2} + c(f)^2 + n^{-1/2}]$$

where $c(f) = \max_{j=1}^3 [\sup_{y \in I} |f^{(j)}(y)|/\inf_{y \in I} f^{j+1}(y)]$.

At the end of this section we shall give an example showing that Theorem 4.3.1 does not hold for $r_i - r_{i-1} = 1$. It is difficult to make a conjecture whether the result holds for $r_i - r_{i-1} = 2$ or $r_i - r_{i-1} = 3$. As we will see in the proof of Theorem 4.3.1 one reason for the restriction $r_i - r_{i-1} \geq 4$ is that the supports of the two joint distributions are unequal.

Theorem 4.3.1 is a slight improvement of Theorem 2.1 in Reiss (1981b) which was proved under the stronger condition that $r_i - r_{i-1} \geq 5$. Therefore, the proof is given in its full length. Another reason for running through all the technical details is to facilitate and to encourage further research work. Theorem 4.3.1 may be of interest as a challenging problem that can only be solved when having a profound knowledge of the distributional properties of order statistics.

Theorem 4.3.1 also serves as a powerful tool to prove various results for order statistics. As an example we mention a result of Section 4.5 stating that several order statistics of i.i.d. exponential r.v.'s are jointly asymptotically normal. By making use of Theorem 4.3.1, this may easily be extended to other r.v.'s. However, one should notice that a stronger result may be achieved by using a method adjusted to the particular problem. Thus, applications of Theorem 4.3.1 will lead to results of a preliminary character which may stimulate further research work. Another application of Theorem 4.3.1 will concern linear combinations of order statistics (see Section 6.2).

PROOF OF THEOREM 4.3.1. Part I. We write $\mu_i = F^{-1}(b_i)$, $f_i = f(\mu_i)$ and, more generally, $f_i^{(j)} = f^{(j)}(\mu_i)$. Denote by Q_0 and Q_1 the distributions of

$$(U_{r_i:n} - b_i)_{i=1}^k \quad \text{and, respectively,} \quad (f_i(X_{r_i:n} - \mu_i))_{i=1}^k,$$

and by g_0 and g_1 the corresponding densities.

From Lemma 3.3.9(i) and Lemma A.3.5 we obtain

$$\sup_B |Q_0(B) - Q_1(B)| \leq \left[2Q_0(A^c) + \int_A \left(-\log \frac{g_1}{g_0} \right) dQ_0 \right]^{1/2} \tag{1}$$

for some Borel set A to be fixed later. The main difficulty of the proof is to obtain a sharp lower bound of $\int_A \log g_1/g_0 \, dQ_0$.

We have

$$g_0 = \kappa \left(\prod_{i=1}^{k+1} \psi_i^{r_i - r_{i-1} - 1} \right) 1_{\{\mathbf{x}:\, 0 < x_1 + b_1 < \cdots < x_k + b_k < 1\}}$$

and

$$g_1 = \kappa \left(\prod_{i=1}^{k} h_i \right) \prod_{i=1}^{k+1} (\delta_i + \psi_i)^{r_i - r_{i-1} - 1} 1_{A_1},$$

where

$$A_1 = \{\mathbf{x}: F(\mu_1 + x_1/f_1) < \cdots < F(\mu_k + x_k/f_k)\}.$$

Moreover, κ is a normalizing constant, $h_i(\mathbf{x}) = f(\mu_i + x_i/f_i)/f_i$, $\psi_i(\mathbf{x}) = x_i - x_{i-1} + (b_i - b_{i-1})$, $\delta_i(\mathbf{x}) = F(\mu_i + x_i/f_i) - F(\mu_{i-1} + x_{i-1}/f_{i-1}) - \psi_i(\mathbf{x})$ for $i = 1, \ldots, k + 1$ [with the convention that $x_0 = x_{k+1} = 0$, $F(\mu_0 + x_0/f_0) = 0$ and $F(\mu_{k+1} + x_{k+1}/f_{k+1}) = 1$]. Thus, for $A \subset A_1$ we have

$$\int_A \left(\log \frac{g_1}{g_0} \right) dQ_0 = \sum_{i=1}^{k} \int_A (\log h_i) \, dQ_0$$
$$+ \sum_{i=1}^{k+1} (r_i - r_{i-1} - 1) \int_A \log\left(1 + \frac{\delta_i}{\psi_i} \right) dQ_0. \tag{2}$$

To obtain an expansion of $\log(1 + \delta_i/\psi_i)$, we introduce the sets

$$A_{2,i} = \{\delta_i/\psi_i \geq -\tfrac{1}{2}\}, \qquad i = 1, \ldots, k + 1.$$

Notice that

$$|\log(1 + \delta_i/\psi_i) - \delta_i/\psi_i| \leq C(\delta_i/\psi_i)^2 \tag{3}$$

on $A_{2,i}$ where, throughout the proof, C denotes a universal constant that is not necessarily the same at each appearance. Moreover, we write

$$A_{3,i} = \{\mathbf{x}: |x_i| \leq 5n^{-1/2}(\log n)\sigma_i\}$$

and

$$A = A_1 \cap \left(\bigcap_{i=1}^{k+1} A_{2,i} \right) \cap \bigcap_{i=1}^{k} A_{3,i}. \tag{4}$$

We shall verify that the following three inequalities hold:

$$\left| \sum_{i=1}^{k} \int_A (\log h_i) \, dQ_0 \right| \leq C[c(f)Q_0(A^c)^{2/3} k/n^{1/2} + (c(f) + c(f)^2)k/n], \tag{5}$$

$$\left| \sum_{i=1}^{k+1} (r_i - r_{i-1} - 1) \int_A \log(1 + \delta_i/\psi_i) \, dQ_0 \right|$$
$$\leq C(c(f) + c(f)^2) \left[\frac{k}{n} + \frac{k}{n^{1/2}} Q_0(A^c)^{7/12} \right], \tag{6}$$

$$Q_0(A^c) \le C\left[\frac{k}{n^3} + c(f)^4 (\log n)^{1/2} \frac{k}{n^2}\right]. \tag{7}$$

The assertion of the theorem is immediate from (1), (2), and (5)–(7).

A Taylor expansion of $\log(f/f_i)$ about μ_i yields

$$|\log h_i(\mathbf{x}) - (f_i^{(1)}/f_i^2)x_i| \le C(c(f) + c(f)^2)x_i^2$$

for $\mathbf{x} \in A_{3,i}$ and $i = 1, \ldots, k$. Since $\int x_i \, dQ_0(\mathbf{x}) = 0$ we obtain

$$\left|\int_A (\log h_i) \, dQ_0\right| \le c(f) \int_{A^c} |x_i| \, dQ_0(\mathbf{x}) + C(c(f) + c(f)^2) \int x_i^2 \, dQ_0(\mathbf{x}) \tag{8}$$

and hence, (5) is immediate from (1.7.4).

Next, we shall prove a lower bound of $\sum_{i=1}^{k+1} (r_i - r_{i-1} - 1) \int_A \log(1 + \delta_i/\psi_i) \, dQ_0$. It is obvious from (3) that

$$\sum_{i=1}^{k+1} (r_i - r_{i-1} - 1) \int_A \log(1 + \delta_i/\psi_i) \, dQ_0 \le C(|\rho_1| + |\rho_2| + \rho_3) \tag{9}$$

with

$$\rho_1 = \sum_{i=1}^{k+1} (r_i - r_{i-1} - 1) \int_A \frac{a_i x_i^2 - a_{i-1} x_{i-1}^2}{\psi_i(\mathbf{x})} \, dQ_0(\mathbf{x}),$$

$$\rho_2 = \sum_{i=1}^{k+1} (r_i - r_{i-1} - 1) \int_A \frac{\delta_i(\mathbf{x}) - (a_i x_i^2 - a_{i-1} x_{i-1}^2)}{\psi_i(\mathbf{x})} \, dQ_0(\mathbf{x}),$$

$$\rho_3 = \sum_{i=1}^{k} (r_i - r_{i-1} - 1) \int_A (\delta_i/\psi_i)^2 \, dQ_0,$$

where the constants a_i are given by $a_i = f_i^{(1)}/2f_i^2$ for $i = 1, \ldots, k$, and $a_0 = a_{k+1} = 0$. From P.1.25 it is easily seen that

$$\rho_1 = -\sum_{i=1}^{k+1} (r_i - r_{i-1} - 1) \int_{A^c} \frac{a_i x_i^2 - a_{i-1} x_{i-1}^2}{\psi_i(\mathbf{x})} \, dQ_0(\mathbf{x}).$$

Some straightforward calculations yield

$$|a_i x_i^2 - a_{i-1} x_{i-1}^2| \le c(f)|x_i^2 - x_{i-1}^2| + (c(f) - c(f)^2)(b_i - b_{i-1})x_{i-1}^2 \tag{10}$$

for every $\mathbf{x}$ and $i = 2, \ldots, k$. Moreover, $\sum_{i=1}^{k+1} (a_i x_i^2 - a_{i-1} x_{i-1}^2) = 0$ and $r_i - r_{i-1} - (n + 1)\psi_i = -(n + 1)(x_i - x_{i-1})$. Combining these relations and applying the Hölder inequality we obtain

$$|\rho_1| \le (Q_0(A^c))^{7/12} \sum_{i=1}^{k+1} \left[\int \psi_i^{-3} \, dQ_0\right]^{1/3} \left(\int \left[(c(f)|x_i - x_{i-1}|(|x_i| + |x_{i-1}|)\right.\right.$$

$$\left.\left. + (c(f) + c(f)^2)(b_i - b_{i-1})x_{i-1}^2)(1 + (n + 1)|x_i - x_{i-1}|)\right]^{12} \, dQ_0\right)^{1/12}.$$

Since $r_i - r_{i-1} \geq 4$ we know that P.1.23 is applicable to $\int \psi_i^{-3} \, dQ_0$ and hence the Hölder inequality, Lemma 3.1.3 and Corollary 1.6.8 yield

$$|\rho_1| \leq C(c(f) + c(f)^2) \frac{k}{n^{1/2}} (Q_0(A^c))^{7/12}. \tag{11}$$

To obtain a sharp upper bound of $|\rho_2|$ one has to utilize some tedious estimates of $|\delta_i(\mathbf{x}) - (a_i x_i^2 - a_{i-1} x_{i-1}^2)|$. A Taylor expansion of $G(y) = F(\mu_i + yx_i/f_i) - F(\mu_{i-1} + yx_{i-1}/f_{i-1})$ about $y = 0$ yields

$$|\delta_i(\mathbf{x}) - (a_i x_i^2 - a_{i-1} x_{i-1}^2)| = \frac{1}{6} \left| f^{(2)}\left(\mu_i + \theta \frac{x_i}{f_i}\right) \frac{x_i^3}{f_i^3} - f^{(2)}\left(\mu_{i-1} + \theta \frac{x_{i-1}}{f_{i-1}}\right) \frac{x_{i-1}^3}{f_{i-1}^3} \right|$$

for every $i = 2, \ldots, k$ and $\mathbf{x} \in A_{3,i} \cap A_{3,i-1}$ where $\theta \in (0, 1)$. Thus, by further Taylor expansions of F^{-1} and of derivatives of F we get

$$|\delta_i(\mathbf{x}) - (a_i x_i^2 - a_{i-1} x_{i-1}^2)|$$
$$\leq C(c(f)|x_i^3 - x_{i-1}^3| + x_{i-1}^3 [c(f)|x_i - x_{i-1}| + (c(f) + c(f)^2)(b_i - b_{i-1})])$$
$$=: \eta_i(\mathbf{x}). \tag{12}$$

For $i = 1$ and $\mathbf{x} \in A_{3,1}$ and, respectively, $i = k + 1$ and $\mathbf{x} \in A_{3,k+1}$ we get

$$|\delta_i(\mathbf{x}) - (a_i x_i^2 - a_{i-1} x_{i-1}^2)| \leq C c(f) |x_i - x_{i-1}|^3 =: \eta_i(\mathbf{x}). \tag{13}$$

Since $\sum_{i=1}^{k+1} [\delta_i(\mathbf{x}) - (a_i x_i^2 - a_{i-1} x_{i-1}^2)] = 0$ we obtain—using again the Hölder inequality and applying (12) and (13)—that

$$|\rho_2| \leq \sum_{i=1}^{k+1} \int [|\eta_i(\mathbf{x})|(1 + (n + 1)|x_i - x_{i-1}|)/\psi_i(\mathbf{x})] \, dQ_0(\mathbf{x})$$

$$\leq \sum_{i=1}^{k+1} \left(\int [\eta_i(\mathbf{x})(1 + (n + 1)|x_i - x_{i-1}|)]^2 \, dQ_0(\mathbf{x}) \right)^{1/2} \left(\int \psi_i^{-2} \, dQ_0 \right)^{1/2}$$

Proceeding as in the proof of (11) we obtain

$$|\rho_2| \leq C(c(f) + c(f)^2)k/n. \tag{14}$$

Moreover, the arguments used to prove (11) and (14) also lead to

$$\rho_3 \leq \sum_{i=1}^{k+1} (r_i - r_{i-1} - 1) \left(\int [\eta_i(\mathbf{x}) + c(f)|x_i^2 - x_{i-1}^2| \right.$$

$$\left. + (c(f) + c(f)^2)(b_i - b_{i-1})x_{i-1}^2]^6 \, dQ_0(\mathbf{x}) \right)^{1/3} \left(\int \psi_i^{-3} \, dQ_0 \right)^{2/3} \tag{15}$$

$$\leq C(c(f) + c(f)^2)k/n.$$

Combining (9), (11), (14), and (15) we obtain (6).

Finally, we prove (7). Applying Lemma 3.1.1 we get

$$Q_0\{\mathbf{x}: |x_i| \geq (50/11)\sigma_i(\log n)/n^{1/2}\} \leq Cn^{-3} \tag{16}$$

for $i = 1, \ldots, k$. Hence

$$Q_0(A_{2,i}^c) \leq Cn^{-3} \tag{17}$$

for $i = 1, \ldots, k$, and in view of Corollary 1.6.8,

$$Q_0\{\mathbf{x}: |x_i - x_{i-1}| \geq 5(b_i - b_{i-1})^{1/2}(\log n)/n^{1/2}\} \leq Cn^{-3} \tag{18}$$

for $i = 2, \ldots, k$. From (10), (11), (13), (17), and (18) we infer that

$$Q_0\{\delta_i \geq -\varepsilon_n\} \geq 1 - Cn^{-3} \tag{19}$$

for $i = 1, \ldots, k+1$ where $\varepsilon_n = c(f)(b_i - b_{i-1})^{1/2}(\log n)^3/n^{1/2}$. Since $r_i - r_{i-1} \geq 4$ we deduce from Lemma 3.1.2 that

$$Q_0\{\psi_i \geq 3\varepsilon_n\} \geq 1 - Cc(f)^4(\log n)^{1/2}/n^2 \tag{20}$$

for $i = 1, \ldots, k+1$. Combining (19) and (20) we get

$$Q_0(A_{3,i}^c) \leq C[n^{-3} + c(f)^4(\log n)^{1/2}/n^2] \tag{21}$$

for $i = 1, \ldots, k+1$. It is immediate that $Q_0(A_1) \geq Q_0(\bigcap_{i=1}^k A_{3,i})$. This together with (17) and (20) yields

$$Q_0(A^c) \leq C[k/n^3 + c(f)^4(\log n)^{1/2}k/n^2]. \tag{22}$$

Thus, (7) holds and the proof is complete. $\qquad \square$

Counterexample

Theorem 4.3.1 was proved under the condition $r_i - r_{i-1} \geq 4$. A counterexample in Reiss (1981b) shows that this result does not hold if $r_i - r_{i-1} = 1$ for $i = 1, 2, \ldots, k$.

EXAMPLE 4.3.2. Let $X_{i:n}$ be the ith order statistic of n i.i.d. standard exponential r.v.'s (with common d.f. G and density g).

Then, if $n^{1/2} = o(k(n))$ and $[nq] + k(n) \leq n$ where $q \in (0, 1)$ is fixed, we obviously have

$$P\{U_{i:n} - U_{i-1:n} > 0 \quad \text{for } i = [nq], \ldots, [nq] + k(n)\} = 1$$

and, with $b_i = i/(n+1)$ and $\mu_i = G^{-1}(b_i)$ it can be verified that

$$\limsup_n P\{g(\mu_i)(X_{i:n} - \mu_i) - g(\mu_{i-1})(X_{i-1:n} - \mu_{i-1}) + (b_i - b_{i-1})$$
$$> 0 \quad \text{for } i = [nq], \ldots, [nq] + k(n)\} < 1.$$

Thus, the remainder term in Theorem 4.3.1 is not of order $O((k/n)^{1/2})$ for the sets

$$B = \{(x_1, x_2, \ldots, x_{k+1}): x_i - x_{i-1} + b_i - b_{i-1} > 0, i = 1, \ldots, k(n)\}.$$

4.4. The Approximate Multivariate Normal Distribution

From Section 4.3 we already know that normalized joint distributions of central order statistics are asymptotically independent of the underlying d.f. F. In Section 4.5 we shall prove that, under appropriate regularity conditions, the joint distributions are approximately normal. In the present section we introduce and study some properties of such normal distributions.

To find these approximate normal distributions it suffices to consider order statistics $U_{r_1:n} \leq U_{r_2:n} \leq \cdots \leq U_{r_k:n}$ of n i.i.d. random variables uniformly distributed on $(0, 1)$. Put $b_i = r_i/(n + 1)$. Then the normalized order statistics

$$(n + 1)^{1/2}(U_{r_i:n} - b_i), \qquad i = 1, \ldots, k,$$

have expectation equal to zero and covariances approximately equal to $b_i(1 - b_j)$ for $i \leq j$. Thus, adequate candidates of approximate joint normal distribution of central order statistics are the k-variate normal distributions $N_{(0,\Sigma)}$ with mean vector zero and covariance matrix $\Sigma = (\sigma_{i,j})$ where $\sigma_{i,j} = b_i(1 - b_j)$ for $1 \leq i \leq j \leq k$. Below the b_i are replaced by arbitrary λ_i.

Representations

Our first aim is to represent $N_{(0,\Sigma)}$ as a distribution induced by the k-variate standard normal distribution $N_{(0,I)}$ where I denotes the unit matrix. Obviously, $N_{(0,I)} = N_{(0,1)}^k$. Given $0 = \lambda_0 < \lambda_1 < \cdots < \lambda_k < 1$ define the linear map T by

$$T(\mathbf{x}) = \left[(1 - \lambda_i) \sum_{j=1}^{i} \left[\frac{\lambda_j - \lambda_{j-1}}{(1 - \lambda_{j-1})(1 - \lambda_j)} \right]^{1/2} x_j \right]_{i=1}^{k}. \qquad (4.4.1)$$

Lemma 4.4.1. $\qquad\qquad TN_{(0,I)} = N_{(0,\Sigma)}$

(that is, $N_{(0,I)}\{T \in B\} = N_{(0,\Sigma)}(B)$ for every Borel set B).

PROOF. Let T also denote the matrix which corresponds to the linear map. The standard formula for normal distributions yields that $TN_{(0,I)}$ has the covariance matrix $H = (\eta_{i,j}) = TT^t$ where T^t is the transposed of T. Thus,

$$\eta_{i,j} = (1 - \lambda_i)(1 - \lambda_j) \sum_{m=1}^{j} \frac{\lambda_m - \lambda_{m-1}}{(1 - \lambda_{m-1})(1 - \lambda_m)} \qquad \text{for } i \geq j.$$

By induction over $j = 1, \ldots, k$ we get

$$\sum_{m=1}^{j} \frac{\lambda_m - \lambda_{m-1}}{(1 - \lambda_{m-1})(1 - \lambda_m)} = \frac{\lambda_j}{(1 - \lambda_j)}$$

and hence $\eta_{i,j} = (1 - \lambda_i)\lambda_j$ for $i \geq j$. Since $\eta_{i,j} = \eta_{j,i}$ the proof is complete. $\qquad\square$

From standard calculus for normal distributions we know that the density $\varphi_{(0,\Sigma)}$ of $N_{(0,\Sigma)}$ is given by

$$\varphi_{(0,\Sigma)}(\mathbf{x}) = [\det \Sigma^{-1}/(2\pi)^k]^{1/2} \exp[-\tfrac{1}{2}\mathbf{x}'\Sigma^{-1}\mathbf{x}] \qquad (4.4.2)$$

where $\mathbf{x} = (x_1,\dots,x_k)'$ and Σ^{-1} is the inverse matrix of Σ. By elementary calculations and by formula (4.4.4) below we get an alternative representation of $\varphi_{(0,\Sigma)}$, namely,

$$\varphi_{(0,\Sigma)}(\mathbf{x}) = \left[(2\pi)^k \prod_{i=1}^{k+1} (\lambda_i - \lambda_{i-1}) \right]^{-1/2} \exp\left[-\frac{1}{2} \sum_{i=1}^{k+1} \frac{(x_i - x_{i-1})^2}{\lambda_i - \lambda_{i-1}} \right] \qquad (4.4.3)$$

where $\lambda_0 = 0$, $\lambda_{k+1} = 1$ and $x_0 = x_{k+1} = 0$.

Lemma 4.4.2. (i) *The matrix* $\Sigma^{-1} = (\alpha_{i,j})$ *is given by*

$$\alpha_{i,i} = \frac{\lambda_{i+1} - \lambda_{i-1}}{(\lambda_{i+1} - \lambda_i)(\lambda_i - \lambda_{i-1})}, \qquad i = 1,\dots,k,$$

and $\alpha_{i,i-1} = \alpha_{i-1,i} = -(\lambda_i - \lambda_{i-1})^{-1}$, $i = 2,\dots,k$, *and* $\alpha_{i,j} = 0$, *otherwise.*

(ii) $\det \Sigma^{-1} = \prod_{i=1}^{k+1} (\lambda_i - \lambda_{i-1})^{-1}.$ $\qquad (4.4.4)$

PROOF. (i) Let T be defined as in (4.4.1). The inverse of T is represented by the matrix $B = (\beta_{i,j})$ given by

$$\beta_{i,i} = \left[\frac{1 - \lambda_{i-1}}{(1 - \lambda_i)(\lambda_i - \lambda_{i-1})} \right]^{1/2}, \qquad i = 1,\dots,k,$$

and

$$\beta_{i,i-1} = -\left[\frac{1 - \lambda_i}{(1 - \lambda_{i-1})(\lambda_i - \lambda_{i-1})} \right]^{1/2}, \qquad i = 2,\dots,k,$$

and $\beta_{i,j} = 0$, otherwise. Notice that $\Sigma^{-1} = B'B = [\sum_{m=1}^k \beta_{m,i}\beta_{m,j}]_{i,j}$ and, thus, $\alpha_{i,i} = \beta_{i,i}^2 + \beta_{i+1,i}^2$, $\alpha_{i,i-1} = \alpha_{i-1,i} = \beta_{i,i}\beta_{i,i-1}$ and $\alpha_{i,j} = 0$, otherwise. The proof of (i) is complete.

(ii) Moreover,

$$\det \Sigma^{-1} = (\det B)^2 = \prod_{i=1}^k \beta_{i,i}^2 = \prod_{i=1}^k (\lambda_i - \lambda_{i-1})^{-1} \prod_{i=1}^k \frac{1 - \lambda_{i-1}}{1 - \lambda_i}$$

$$= \prod_{i=1}^{k+1} (\lambda_i - \lambda_{i-1})^{-1}. \qquad \qquad \square$$

Moments

Recall that the absolute moments of the standard normal distribution $N_{(0,1)}$ are given by

$$\int |x|^j \, dN_{(0,1)}(x) = \frac{1 \cdot 3 \cdot 5 \cdot \ldots \cdot (j-1)}{(2^j/\pi)^{1/2}((j-1)/2)!} \quad \text{if} \quad \begin{array}{l} j \text{ even} \\ j \text{ odd} \end{array} \tag{4.4.5}$$

for $j = 1, 2, \ldots$.

Since $N_{(0, C\Sigma C^t)}$ is the normal distribution induced by $N_{(0, \Sigma)}$ and the map $\mathbf{x} \to C\mathbf{x}$ where C is a m,k-matrix with rank m we know that the distribution induced by $N_{(0, \Sigma)}$ and the map $\mathbf{x} \to x_i - x_{i-1}$ is the univariate normal distribution $N_{(0, (\lambda_i - \lambda_{i-1})(1 - (\lambda_i - \lambda_{i-1})))}$.

This together with (4.4.5) implies that

$$\int |x_i - x_{i-1}|^j \, dN_{(0,\Sigma)}(\mathbf{x})$$
$$= \frac{1 \cdot 3 \cdot 5 \cdot \ldots \cdot (j-1) [(\lambda_i - \lambda_{i-1})(1 - (\lambda_i - \lambda_{i-1}))]^{j/2}}{(2^j/\pi)^{1/2}((j-1)/2)! [(\lambda_i - \lambda_{i-1})(1 - (\lambda_i - \lambda_{i-1}))]^{j/2}} \quad \text{if} \quad \begin{array}{l} j \text{ even} \\ j \text{ odd.} \end{array} \tag{4.4.6}$$

Further, by applying Lemma 4.4.1, we obtain for $i = 2, \ldots, k - 1$,

$$\int x_i^2 x_{i-1} \, dN_{(0,\Sigma)}(\mathbf{x}) = \int x_i x_{i-1}^2 \, dN_{(0,\Sigma)}(\mathbf{x}) = 0. \tag{4.4.7}$$

4.5. Asymptotic Normality and Expansions of Joint Distributions

In the particular case of exponential r.v.'s we know that spacings are independent so that it will be easy to deduce the asymptotic normality and an expansion of the joint distribution of several central order statistics from the corresponding expansion for a single order statistic.

In a second step the result will be extended to a larger class of order statistics by using the transformation technique.

We will use the abbreviations of Section 4.4: Given positive integers n, k, and r_i with $1 \le r_1 < r_2 < \cdots < r_k \le n$, put $b_i = r_i/(n+1)$ and $\sigma_{i,j} = b_i(1 - b_j)$ for $1 \le i \le j \le k$. Moreover, denote by $N_{(0,\Sigma)}$ the k-variate normal distribution with mean vector zero and covariance matrix $\Sigma = (\sigma_{i,j})$. Again, the unit matrix is denoted by I.

Normal Approximation: Exponential R.V.'s

First let us consider the case of order statistics from exponential r.v.'s. Before treating the expansion of length two we shall discuss the result and the proof in connection with the simpler normal approximation.

Let $X_{i:n}$ be the ith order statistic of n i.i.d. standard exponential r.v.'s. Denote by P_n the joint distribution of

$$(n+1)^{1/2} g(G^{-1}(b_i))(X_{r_i:n} - G^{-1}(b_i)), \qquad i = 1, \ldots, k, \tag{4.5.1}$$

where G is the standard exponential d.f. with density g. Moreover, $\| \ \|$ denotes again the variational distance.

Theorem 4.5.1. *For all positive integers k and r_i with $0 = r_0 < r_1 < r_2 < \cdots < r_k < r_{k+1} = n + 1$ the following inequality holds:*

$$\| P_n - N_{(0,\Sigma)} \| \leq C \exp(C\rho_n)\rho_n^{1/2} \qquad (4.5.2)$$

where $C = \max(1, 2C_2)$, C_2 is the constant in Theorem 4.2.4 for $m = 2$, and ρ_n is defined by

$$\rho_n = 2 \sum_{i=1}^{k+1} (r_i - r_{i-1})^{-1}. \qquad (4.5.3)$$

Since $\sum_{i=1}^{k+1} (r_i - r_{i-1})/(n + 1) = 1$ we infer from Jensen's inequality (see P.3.9) that

$$\rho_n \geq 2k^2/n$$

which shows that $N_{(0,\Sigma)}$ will provide an accurate approximation to P_n only if the number of order statistics under consideration is bounded away from $n^{1/2}$. From the expansion of length 2 we shall learn that the bound in (4.5.2) is sharp.

Next we make some comments about the proof of Theorem 4.5.1. Notice that the asymptotic normality of several order statistics holds if the corresponding spacings have this property. Let Q_n denote the joint distribution of the normalized spacings

$$\left(\frac{(n + 1)(1 - b_{i-1})(1 - b_i)}{b_i - b_{i-1}} \right)^{1/2} (X_{r_i:n} - X_{r_{i-1}:n} - (G^{-1}(b_i) - G^{-1}(b_{i-1}))) \qquad (4.5.4)$$

for $i = 1, \ldots, k$ (with the convention that $b_0 = 0$ and $G^{-1}(b_0) = 0$).

Denote again by T the map in (4.4.1) which transforms $N_{(0,I)}$ to $N_{(0,\Sigma)}$ [that is, $TN_{(0,I)} = N_{(0,\Sigma)}$]. Since $G^{-1}(b_i) = -\log(1 - b_i)$ and hence $g(G^{-1}(b_i)) = 1 - b_i$ it is easy to see that

$$TQ_n = P_n.$$

Therefore,

$$\| P_n - N_{(0,\Sigma)} \| = \| Q_n - N_{(0,I)} \|. \qquad (4.5.5)$$

On the right-hand side of (4.5.5) one has to calculate the variational distance of the two product measures $Q_n := \times_{i=1}^{k} Q_{n,i}$ and $N_{(0,I)} = N_{(0,1)}^k$ where $Q_{n,i}$ is the distribution of the ith spacing as given in (4.5.4).

From Lemma 1.4.3 we know that spacings of exponential r.v.'s are distributed like order statistics of exponential r.v.'s. Since $G^{-1}(b_i) - G^{-1}(b_{i-1}) = G^{-1}((r_i - r_{i-1})/(n - r_{i-1} + 1))$ we obtain that $Q_{n,i}$ is the distribution of the normalized order statistic

$$\frac{(m_i + 1)^{3/2} g(G^{-1}(s_i/(m_i + 1)))}{(s_i(m_i - s_i + 1))^{1/2}} (X_{s_i:m_i} - G^{-1}(s_i/(m_i + 1))) \qquad (4.5.6)$$

where $m_i = n - r_{i-1}$ and $s_i = r_i - r_{i-1}$.

Section 3.3 provides the inequalities $\|Q_n - N_{(0,I)}\| \le \sum_{i=1}^{k} \|Q_{n,i} - N_{(0,1)}\|$ as well as

$$\|Q_n - N_{(0,I)}\| \le \left(\sum_{i=1}^{k} H(Q_{n,i}, N_{(0,1)})^2 \right)^{1/2}$$

where H denotes the Hellinger distance. The first inequality and upper bounds of $\|Q_{n,i} - N_{(0,1)}\|$, $i = 1, \ldots, k$ (compare with Corollary 4.2.7) lead to an inaccurate upper bound of $\|Q_n - N_{(0,I)}\|$. The second inequality is not applicable since a bound of the Hellinger distance between $Q_{n,i}$ and $N_{(0,1)}$ is not at our disposal. The way out of this dilemma will be the use of an expansion of length two.

Expansion of Length Two: Exponential R.V.'s

To simplify our notation we shall only establish an expansion of length two. Expansions of length m can be proved by the same method.

Theorem 4.5.2. *Let* C, $X_{i:n}$, r_i, P_n *and* ρ_n *be as in Theorem 4.5.1. Then, the following inequality holds:*

$$\sup_{B} \left| P_n(B) - \int_{B} (1 + L_{\mathbf{r},n}) \, dN_{(0,\Sigma)} \right| \le C \exp(C\rho_n)\rho_n \qquad (4.5.7)$$

where $L_{\mathbf{r},n}$ *is the polynomial defined by*

$$L_{\mathbf{r},n}(\mathbf{x}) = \sum_{i=1}^{k} L_{1,r_i-r_{i-1},n-r_{i-1}}(x_i/\gamma_{i,i} - x_{i-1}/\gamma_{i-1,i})$$

with $L_{1,r,n}$ *defined as in Corollary 4.2.7,* $x_0 = 0$ *and*

$$\gamma_{i,j} = (1 - b_i)[(b_j - b_{j-1})/(1 - b_{j-1})(1 - b_j)]^{1/2}.$$

PROOF. From (4.5.6) and Corollary 4.2.7 it is immediate that

$$\sup_{B} \left| Q_{n,i}(B) - \int_{B} (1 + L_{1,r_i-r_{i-1},n-r_i}) \, dN_{(0,1)} \right|$$

$$\le C_2 \frac{n - r_{i-1}}{(r_i - r_{i-1})(n - r_i + 1)} =: C_2 \delta_i. \qquad (1)$$

The bound for the variational distance between product measures via the variational distance between the single components (compare with Corollary

A.3.4)) yields

$$\sup_B \left| \left(\bigtimes_{i=1}^k Q_{n,i} \right)(B) - \int_B \prod_{i=1}^k (1 + L_{1,r_i-r_{i-1},n-r_{i-1}}(x_i))\, dN_{(0,1)}^k(\mathbf{x}) \right|$$

$$\le C_2 \exp\left[2C_2 \sum_{i=1}^k \delta_i \right] \sum_{i=1}^k \delta_i. \tag{2}$$

Next we verify that the integral in (2) can be replaced by that in (4.5.7). Lemma A.3.6, applied to $g_i = L_{1,r_i-r_{i-1},n-r_{i-1}}$, yields

$$\sup_B \left| \int_B \prod_{i=1}^k [1 + L_{1,r_i-r_{i-1},n-r_{i-1}}(x_i)]\, dN_{(0,1)}^k(\mathbf{x}) \right.$$

$$\left. - \int_B \left[1 + \sum_{i=1}^k L_{1,r_i-r_{i-1},n-r_{i-1}}(x_i) \right] dN_{(0,1)}^k(\mathbf{x}) \right| \tag{4.5.8}$$

$$\le 8^{-1/2} \exp\left[2^{-1} \sum_{i=1}^k \int L_{1,r_i-r_{i-1},n-r_{i-1}}^2\, dN_{(0,1)} \right] \sum_{i=1}^k \int L_{1,r_i-r_{i-1},n-r_{i-1}}^2\, dN_{(0,1)}$$

$$\le 18^{-1/2} \exp\left[3^{-1} \sum_{i=1}^k \delta_i \right] \sum_{i=1}^k \delta_i$$

where the last step is immediate from (4.2.7).

Check that $\sum_{i=1}^k \delta_i \le \rho_n$. Combining (2) and (4.5.8) we obtain

$$\sup_B \left| \left(\bigtimes_{i=1}^k Q_{n,i} \right)(B) - \int_B \left[1 + \sum_{i=1}^k L_{1,r_i-r_{i-1},n-r_{i-1}}(x_i) \right] dN_{(0,1)}^k(\mathbf{x}) \right|$$

$$\le C_2 \exp[2C_2\rho_n]\rho_n + 18^{-1/2} \exp[3^{-1}\rho_n]\rho_n \le C \exp(C\rho_n)\rho_n. \tag{4.5.9}$$

Now, the transformation, as explained in (4.5.5), yields the desired inequality (4.5.7). For this purpose apply the transformation theorem for densities. Note that the inverse S of T is given by

$$S(\mathbf{x}) = [x_i/\gamma_{i,i} - x_{i-1}/\gamma_{i-1,i}]_{i=1}^k. \qquad \square$$

From (4.5.9) we also deduce for the normalized, joint distribution P_n of order statistics that

$$\|P_n - N_{(0,\Sigma)}\| = 2^{-1} \int \left| \sum_{i=1}^k L_{1,r_i-r_{i-1},n-r_{i-1}}(x_i) \right| dN_{(0,1)}^k(\mathbf{x}) + O(\rho_n)$$

$$\le 2^{-1}\rho_n^{1/2} + O(\rho_n) \tag{4.5.10}$$

where the last inequality follows by means of the Schwarz inequality.

Notice that (4.5.10) is equivalent to (4.5.2) as far as the order of the normal approximation is concerned. However, to prove (4.5.2) with the constant as stated there one has to utilize a slight modification of the proof of Theorem 4.5.2.

PROOF OF THEOREM 4.5.1. Applying Lemma A.3.6 again we obtain

$$\sup_B \left| \int_B \prod_{i=1}^k [1 + L_{1, r_i - r_{i-1}, n - r_{i-1}}(x_i)] \, dN_{(0,1)}^k(\mathbf{x}) - N_{(0,1)}^k(B) \right|$$

$$\le \exp[3^{-1} \rho_n] (\rho_n/6)^{1/2}$$

(4.5.8')

showing that (4.5.2) can be proved in the same way as (4.5.7) by applying (4.5.8') in place of (4.5.8).   □

Normal Approximation: General Case

Hereafter, let P_n denote the joint distribution of the normalized order statistics

$$(n + 1)^{1/2} f(F^{-1}(b_i))(X_{r_i : n} - F^{-1}(b_i)), \qquad i = 1, \dots, k, \qquad (4.5.11)$$

where $X_{i:n}$ is the ith order statistics of n i.i.d. random variables with common d.f. F and density f, and $b_i = r_i/(n + 1)$. Recall that the covariance matrix Σ is defined by $\sigma_{i,j} = b_i(1 - b_j)$ for $1 \le i \le j \le k$.

From Theorem 4.3.1 and 4.5.1 it is easily seen that under certain regularity conditions,

$$\|P_n - N_{(\mathbf{0}, \Sigma)}\| = O(\rho_n^{1/2}) \qquad (4.5.12)$$

with ρ_n as in (4.5.3). The crucial point is that the underlying density is assumed to possess three bounded derivatives. The aim of the following considerations is to show that (4.5.12) holds if f has two bounded derivatives. The bound $O(\rho_n^{1/2})$ is sharp as far as the normal approximation is concerned, however, $\rho_n^{1/2}$ is of a larger order than the upper bound in Theorem 4.3.1.

Theorem 4.5.3. *Denote by P_n the joint distribution of the normalized order statistics in (4.5.11). Assume that the underlying density f has two derivatives on the intervals $I_i = (F^{-1}(b_i) - \varepsilon_i, F^{-1}(b_i) - \varepsilon_i)$, $i = 1, \dots, k$, where $\varepsilon_i = 5[\sigma_{i,i} \log(n)/(n + 1)]^{1/2}/f(F^{-1}(b_i))$. Moreover, assume that $\min(b_1, 1 - b_k) \ge 10 \log(n)/(n + 1)$.*

Then there is a universal constant $C > 0$ such that

$$\|P_n - N_{(\mathbf{0}, \Sigma)}\| \le C(1 + d(f))\rho_n^{1/2}$$

where $d(f) = \max_{j=1}^2 \max_{i=1}^k (\sup_{y \in I_i} |f^{(j)}(y)|/\inf_{y \in I_i} f^{j+1}(y))$.

PROOF. In the first part of the proof we deal with the special case of order statistics $U_{r:n}$ of n i.i.d. random variables with uniform distribution on $(0, 1)$. In this case, an application of Theorem 4.3.1 would yield a result which is only slightly weaker than that stated above. The present method has the advantage of being simpler than that of Theorem 4.3.1 and, moreover, it will also be applicable in the second part.

I. Let Q_n denote the joint distribution of normalized order statistics $X_{r_1:n}, \ldots, X_{r_k:n}$ of standard exponential r.v.'s with common d.f. G and density g. Write $g_i = g(G^{-1}(b_i))$. Denote by Q_n^* the joint distribution of $(n+1)^{1/2}(U_{r_j:n} - b_i)$, $i = 1, \ldots, k$. From Corollary 1.2.6 it is easily seen that

$$Q_n^* = TQ_n \tag{1}$$

where $T(\mathbf{x}) = (T_1(x_1), \ldots, T_k(x_k))$ and

$$T_i(x_i) = (n+1)^{1/2}\left(G\left(G^{-1}(b_i) + \frac{x_i}{(n+1)^{1/2}g_i} \right) - b_i \right) \tag{2}$$

for every $\mathbf{x}$ such that $G^{-1}(b_i) + x_i/((n+1)^{1/2}g_i) > 0$, $i = 1, \ldots, k$.
    Theorem 4.5.1 and (1) yield

$$\|Q_n^* - N_{(\mathbf{0},\Sigma)}\| \le \|TQ_n - TN_{(\mathbf{0},\Sigma)}\| + \|TN_{(\mathbf{0},\Sigma)} - N_{(\mathbf{0},\Sigma)}\|$$
$$\le C\rho_n^{1/2} + \|TN_{(\mathbf{0},\Sigma)} - N_{(\mathbf{0},\Sigma)}\|$$

where, throughout, C denotes a universal constant that will not be the same at each appearance. Thus, it remains to prove that

$$\|TN_{(\mathbf{0},\Sigma)} - N_{(\mathbf{0},\Sigma)}\| \le C\rho_n^{1/2}. \tag{3}$$

The inverse S of T is given by $S(\mathbf{x}) = (S_1(x_1), \ldots, S_k(x_k))$ where

$$S_i(x_i) = (n+1)^{1/2}g_i(G^{-1}(b_i + x_i/(n+1)^{1/2}) - G^{-1}(b_i)), \qquad i = 1, \ldots, k, \tag{4}$$

for $\mathbf{x}$ with $0 < b_i + x_i/(n+1)^{1/2} < 1$. Inequality (3) holds if

$$\|N_{(\mathbf{0},\Sigma)} - SN_{(\mathbf{0},\Sigma)}\| \le C\rho_n^{1/2}. \tag{5}$$

We prefer to prove (5) instead of (3) since this is the inequality that also has to be verified in the second part of the proof with G replaced by F.
    Denote by N_T and N_S the restrictions of $N_{(\mathbf{0},\Sigma)}$ to the domains D_T of T and D_S of S. Check that

$$\|TN_{(\mathbf{0},\Sigma)} - N_{(\mathbf{0},\Sigma)}\| \le \|(T \circ S \circ T)N_{(\mathbf{0},\Sigma)} - (T \circ S)N_{(\mathbf{0},\Sigma)}\| + \|N_S - N_{(\mathbf{0},\Sigma)}\|$$
$$\le \|N_{(\mathbf{0},\Sigma)} - SN_{(\mathbf{0},\Sigma)}\| + N_{(\mathbf{0},\Sigma)}(D_T^c) + N_{(\mathbf{0},\Sigma)}(D_S^c)$$

which shows that (5) implies (3) since

$$\max(N_{(\mathbf{0},\Sigma)}(D_T^c), N_{(\mathbf{0},\Sigma)}(D_S^c)) \le C\rho_n. \tag{6}$$

(6) in conjunction with (A.3.5) yields

$$\|N_{(\mathbf{0},\Sigma)} - SN_{(\mathbf{0},\Sigma)}\| \le C\rho_n + \left[2N_{(\mathbf{0},\Sigma)}(B^c) + \int_B (-\log(f_1/f_0))\, dN_{(\mathbf{0},\Sigma)} \right]^{1/2} \tag{7}$$

for sets B in the domain of T, and f_0, f_1 being the densities of $N_{(\mathbf{0},\Sigma)}$ and $SN_{(\mathbf{0},\Sigma)}$. Applying the transformation theorem for densities (1.4.4) we obtain

$$-\log(f_1/f_0)(\mathbf{x}) = -\sum_{i=1}^{k} \log T_i'(x_i) + 2^{-1} \sum_{i=1}^{k+1} \frac{\delta_i(\mathbf{x})(x_i - x_{i-1}) + \delta_i^2(\mathbf{x})/2}{b_i - b_{i-1}}, \qquad (8)$$

$\mathbf{x} \in B$, where

$$\delta_i(\mathbf{x}) = (T_i(x_i) - x_i) - (T_{i-1}(x_{i-1}) - x_{i-1})$$

(with the convention that $T_{k+1}(x_{k+1}) = T_0(x_0) = x_{k+1} = x_0 = 0$ and $b_{k+1} = 1$, $b_0 = 0$).

Check that

$$-\log T_i'(x_i) = x_i/(n + 1)^{1/2}(1 - b_i) \qquad (9)$$

and, for $x_i \geq -(n + 1)^{1/2}\sigma_{i,i}$,

$$\left| T_i(x_i) - x_i + \frac{x_i^2}{2(n + 1)^{1/2}(1 - b_i)} \right| \leq \frac{|x_i|^3}{2(n + 1)(1 - b_i)^2}. \qquad (10)$$

Define

$$B = \{\mathbf{x}: x_i > -(10(\log n)\sigma_{i,i})^{1/2}, i = 1, \ldots, k\}.$$

Applying the inequality $1 - \Phi(x) \leq \varphi(x)/x$ we obtain

$$N_{(\mathbf{0},\Sigma)}(B^c) \leq n^{-4}. \qquad (11)$$

The condition $\min(b_1, 1 - b_k) \geq 10 \log(n)/(n + 1)$ yields $B \subset D_T$ and (10) holds for $\mathbf{x} \in B$ for $i = 1, \ldots, k$. Since

$$\int x_i \, dN_{(\mathbf{0},\Sigma)}(\mathbf{x}) = 0, \qquad i = 1, \ldots, k, \qquad (12)$$

we obtain, by applying (9) and the Schwarz inequality, that

$$\int_B \left(-\sum_{i=1}^{k} \log T_i'(x_i)\right) dN_{(\mathbf{0},\Sigma)}(\mathbf{x}) \leq \sum_{i=1}^{k} \int_{B^c} \frac{|x_i|}{(n + 1)^{1/2}(1 - b_i)} dN_{(\mathbf{0},\Sigma)}(\mathbf{x})$$

$$\leq Cn^{-1}. \qquad (13)$$

Notice that according to (4.4.7),

$$\sum_{i=1}^{k} \int (x_i^2 - x_{i-1}^2)(x_i - x_{i-1}) \, dN_{(\mathbf{0},\Sigma)}(\mathbf{x}) = 0, \qquad (14)$$

and hence, applying (4.4.5) and (4.4.6), we obtain by means of some straightforward calculations that

$$\int_B \left(\sum_{i=1}^{k+1} \frac{\delta_i(\mathbf{x})(x_i - x_{i-1}) + \delta_i^2(\mathbf{x})/2}{b_i - b_{i-1}}\right) dN_{(\mathbf{0},\Sigma)}(\mathbf{x}) \leq C\rho_n. \qquad (15)$$

Combining (11), (13), and (15) we see that the assertion of Part I holds.

II. Notice that $P_n = SQ_n^*$ where S is defined as in (4) with G and g_i replaced by F and $f(F^{-1}(b_i))$. Using Taylor expansions of $\log T_i'(x_i)$ and $T_i(x_i)$ the proof of this part runs along the lines of Part I. $\square$

Final Remarks

In Reiss (1981a) one can also find expansions of length $m > 2$ for the joint distribution of central order statistics of exponential r.v.'s. Starting with this special case, one may derive expansions in case of r.v.'s with sufficiently smooth d.f. by using the method as adopted in Reiss (1975a); that is, one has to expand the densities and to integrate the densities over Borel sets in a more direct way.

4.6. Expansions of Distribution Functions of Order Statistics

In Sections 4.2 and 4.5, expansions of distributions of central order statistics were established which hold w.r.t. the variational distance. These expansions can be represented by means of polynomials that are densities w.r.t. the standard normal distribution.

Expansions for d.f.'s can be written in a way which is more adjusted to d.f.'s The results for d.f.'s of order statistics hold under conditions which are weaker than those required for approximations in the strong sense. Along with the reformulation of the results of Section 4.2 we shall study expansions of d.f.'s of order statistics under conditions that hold for order statistics of discrete r.v.'s.

Write again

$$a_{r,n}^2 = r(n - r + 1)/(n + 1)^3 \quad \text{and} \quad b_{r,n} = r/(n + 1).$$

Continuous D.F.'s

First, the results of Section 4.2 will be rewritten in terms of d.f.'s.

Corollary 4.6.1. *Under the conditions of Theorem 4.2.4 there exist polynomials $S_{i,r,n}$ of degree $\leq 3i - 1$ such that*

$$\sup_t \left| P\{a_{r,n}^{-1} f(F^{-1}(b_{r,n}))(X_{r:n} - F^{-1}(b_{r,n})) \leq t\} - \left(\Phi(t) + \varphi(t) \sum_{i=1}^{m-1} S_{i,r,n}(t) \right) \right|$$

$$\leq C_m \left[(n/r(n-r))^{m/2} + \max_{j=1}^{m} |\alpha_{j,r,n}|^{m/j} \right] \tag{4.6.1}$$

where $\alpha_{j,r,n}$ are the terms in Theorem 4.2.4.

PROOF. Apply Lemma 3.2.6. $\qquad\qquad\qquad\qquad\qquad\qquad\qquad\qquad$ □

Let us note the explicit form of $S_{1,r,n}$ and $S_{2,r,n}$. We have

$$(\varphi S_{i,r,n})' = \varphi L_{i,r,n}$$

with $L_{i,r,n}$ as in Addendum 4.2.5. Moreover,

$$S_{1,r,n}(t) = \frac{n - 2r + 1}{3[r(n - r + 1)(n + 1)]^{1/2}}(1 - t^2) + \alpha_{1,r,n}t^2/2 \qquad (4.6.2)$$

and

$$S_{2,r,n}(t) = \frac{1}{r(n - r + 1)(n + 1)}[-(n - 2r + 1)^2(15t + 5t^3 + t^5)/18$$

$$+ [7(n - 2r + 1)^2 + 3r(n - r + 1)](3t + t^3)/12 + (n - r + 1)^2 t]$$

$$+ \alpha_{1,r,n}\frac{t}{2}L_{1,r,n}(t) - \alpha_{1,r,n}^2\frac{t^5}{8} + \alpha_{2,r,n}\frac{t^3}{6} \qquad (4.6.3)$$

with $\alpha_{j,r,n}$ as in Theorem 4.2.4 and $L_{1,r,n}$ as in (4.2.2).

EXAMPLE 4.6.2. We have

$$\sup_t \left| P\left\{a_{r,n}^{-1}\left(U_{r:n} - \frac{r}{n + 1}\right) \le t\right\} - \left(\Phi(t) + \varphi(t)\sum_{i=1}^{m-1} S_{i,r,n}(t)\right)\right|$$

$$\le C_m \left(\frac{n}{r(n - r)}\right)^{m/2} \qquad (4.6.4)$$

where $S_{i,r,n}$ are the polynomials of Corollary 4.6.1 with $\alpha_{i,r,n} = 0$.

Discrete D.F.'s

The conditions of Theorem 4.2.4 exclude discrete d.f.'s F. The key idea of the following is to approximate the d.f. F (which may be discrete) by some function G which fulfills an appropriate Taylor expansion.

As an example we shall treat the case of d.f.'s F that permit an Edgeworth expansion (like binomial d.f.'s).

We start with a technical lemma.

Lemma 4.6.3. *Let $X_{i:n}$ be the order statistics of n i.i.d. random variables with common d.f. F.*

Let G be a function and u a fixed real number such that for all reals y,

$$\left| G(u + y) - G(u) - \sum_{i=1}^m \frac{c_i}{i!}y^i \right| \le \frac{c_{m+1}}{(m + 1)!}|y|^{m+1}. \qquad (4.6.5)$$

Then, if $c_1 > 0$ there exists a universal constant $C_m > 0$ and polynomials $S_{i,r,n}$ of degree $\leq 3i - 1$ such that for all reals t the following inequality holds:

$$\left| P\{a_{r,n}^{-1}c_1(X_{r:n} - u) \leq t\} - \left(\Phi(t) + \varphi(t) \sum_{i=1}^{m-1} S_{i,r,n}(t) \right) \right|$$

$$\leq C_m \left[\left(\frac{n}{r(n-r+1)} \right)^{m/2} + a_{r,n}^m \max_{j=1}^{m} (c_{j+1}/c_1^{j+1})^{m/j} \right. \tag{4.6.6}$$

$$\left. + a_{r,n}^{-1}(|F(u + ta_{r,n}/c_1) - G(u + ta_{r,n}/c_1)| + |G(u) - b_{r,n}|) \right].$$

PROOF. Writing $x = u + ta_{r,n}/c_1$ we get

$$P\{a_{r,n}^{-1}c_1(X_{r:n} - u) \leq t\} = P\{a_{r,n}^{-1}(U_{r:n} - b_{r,n}) \leq a_{r,n}^{-1}(F(x) - b_{r,n})\}.$$

Denote by $\tilde{S}_{i,r,n}$ the polynomials of Example 4.6.2. Since

$$a_{r,n}^{-1}(F(x) - b_{r,n}) = a_{r,n}^{-1}(F(x) - G(x)) + V(t) + a_{r,n}^{-1}(G(u) - b_{r,n}),$$

with $V(t) = a_{r,n}^{-1}(G(x) - G(u))$, it is immediate from Example 4.6.2 that

$$\left| P\{a_{r,n}^{-1}c_1(X_{r:n} - u) \leq t\} - \left[\Phi(V(t)) + \varphi(V(t)) \sum_{i=1}^{m-1} \tilde{S}_{i,r,n}(V(t)) \right] \right|$$

$$\leq C_m \left[\left(\frac{n}{r(n-r+1)} \right)^{m/2} + a_{r,n}^{-1}(|F(x) - G(x)| + |G(u) - b_{r,n}|) \right].$$

Using condition (4.6.5) we obtain an expansion of $V(t)$ of length m, namely,

$$V(t) = t + \sum_{i=2}^{m} \frac{c_i a_{r,n}^{i-1}}{i! c_1^i} t^i + \theta_m(t) \frac{c_{m+1} a_{r,n}^m}{(m+1)! c_1^{m+1}} |t|^{m+1}$$

where $|\theta_m(t)| \leq 1$. Now arguments analogous to those of the proof to Theorem 4.2.4 lead to (4.6.6). $\qquad\square$

The polynomials in Lemma 4.6.3 are of the same form as those in Corollary 4.6.1 with $\alpha_{j,r,n}$ replaced by $a_{r,n}^j c_{j+1}/c_1^{j+1}$.

Next, Lemma 4.6.3 will be specialized to d.f.'s $F \equiv F_N$ permitting an Edgeworth expansion $G \equiv G_{M,N}$ of the form

$$G_{M,N}(t) = \Phi(t) + \varphi(t) \sum_{i=1}^{M-1} N^{-i/2} Q_i(t)$$

where M and N are positive integers, and Q_i is a polynomial for $i = 1, \ldots, M - 1$. Let us assume that

$$|F_N(t) - G_{M,N}(t)| \leq C_M N^{-M/2} \tag{4.6.7}$$

uniformly over $t \in I$ where I will be specified below.

If F_N stems from a N-fold convolution, typically one has the following two cases:

(i) I is the real line if the Cramér–von Mises condition holds,

(ii) $I = \{y + kh: k \text{ integer}\}$ where y and $h > 0$ are fixed.

Moreover, define an "inverse" $G_{M,N}^*$ of $G_{M,N}$ by

$$G_{M,N}^* = \Phi^{-1} + \sum_{i=1}^{M-1} N^{-i/2} Q_i^*(\Phi^{-1})$$

where the Q_i^* are the polynomials as described in Pfanzagl (1973c), Lemma 7.

We note that

$$Q_1^* = -Q_1$$

and

$$Q_2^*(t) = Q_1(t)Q_1'(t) - \tfrac{1}{2}Q_1(t)^2 - Q_2(t).$$

$$\tag{4.6.8}$$

Since $G_{M,N}$ is an approximation to F_N we know that $G_{M,N}^*$ is an approximation to F_N^{-1}. As an application of Lemma 4.6.3 to $F \equiv F_N$, $G \equiv G_{M,N}$, and $u = G_{M,N}^*(b_{r,n})$ we obtain the following

Corollary 4.6.4. *Under condition* (4.6.7) *there exists* $C_{m,M} > 0$ *such that for every positive integer* n, $r \in \{1, \dots, n\}$ *and* $t \in I$:

$$\left| P\{X_{r:n} \le t\} - \left(\Phi + \varphi \sum_{i=1}^{m-1} S_{i,r,n} \right)(s_M(t)) \right|$$

$$\le C_{m,M}[a_{r,n}^{-1} N^{-M/2} + a_{r,n}^m \varphi(\Phi^{-1}(b_{r,n}))^{-2m}]$$

$$\tag{4.6.9}$$

where

$$s_M(t) = a_{r,n}^{-1} G_{M,N}'[G_{M,N}^*(b_{r,n})](t - G_{M,N}^*(b_{r,n}))$$

and the $S_{i,r,n}$ *are the polynomials of Lemma 4.6.3 with* $c_i = G_{M,N}^{(i)}(G_{M,N}^*(b_{r,n}))$.

PROOF. To make Lemma 4.6.3 applicable one has to verify that

$$G_{M,N}(G_{M,N}^*(b_{r,n})) = b_{r,n} + O(N^{-m/2}). \tag{1}$$

It suffices to prove that (1) holds uniformly over all r and n such that $|\Phi^{-1}(b_{r,n})| = O(\log N)$. A standard technique [see Pfanzagl (1973c), page 1016] yields

$$G_{M,N}(\tilde{G}_{M,N}(t)) = \Phi(t) + O(N^{-m/2}) \tag{2}$$

uniformly over $|t| = O(\log N)$ where $\tilde{G}_{M,N}(t) = t + \sum_{i=1}^{M-1} N^{-i/2} Q_i^*(t)$. Thus, (1) is immediate from (2) applied to $t = \Phi^{-1}(b_{r,n})$. $\qquad\square$

To exemplify the usefulness of Corollary 4.6.4 we study the d.f. of an order statistic $X_{r:n}$ of n i.i.d. binomial r.v.'s with parameters N and $p \in (0, 1)$. It is clear that

$$P\{X_{r:n} \le t\} = \sum_{i=r}^{n} \binom{n}{i} F_N^i(t)(1 - F_N(t))^{n-i}$$

where

$$F_N(t) = \sum_{k=0}^{[t]} \binom{N}{k} p^k (1 - p)^{N-k}$$

with [ ] denoting the integer function. Moreover, $P\{X_{r:n} \leq t\} = P\{X_{r:n} \leq [t]\}$ so that $P\{X_{r:n} \leq t\}$ has to be evaluated at $t \in \{0, \ldots, N\}$ only.

As an approximation to the normalized version of F_N we use the standard normal d.f. Φ and the Edgeworth expansion $\Phi + N^{-1/2} \varphi Q_1$ of length 2 where (see Bhattacharya and Rao (1976), Theorem 23.1)

$$Q_1(t) = [(2p - 1)t^2 + (4 - 2p)]/6(p(1 - p))^{1/2}.$$

Table 4.6.1. Maximum Absolute Deviation of Exact Values and Expansions

	$p = .2$ $N = n$ $r = [n/4]$		$p = .5$ $N = n$ $r = [n/2]$		$p = .2$ $N = [n^{4/3}]$ $r = [n/4]$		$p = .5$ $N = [n^{4/3}]$ $r = [n/2]$	
	(m, M)							
n	$(1, 1)$	$(2, 2)$	$(1, 1)$	$(2, 2)$	$(1, 1)$	$(2, 2)$	$(1, 1)$	$(2, 2)$
20	.33	.01	.35	.01	.29	.007	.27	.006
80	.38	.002	.32	.003	.22	.002	.20	.0028
200	.42	.0001	.31	.0001	.20	.0007	.16	.0005

Table 4.6.1 presents a numerical comparison of the approximations in Corollary 4.6.4 in the special cases of $(m, M) = (1, 1)$ and $(m, M) = (2, 2)$. Thus, if $(m, M) = (1, 1)$ we compute the maximum value of

$$\left| P\{X_{r:n} \leq k\} - \Phi\left[a_{r,n}^{-1} \varphi(\Phi^{-1}(b_{r,n})) \left(\frac{k - Np}{N^{1/2}(p(1 - p))^{1/2}} - \Phi^{-1}(b_{r,n}) \right) \right] \right|$$

over $k = 0, \ldots, N$.

4.7. Local Limit Theorems and Moderate Deviations

In Section 4.2 we proved expansions of distributions of single order statistics uniformly over the Borel sets. The main technical tool was an expansion of one factor of the density (compare with the proof to (4.7.2)). The expansion of the density was not given explicitly to concentrate our attention on the result of statistical relevance, namely, the expansion of distributions.

The final section of this chapter is the proper place to give some explicit formulas for expansions of densities with an error bound that is nonuniform in x. By integration we shall also get inequalities which are relevant for probabilities of moderate deviation.

Let again

$$a_{r,n}^2 = r(n - r + 1)/(n + 1)^3 \quad \text{and} \quad b_{r,n} = r/(n + 1).$$

Denote again by $U_{r:n}$ the rth order statistic of n i.i.d. $(0, 1)$-uniformly distributed r.v.'s. From Lemma 3.1.1 we obtain

$$P\{a_{r,n}^{-1}|U_{r:n} - b_{r,n}| \geq \varepsilon\} \leq 2\exp\left(-\frac{\varepsilon^2}{3[1 + n^{-1} + \varepsilon/(a_{r,n}n)]}\right), \qquad \varepsilon > 0. \tag{4.7.1}$$

A refinement of this result will be obtained in the second part of this section.

Local Limit Theorems

Denote by $g_{r,n}$ the density of

$$a_{r,n}^{-1}(U_{r:n} - b_{r,n})$$

and by Φ and φ the standard normal d.f. and density. The most simple "local limit theorem" is given by the inequality

$$|g_{r,n}(x) - \varphi(x)| \leq C\varphi(x)\left(\frac{n}{r(n - r + 1)}\right)^{1/2}(1 + |x|^3) \tag{4.7.2}$$

which holds for

$$x \in A(r, n) := \{x : |x| \leq (r(n - r)/n)^{1/6}\} \tag{4.7.3}$$

where the constant $C > 0$ is independent of x.

To prove (4.7.2) let us follow the lines of the proof to Theorem 4.2.1. The density $g_{r,n}$ of $a_{r,n}^{-1}(U_{r:n} - b_{r,n})$ is written as $\rho_{r,n}h_{r,n}$ where $\rho_{r,n}$ is a normalizing constant. From the proof of Theorem 4.2.1(1) we know that

$$|\exp(x^2/2)h_{r,n}(x) - 1| \leq C[n/r(n - r + 1)]^{1/2}(|x| + |x|^3) \tag{4.7.4}$$

for $x \in A(r, n)$.

We also need an expansion of the factor $\rho_{r,n}$. By integration over an interval B we get uniformly in r and n that

$$\rho_{r,n} = \int_B g_{r,n}(x)\,dx \Big/ \int_B h_{r,n}(x)\,dx$$

$$= P\{a_{r,n}^{-1}(U_{r:n} - b_{r,n}) \in B\}/[(2\pi)^{1/2}N_{(0,1)}(B) + O((n/r(n - r))^{1/2}] \tag{4.7.5}$$

$$= (2\pi)^{-1/2} + O((n/r(n - r))^{1/2})$$

where the final step is immediate by specifying $B = \{x : |x| \leq \log(r(n - r)/n)\}$ and applying (4.7.1) to $\varepsilon = \log(r(n - r)/n)$.

An expansion of length m can be established in the same way. For some constant $C_m > 0$ we get

$$\left| g_{r,n}(x) - \varphi(x)\left(1 + \sum_{i=1}^{m-1} L_{i,r,n}(x)\right)\right| \le C_m \varphi(x)\left(\frac{n}{r(n-r+1)}\right)^{m/2}(1+|x|^{3m})$$

$$(4.7.6)$$

for $x \in A(r,n)$ with polynomials $L_{i,r,n}$ as given in Theorem 4.2.1.

In analogy to Theorem 4.2.4 we also establish an expansion of the density of the normalized rth order statistic under the condition that the underlying d.f. has $m + 1$ derivatives.

Theorem 4.7.1. *For some $r \in \{1,\ldots,n\}$ let $X_{r:n}$ be the rth order statistic of n i.i.d. random variables with common d.f. F and density f. Assume that $f(F^{-1}(b_{r.n})) > 0$ and that the function $S_{r,n}$ defined by*

$$S_{r,n}(x) = a_{r,n}^{-1}(F[F^{-1}(b_{r,n}) + xa_{r,n}/f(F^{-1}(b_{r,n}))] - b_{r,n})$$

has $m + 1$ derivatives on the interval $I_{r,n} := \{x : |x| \le c_{r,n}\}$ where $\log(r(n-r)/n) \le c_{r,n} \le (r(n-r)/n)^{1/6}/2$. Denote by $f_{r,n}$ the density of

$$a_{r,n}^{-1} f(F^{-1}(b_{r.n}))(X_{r:n} - F^{-1}(b_{r,n})).$$

Then there exists a constant $C_m > 0$ (only depending on m) such that

$$\left| f_{r,n}(x) - \varphi(x)\left(1 + \sum_{i=1}^{m-1} L_{i,r,n}\right)\right|$$

$$(4.7.7)$$

$$\le C_m \varphi(x)(1+|x|^{3m})\left[(n/r(n-r))^{m/2} + \max_{j=1}^{m} |\alpha_{j,r,n}|^{m/j}\right]$$

for $x \in I_{r,n}$ with polynomials $L_{i,r,n}$ as given in Theorem 4.2.4. Moreover, $\alpha_{j,r,n} = S_{r,n}^{(j+1)}(0), j = 1,\ldots, m-1$, and $\alpha_{m,r,n} = \sup\{|S_{r,n}^{(m+1)}(x)| : x \in I_{r,n}\}$.

PROOF. We give a short sketch of the proof. Check that

$$f_{r,n} = S_{r,n}' g_{r,n}(S_{r,n})$$

with $g_{r,n}$ as above. Applying (4.7.6) we obtain

$$\left| f_{r,n} - S_{r,n}'\varphi(S_{r,n})\left(1 + \sum_{i=1}^{m-1} L_{i,r,n}(S_{r,n})\right)\right|$$

$$\le C_m |S_{r,n}'|\varphi(S_{r,n})(n/r(n-r))^{m/2}(1+|S_{r,n}|^{3m})$$

with polynomials $L_{i,r,n}$ as given in (4.7.6). Now, using Taylor expansions of $S_{r,n}'$ and $S_{r,n}$ about zero and of φ about x we obtain the desired result by arranging the terms in the appropriate order. □

Moderate Deviations

We shall only study a simple application of (4.7.1). It will be shown that the right-hand side of (4.7.1) can be replaced by a term $C \exp(-\varepsilon^2/2)/\varepsilon$ for certain ε.

Lemma 4.7.2. *For some constant $C > 0$,*

(i)
$$|P\{a_{r,n}^{-1}(U_{r:n} - b_{r,n}) \in B\} - N_{(0,1)}(B)|$$

$$\leq C\left(\frac{n}{r(n-r+1)}\right)^{1/2} \int_B (1 + |x|^3)\varphi(x)\,dx$$

for every Borel set $B \subset A(r, n)$ [defined in (4.7.3)].
(ii) *Moreover,*

$$P\{a_{r,n}^{-1}|U_{r:n} - b_{r,n}| \geq \varepsilon\} \leq C\exp(-\varepsilon^2/2)/\varepsilon$$

if $\varepsilon \leq (r(n-r+1)/n)^{1/6}/2$.

PROOF. (i) is immediate from (4.7.2) by integrating over B.
(ii) follows from (4.7.1) and (i). Put $d = (r(n-r+1)/n)^{1/6}$. We get

$$P\{a_{r,n}^{-1}|U_{r:n} - b_{r,n}| \geq \varepsilon\}$$

$$= P\{a_{r,n}^{-1}|U_{r:n} - b_{r,n}| \geq d\} + P\{\varepsilon \leq a_{r,n}^{-1}|U_{r:n} - b_{r,n}| \leq d\}$$

$$\leq 2\exp\left(-\frac{d^2}{3[1 + n^{-1} + d/(a_{r,n}n)]}\right)$$

$$+ C\left((1 - \Phi(\varepsilon)) + \left(\frac{n}{r(n-r+1)}\right)^{1/2}\int_\varepsilon^\infty |x|^3\,\varphi(x)\,dx\right)$$

$$\leq C\exp(-\varepsilon^2/2)/\varepsilon$$

where the final step is immediate from (3.2.3) and (3.2.12).   □

P.4. Problems and Supplements

1. (Asymptotic d.f.'s of central order statistics)
   (i) Let $r(n) \in \{1, \ldots, n\}$ be such that $n^{1/2}(r(n)/n - q) \to 0$, $n \to \infty$, for some $q \in (0, 1)$. The possible nondegenerate limiting d.f.'s of the sequence of order statistics $X_{r(n):n}$ of i.i.d. r.v.'s are of the following type:

$$H_{1,\alpha}(x) = \begin{cases} 0 & x < 0, \\ \Phi(x^\alpha) & x \geq 0, \end{cases} \quad \text{if}$$

$$H_{2,\alpha}(x) = \begin{cases} \Phi(-(-x)^\alpha) & x < 0, \\ 1 & x \geq 0, \end{cases} \quad \text{if}$$

$$H_{3,\alpha,\sigma}(x) = H_{1,\alpha}(x/\sigma)1_{[0,\infty)}(x) + H_{2,\alpha}(x)1_{(-\infty,0)}(x),$$

$$H_4 = (1_{[-1,\infty)} + 1_{[1,\infty)})/2$$

where $\alpha, \sigma > 0$.

(Smirnov, 1949)

   (ii) There exists an absolutely continuous d.f. F such that for every $q \in [0, 1]$ and every d.f. H there exists $r(n)$ with $r(n)/n \to q$ and $\min(r(n), n - r(n)) \to \infty$ as

$n \to \infty$ having the following property: Let $X_{r(n):n}$ denote the $r(n)$th order statistic of n i.i.d. random variables with common d.f. F. Then, the d.f. of $a_n^{-1}(X_{r(n):n} - b_n)$ converges weakly to H for certain $a_n > 0$ and b_n.

(Balkema and de Haan, 1978b)

(iii) The set of all d.f.'s F such that (ii) holds is dense in the set of d.f.'s w.r.t. the topology of weak convergence.

(Balkema and de Haan, 1978b)

(iv) Let $X_1, X_2, X_3, \ldots$ be a stationary, standard normal sequence with covariances $\tau(n) = EX_1 X_{n+1}$ satisfying the condition $\sum_{i=1}^{\infty} |\tau(n)| < \infty$. Let $r(n) \in \{1, \ldots, n\}$ be such that $r(n)/n \to \lambda$, $n \to \infty$, where $0 < \lambda < 1$. Denote by $X_{r(n):n}$ the $r(n)$th order statistic of $X_1, \ldots, X_n$. Then, for every x,

$$P\left\{ \frac{n^{1/2} \varphi(\Phi^{-1}(\lambda))}{(\lambda(1 - \lambda))^{1/2}} \left(X_{r(n):n} - \Phi^{-1}\left(\frac{r(n)}{n} \right) \right) \le x \right\}$$

$$\to \Phi\left[x \Bigg/ \left(1 + 2\lambda^{-1} \sum_{n=1}^{\infty} \rho(\Phi^{-1}(\lambda), \tau(n)) \right)^{1/2} \right], \qquad n \to \infty,$$

where

$$\rho(u, r) = (2\pi)^{-1} \int_0^r (1 - z^2)^{1/2} e^{-u^2/(1+z)} \, dz.$$

(Rootzén, 1985)

2. Let again $N_{(\mu, \Sigma)}$ be a k-variate normal distribution with mean vector μ and covariance matrix $\Sigma = (\sigma_{i,j})$. Moreover, let I denote the unit matrix.

   (i) Prove that

$$\|N_{(0, \Sigma)} - N_{(0, I)}\| \le 2^{-1/2} \left[\sum_{i=1}^{k} (\sigma_{i,i} - 1) - \log(\det(\Sigma)) \right]^{1/2}.$$

(Hint: Apply (4.4.2) and an inequality involving the Kullback–Leibler distance.]

   (ii) If Σ is a diagonal matrix then (i) yields

$$\|N_{(0, \Sigma)} - N_{(0, I)}\| \le 2^{-1/2} \left[\sum_{i=1}^{k} (\sigma_{i,i} - 1)^2 / \sigma_{ii} \right]^{1/2}.$$

   (iii) Alternatively,

$$\|N_{(0, \Sigma)} - N_{(0, I)}\| \le k 2^{k+1} \|\Sigma - I\|_2,$$

where $\|\cdot\|_2$ denotes the Euclidean norm.

(Pfanzagl, 1973b, Lemma 12)

   (iv) Denote again by K the Kullback–Leibler distance. Prove that

$$K(N_{(\mu, I)}, N_{(0, I)}) = 2^{-1} \|\mu\|_2^2.$$

   (v) Prove that

$$\|N_{(\mu_1, I)} - N_{(\mu_2, I)}\| \le 2^{-1/2} \|\mu_1 - \mu_2\|_2.$$

3. Let $N_{(0, \Sigma)}$ be the k-variate normal distribution given in Lemma 4.4.1. Define the linear map S by

$$S(\mathbf{x}) = \left(\frac{x_i - x_{i-1}}{(\lambda_i - \lambda_{i-1})^{1/2}} - (1 + (1 - \lambda_k)^{-1/2}) \frac{(\lambda_i - \lambda_{i-1})^{1/2}}{\lambda_k} x_k \right)^k_{i=1}.$$

Then, with I denoting the unit matrix, we have

$$SN_{(\mathbf{0}, \Sigma)} = N_{(\mathbf{0}, I)}.$$

(Reiss, 1975a)

4. (Spacings)
Given $1 \le r_1 < \cdots < r_k \le n$ put again $\lambda_i = r_i/(n+1)$, $\sigma_{i,j} = \lambda_i(1 - \lambda_j)$ for $1 \le i \le j \le k$, and $f_i = F'(F^{-1}(\lambda_i))$. Moreover, we introduce

$$a_i^2 = \sigma_{i-1,i-1}/f_{i-1}^2 - 2\sigma_{i-1,i}/(f_{i-1} f_i) + \sigma_{i,i}/f_i^2$$

for $i = 1, \ldots, k$ (with the convention that $a_1^2 = \sigma_{1,1}/f_1^2$).

Let $X_{i:n}$ be the order statistics of n i.i.d. random variables with common d.f. F. Denote by Q_n the joint distribution of the normalized spacings

$$(n+1)^{1/2} a_i^{-1} [X_{r_i:n} - X_{r_{i-1}:n} - (F^{-1}(\lambda_i) - F^{-1}(\lambda_{i-1}))], \qquad i = 1, \ldots, k,$$

and by P_n the joint distribution of the normalized order statistics

$$(n+1)^{1/2} f_i [X_{r_i:n} - F^{-1}(\lambda_i)], \qquad i = 1, \ldots, k.$$

After this long introduction we can offer some simple problems.
(i) Show that

$$\|Q_n - N_{(\mathbf{0}, I)}\| \le \|P_n - N_{(\mathbf{0}, \Sigma)}\| + \Delta^{1/2}$$

where I is the unit matrix, $\Sigma = (\sigma_{i,j})$ and

$$\Delta = 1 - (1 - \lambda_k)^{1/2} \sum_{i=1}^k (\lambda_i - \lambda_{i-1})^{1/2}/(a_i f_i).$$

(ii) $\Delta = 0$ if $k = 1$.
(iii) If F is the uniform d.f. on $(0, 1)$ then

$$\Delta = \lambda_k^2,$$

and as one could expect

$$a_i^{-2} = (\lambda_i - \lambda_{i-1})(1 - (\lambda_i - \lambda_{i-1})).$$

5. (Asymptotic expansions centered at $F^{-1}(q)$)
Let $q \in (0, 1)$ be fixed. Assume that the d.f. F has $m + 1$ bounded derivatives on a neighborhood of $F^{-1}(q)$, and that $f(F^{-1}(q)) > 0$ where $f = F'$. Moreover, assume that $(r(n)/n - q) = O(n^{-1})$. Put $\sigma^2 = q(1 - q)$. Then there exist polynomials $S_{i,n}$ of degree $\le 3i - 1$ (having coefficients uniformly bounded over n) such that

(i) $$\sup_B \left| P \left\{ \frac{n^{1/2} f(F^{-1}(q))}{\sigma} (X_{r(n):n} - F^{-1}(q)) \in B \right\} - \int_B dG_{r(n),n} \right| = O(n^{-m/2})$$

where

$$G_{r(n),n} = \Phi + \varphi \sum_{i=1}^{m-1} n^{-i/2} S_{i,n}.$$

In particular,

$$S_{1,n}(t) = \left[\frac{2q-1}{3\sigma} + \frac{\sigma f'(F^{-1}(q))}{2f(F^{-1}(q))^2}\right]t^2 + \left[\frac{-q+nq-r(n)+1}{\sigma} + \frac{2(2q-1)}{3\sigma}\right].$$

(ii) If the condition $(r(n)/n - q) = O(n^{-1})$ is replaced by $(r(n)/n - q) = o(n^{-1/2})$ then (i) holds for $m = 2$ with $O(n^{-1})$ replaced by $o(n^{-1/2})$.

(iii) Formulate weaker conditions under which (i) holds uniformly over intervals.

(iv) Denote by $f_{r(n),n}$ the density of the normalized distribution of $X_{r(n):n}$ in (i), and put $g_{r(n),n} = G'_{r(n),n}$. Show that

$$|f_{r(n),n}(x) - g_{r(n),n}(x)| = O(n^{-m/2}\varphi(x)(1 + |x|^{3m}))$$

uniformly over $x \in [-\log n, \log n]$.

6. (Asymptotic independence)

Given n i.n.n.i.d. random variables with d.f.'s $F_1, \ldots, F_n$ we have

$$P\{X_{1:n} \le x, X_{n:n} \le x\} - P\{X_{1:n} \le x\}P\{X_{n:n} \le x\}$$

$$= \left[\prod_{i=1}^{n} (F_i(y)(1 - F_i(x)))\right] - \left[\prod_{i=1}^{n} (F_i(y) - F_i(x))\right].$$

(Walsh, 1969)

Bibliographical Notes

Laplace (1818) derived the asymptotic normality of sample medians. He computed the density of the sample median (within a more general framework) and proved a limit theorem for the pointwise convergence of the densities. For a discussion of this result and applications we refer to Stigler (1973). This method was also used by Smirnov (1935) to obtain the asymptotic normality of central order statistics in greater generality. Other approaches reduce the problem to an application of the central limit theorem (that includes as a special case the asymptotic normality of binomial r.v.'s). The reduction is achieved either by means of the representations given in Section 1.6 (Cramér, 1946, and Rényi, 1953), the equality in (1.1.8) (Smirnov, 1949, van der Vaart, 1961, and Iglehart, 1976), or the Bahadur approximation (Sen, 1968).

The problem of charaterizing the possible limiting d.f.'s of central order statistics was dealt with by Smirnov (1949) (see P.4.1(i)) and Balkema and de Haan (1978a, b). If no regularity conditions are supposed, every d.f. is a limiting d.f. of central order statistics (see P.4.1(ii)).

An interesting problem, not treated in the book, occurs if the value of the underlying density at the q-quantile is equal to zero or if the q-quantile is not unique; in this context we refer to the articles of Feldman and Tucker (1966), Kiefer (1969b), Umbach (1981), and Landers and Rogge (1985) for important contributions.

A bound for the accuracy of the normal approximation to the d.f. of a single order statistic was established by Reiss (1974a) (where the terms of the error bound are given explicitly), Egorov and Nevzorov (1976), and Englund (1980).

Expansions of distributions of sample quantiles were established in Reiss (1976). There it was merely assumed that the underlying d.f. F has derivatives on $(F^{-1}(q) - \varepsilon, F^{-1}(q)]$ and $(F^{-1}(q) \, F^{-1}(q) + \varepsilon)$ for some $\varepsilon > 0$. If the left and right derivative of F at $F^{-1}(q)$ are unequal, then the leading term of the expansion is a certain mixture of normal distributions (compare this with P.4.1(i)). In this context, we also refer to Weiss (1969c) who proved a limit theorem under such conditions.

Puri and Ralescu (1986) studied order statistics of a non-random sample size n and a random index which converges to $q \in (0, 1)$ in probability. Among others, the asymptotic normality and a Berry–Esséen type theorem is proved. A result concerning sample quantiles with random sample sizes related to that for maxima (see P.5.11(i)) does not seem to exist in literature.

The problem of asymptotic independence between different groups of order statistics provides an excellent example where a joint treatment of extreme and central order statistics is preferable. The asymptotic independence of lower and upper extremes was first observed by Gumbel (1946). A precise characterization of the conditions that guarantee the asymptotic independence is due to Rossberg (1965, 1967). The corresponding result in the strong sense (that is, approximation w.r.t. the variational distance) was proved by Ikeda (1963) and Ikeda and Matsunawa (1970). In the i.n.n.i.d. case, Walsh (1969) proved the asymptotic independence of sample minimum and sample maximum under the condition that one or several d.f.'s do not dominate the other d.f.'s.

First investigations concerning the accuracy of the asymptotic results were made by Walsh (1970). Sharp bounds of the variational distance in case of extremes were established by Falk and Kohne (1986). Tiago de Oliveira (1961), Rosengard (1962), Rossberg (1965), and Ikeda and Matsunawa (1970) proved independence results that include central order statistics and sample means. The sharp inequalities in Section 4.2 concerning extreme and central order statistics are taken from Falk and Reiss (1988).

The asymptotic independence of ratios of consecutive order statistics was proved by Lamperti (1964) and Dwass (1966); a corresponding result holds for spacings. Smid and Stam (1975) showed that the condition, sufficient for this result, is also necessary.

In Lemma 4.4.3 an upper bound of the distance between the normal distribution $N_{(0, \Sigma)}$ and a distribution induced by $N_{(0, \Sigma)}$ and a function close to the identity is computed. For related results we refer to Pfanzagl [1973a, Lemma 1] and Bhattacharya and Gosh [1978, Theorem 1]. These results are formulated in terms of sequences of arbitrary normal distributions of a fixed dimension and therefore not applicable for our purposes. The normal comparison lemma (see e.g. Leadbetter et al. (1983), Theorem 4.2.1) is related to this.

For $r(i) = r(i, n)$, $i = 1, \ldots, k$, satisfying the condition $r(i, n) \to q_i$, $n \to \infty$, where $0 < q_1 < \cdots < q_k < 1$, the weak convergence of the standardized joint distributions of order statistics $X_{r(i):n}$ to the normal distribution $N_{(0, \Sigma)}$ was proved by Smirnov (1935, 1944), Kendall (1940), and Mosteller (1946).

The normal distributions $N_{(0, \Sigma)}$ are the finite dimensional marginals of the "Brownian Bridge" W^0 which is a special Gaussian process with mean function zero and covariance function $E \, W^0(q) W^0(p) = q(1 - p)$ for $0 \le q \le p \le 1$. The sample quantile process

$$n^{1/2}(F_n^{-1}(q) - q), \qquad q \in [0, 1],$$

here given for $(0, 1)$-uniformly distributed r.v.'s, converges to W^0 in distribution. Thus, the result for order statistics describes the weak convergence of the finite dimensional marginals of the quantile process. For a short discussion of this subject we refer to Serfling (1980). In view of the technique which is needed to rigorously investigate the weak convergence of the quantile process, a detailed study has to be done in conjunction with empirical processes in general (see e.g. M. Csörgő and P. Révész (1981) and G.R. Shorack and J.A. Wellner (1986)). The invariance principle for the sample quantile process provides a powerful tool to establish limit theorems (in the weak sense) for functionals of the sample quantile process, however, one cannot indicate the rate at which the limit theorems are valid. For statistical applications of the quantile process we refer to M. Csörgő (1983) and Shorack and Wellner (1986).

Weiss (1969b) studied the normal approximation of joint distributions of central order statistics w.r.t. the variational distance under the condition that $k = k(n)$ is of order $O(n^{1/4})$. Ikeda and Matsunawa (1972) and Weiss (1973) obtained corresponding results under the weaker condition that $k(n)$ is of order $O(n^{1/3})$. Reiss (1975a) established the asymptotic normality with a bound of order $O(\sum_{i=1}^{k+1} (r_i - r_{i-1})^{-1})^{1/2}$ for the remainder term. We also refer to Reiss (1975a) for an expansion of the joint distribution of central order statistics (see Section 4.5 for an expansion of length two in the special case of exponential r.v.'s). Other notable articles pertaining to this are those of Matsunawa (1975), Weiss (1979a), and Ikeda and Nonaka (1983).

An approximation to the multinomial distribution, with an increasing number of cells as the sample size tends to infinity, by means of the distribution of certain rounded-off normal r.v.'s may be found in Weiss (1976); this method seems to be superior to a more direct approximation by means of a normal distribution as pointed out by Weiss (1978).

The expansions of d.f.'s of order statistics in Section 4.6, taken from Nowak and Reiss (1983), are refinements of those given by Ivchenko (1971, 1974). Ivchenko also considers the multivariate case. In conjunction with this, we mention the article of Kolchin (1980), who established corresponding results for extremes.

CHAPTER 5

Approximations to Distributions of Extremes

The nondegenerate limiting d.f.'s of sample maxima $X_{n:n}$ are the Fréchet d.f.'s $G_{1,\alpha}$, Weibull d.f.'s $G_{2,\alpha}$, and the Gumbel d.f. G_3. Thus, with regard to the variety of limiting d.f.'s the situation of the present chapter turns out to be more complex than that of the preceding chapter, where weak regularity conditions guarantee the asymptotic normality of the order statistics.

As stated in (1.3.11) the limiting d.f.'s are max-stable, that is, for $G \in \{G_{1,\alpha}, G_{2,\alpha}, G_3 : \alpha > 0\}$ we find $c_n > 0$ and reals d_n such that

$$G^n(d_n + xc_n) = G(x).$$

Another interesting class of d.f.'s is that of the generalized Pareto d.f.'s $W \in \{W_{1,\alpha}, W_{2,\alpha}, W_3 : \alpha > 0\}$ as introduced in (1.6.11). These d.f.'s can also be used as a starting point when investigating distributional properties of sample maxima.

Given $G \in \{G_{1,\alpha}, G_{2,\alpha}, G_3 : \alpha > 0\}$ we obtain the associated generalized Pareto d.f. W by restricting the function $\Psi = 1 + \log G$ to certain intervals. The generalized Pareto d.f. W has the property

$$W^n(d_n + xc_n) = G(x) + O(n^{-1})$$

where c_n and d_n are the constants for which $G^n(d_n + xc_n) = G(x)$ holds. The class of generalized Pareto d.f.'s includes as special cases Pareto d.f.'s, uniform d.f.'s, and exponential d.f.'s.

An introduction to our particular point of view for the treatment of extremes will be given in Section 5.1. This section also includes results for the kth largest order statistic.

In Section 5.2 we shall establish bounds for the remainder terms in the limit theorems for sample maxima. In view of statistical applications the distance between the exact and limiting distributions will be measured w.r.t. the Hellinger distance.

In Section 5.3 some preparations are made for the study of the joint distribution of the k largest order statistics; it is shown that there is a close connection between the limiting distributions of the kth largest order statistic $X_{n-k+1:n}$ and the k largest order statistics

$$X_{n-k+1:n} \leq X_{n-k+2:n} \leq \cdots \leq X_{n:n}.$$

Higher order approximations in case of extremes of generalized Pareto r.v.'s are studied in Section 5.4. The accuracy of the approximations to the distribution of the kth largest order statistics and the joint distribution of extreme order statistics is dealt with in Section 5.5.

Finally, in Section 5.6, we shall make some remarks about the connection between extreme order statistics, empirical point processes, and certain Poisson processes.

5.1. Asymptotic Distributions of Extreme Sequences

In this section we shall examine the weak convergence of distributions of extreme order statistics. Moreover, it will be indicated that the strong convergence—that is the convergence w.r.t. the variational distance—holds under the well-known von Mises conditions.

Let $X_{1:n} \leq X_{2:n} \leq \cdots \leq X_{n:n}$ be the order statistics of n i.i.d. random variables with common d.f. F. A nondegenerate limiting d.f. of the sample maximum $X_{n:n}$ has to be—as already pointed out in Section 1.3—one of the Fréchet, Weibull, or Gumbel d.f.'s; that is, if there exist constants $a_n > 0$ and reals b_n such that

$$F^n(b_n + xa_n) \to G(x), \qquad n \to \infty, \tag{5.1.1}$$

for every continuity point of the nondegenerate limiting d.f. G then G has to be of the type $G_{1,\alpha}$, $G_{2,\alpha}$, G_3 for some $\alpha > 0$.

Recall that $G_{1,\alpha}(x) = \exp(-x^{-\alpha})$ for $x > 0$, $G_{2,\alpha}(x) = \exp(-(-x)^\alpha)$ for $x < 0$, and $G_3(x) = \exp(-e^{-x})$ for every x.

Graphical Representation of Extreme Value Densities

The densities $g_{i,\alpha}$ of $G_{i,\alpha}$ are given by

$$g_{1,\alpha}(x) = \alpha x^{-(1+\alpha)} \exp(-x^{-\alpha}), \qquad 0 < x,$$

$$g_{2,\alpha}(x) = \alpha(-x)^{\alpha-1} \exp(-(-x)^\alpha), \qquad x < 0,$$

$$g_3(x) = e^{-x} \exp(-e^{-x}).$$

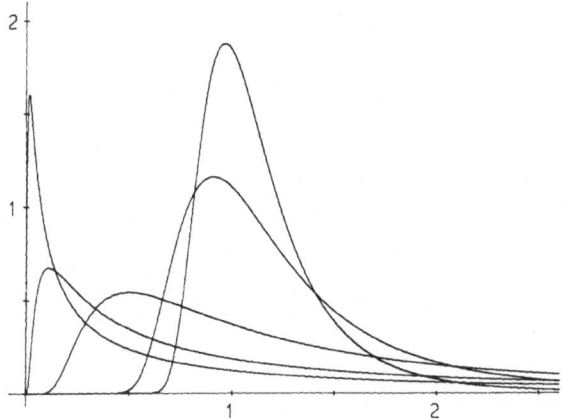

Figure 5.1.1. Fréchet densities $g_{1,\alpha}$ with parameters $\alpha = 0.33, 0.5, 1, 3, 5$; the mode increases as α increases.

Fréchet Densities

Figure 5.1.1 is misleading so far as one density seems to have a pole at zero. A closer look shows that this is not the case. Moreover, from the definition of $g_{1,\alpha}$ it is evident that every Fréchet density is infinitely often differentiable. For $\alpha = 5$ the density already looks like a Gumbel density (compare with Figure 1.3.1).

The density $g_{1,\alpha}$ is unimodal with mode

$$m(1, \alpha) = (\alpha/(1 + \alpha))^{1/\alpha}.$$

It is easy to verify that

$$m(1, \alpha) \to 0, \qquad g_{1,\alpha}(m(1, \alpha)) \to \infty, \qquad \text{as } \alpha \to 0$$

and

$$m(1, \alpha) \to 1, \qquad g_{1,\alpha}(m(1, \alpha)) \to \infty, \qquad \text{as } \alpha \to \infty.$$

Weibull Densities

The "negative" standard exponential density $g_{2,1}$ possesses a central position within the family of Weibull densities. The Weibull densities are again unimodal. From the visual as well as statistical point of view the most significant characteristic of a Weibull density $g_{2,\alpha}$ is its behavior at zero (Figure 5.1.2). Notice that

$$g_{2,\alpha}(x) \sim \alpha(-x)^{\alpha-1}, \qquad x \uparrow 0.$$

One may distinguish between five different classes of Weibull densities as far as the behavior at zero is concerned:

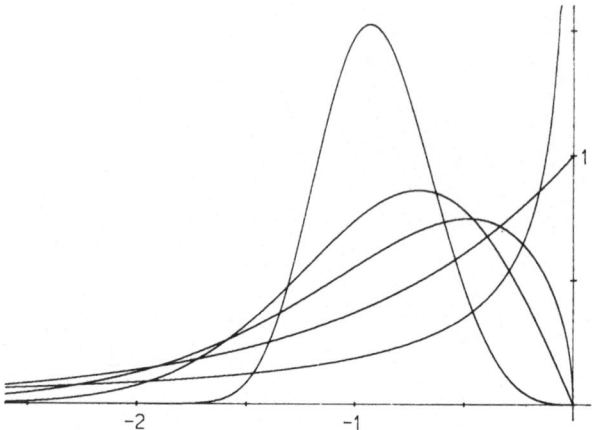

Figure 5.1.2. Weibull densities $g_{2,\alpha}$ with parameters $\alpha = 0.5, 1, 1.5, 2, 4$; the mode decreases as α increases.

$\alpha \in (0, 1)$: pole
$\alpha = 1$: jump
$\alpha \in (1, 2)$: continuous, not differentiable from the left at zero
$\alpha = 2$: differentiable from the left at zero
$\alpha > 2$: differentiable at zero.

If $\alpha > 1$ then the mode of $g_{2,\alpha}$ is equal to

$$m(2, \alpha) = -((\alpha - 1)/\alpha)^{1/\alpha} < 0.$$

Moreover,

$$m(2, \alpha) \to 0, \qquad g_{2,\alpha}(m(2, \alpha)) \to 1, \quad \text{as } \alpha \to 1,$$

and

$$m(2, \alpha) \to 1, \qquad g_{2,\alpha}(m(2, \alpha)) \to \infty, \quad \text{as } \alpha \to \infty.$$

Gumbel Density

The Gumbel density $g_3(x) = e^{-x}\exp(-e^{-x})$ approximately behaves like the standard exponential density e^{-x} as $x \to \infty$. The mode of g_3 is equal to zero. For the graph of g_3 we refer to Figure 1.3.1.

Weak Domains of Attraction

If (5.1.1) holds then F is said to belong to the weak domain of attraction of G. We shall discuss some conditions imposed on F which guarantee the weak convergence of upper extremes.

As mentioned above, $c_n^{-1}(X_{n:n} - d_n)$ has the d.f. $G_{i,\alpha}$ if $F = G_{i,\alpha}$ and if the constants are appropriately chosen. Thus e.g. the sample maximum $X_{n:n}$ of the negative exponential d.f. $G_{2,1}$ may serve as a starting point for the study of asymptotic distributions of sample maxima. However, to extend such a result one has to use the transformation technique (or some equivalent more direct method) so that it can be preferable to work with the sample maximum $U_{n:n}$ or $V_{n:n}$ of n i.i.d. random variables uniformly distributed on $(0, 1)$ or, respectively, $(-1, 0)$. In this case the limiting d.f. will again be $G_{2,1}$. Recall that the uniform distribution on $(-1, 0)$ is the generalized Pareto distribution $W_{2,1}$. As pointed out in (1.3.14) we have

$$P\{n(U_{n:n} - 1) \le x\} = P\{nV_{n:n} \le x\} \to G_{2,1}(x), \qquad n \to \infty, \quad (5.1.2)$$

for every x.

(5.1.2) and Corollary 1.2.7 imply that

$$F^n(b_n + xa_n) = G_{2,1}(n(F(b_n + xa_n) - 1)) + o(1), \qquad n \to \infty, \quad (5.1.3)$$

for every x. Moreover, for $G \in \{G_{1,\alpha}, G_{2,\alpha}, G_3 : \alpha > 0\}$ we may write

$$G = G_{2,1}(\log G) \quad \text{on} \quad (\alpha(G), \omega(G)).$$

This yields

$$F^n(b_n + xa_n) \to G(x), \qquad n \to \infty, \text{ for every } x,$$

if, and only if, $\qquad\qquad\qquad\qquad\qquad\qquad\qquad\qquad\qquad\qquad\qquad\qquad$ (5.1.4)

$$n(1 - F(b_n + xa_n)) \to -\log G(x) =: 1 - \Psi(x), \qquad n \to \infty,$$

for every $x \in (\alpha(G), \omega(G))$.

This well-known equivalence is one of the basic tools to establish necessary and sufficient conditions for the weak convergence of extremes. These conditions [due to Gnedenko (1943) and de Haan (1970)] in their elegance and completeness can be regarded as a corner stone in the classical extreme value theory.

A d.f. F belongs to the weak domain of attraction of an extreme value d.f. $G_{i,\alpha}$ if, and only if, one of the following conditions holds:

$(1, \alpha)$: $\omega(F) = \infty$, $\displaystyle\lim_{t \to \infty} [1 - F(tx)]/[1 - F(t)] = x^{-\alpha}$, $x > 0$;   (5.1.5)

$(2, \alpha)$: $\omega(F) < \infty$, $\displaystyle\lim_{t \downarrow 0} [1 - F(\omega(F) + xt)]/[1 - F(\omega(F) - t)]$

$$= (-x)^\alpha, \quad x < 0; \qquad\qquad\qquad\qquad (5.1.6)$$

(3): $\displaystyle\lim_{t \uparrow \omega(F)} [1 - F(t + xg(t))]/[1 - F(t)] = e^{-x}$, $-\infty < x < \infty$, (5.1.7)

where $g(t) = \int_t^{\omega(F)} (1 - F(y)) \, dy/(1 - F(t))$.

Moreover the constants a_n and b_n can be chosen in the following way:

$$(1, \alpha): \qquad b_n^* = 0, \qquad\qquad\qquad a_n^* = F^{-1}(1 - 1/n); \qquad\qquad (5.1.8)$$

$$(2, \alpha): \qquad b_n^* = \omega(F), \qquad\qquad a_n^* = \omega(F) - F^{-1}(1 - 1/n); \qquad (5.1.9)$$

$$(3): \qquad b_n^* = F^{-1}(1 - 1/n), \qquad a_n^* = g(b_n^*) \qquad\qquad\qquad (5.1.10)$$

where g is defined in (5.1.7).

It is well known that the weak convergence to the limiting d.f. G holds for other choices of constants a_n and b_n if, and only if,

$$a_n/a_n^* \to 1 \quad \text{and} \quad a_n^{-1}(b_n - b_n^*) \to 0 \text{ as } n \to \infty. \qquad (5.1.11)$$

For a well-known extension of this result we refer to P.5.3.

Tail Equivalence of D.F.'s

Further insight into the property that a d.f. belongs to the weak domain of attraction of $G = G_{i,\alpha}$ may be gained by conditions that are more closely related to (5.1.4). Observe that the two statements in (5.1.4) are equivalent to

$$n(1 - F(b_n + xa_n))/[1 - \Psi(x)] \to 1, \qquad n \to \infty, \qquad (5.1.12)$$

for every $x \in (\alpha(G), \omega(G))$ where $\Psi = 1 + \log G$.

Recall that the restriction of Ψ to an appropriate interval is a generalized Pareto d.f. $W \in \{W_{1,\alpha}, W_{2,\alpha}, W_3: \alpha > 0\}$.

Theorem 5.1.1. *Let $G = G_{i,\alpha}$ and $W = W_{i,\alpha}$ for some $i \in \{1, 2, 3\}$ and $\alpha > 0$. Then the following three statements are equivalent:*

(i) $F^n(b_n + xa_n) \to G(x), \qquad n \to \infty$, *for every x,* $\qquad\qquad (5.1.13)$

(ii) $(1 - F(b_n + xa_n))/(1 - W(d_n + xc_n)) \to 1, \qquad n \to \infty, \qquad (5.1.14)$

(iii) $(1 - F(b_n + xa_n))/(1 - G(d_n + xc_n)) \to 1, \qquad n \to \infty, \qquad (5.1.15)$

where (5.1.14) and (5.1.15) have to hold for every $x \in (\alpha(G), \omega(G))$.

Moreover, $d_n = 0$ if $i = 1, 2$, $d_n = \log n$ if $i = 3$, $c_n = n^{1/\alpha}$ if $i = 1$, $c_n = n^{-1/\alpha}$ if $i = 2$, and $c_n = 1$ if $i = 3$.

PROOF. The equivalence of (5.1.13) and (5.1.14) is immediate from (5.1.12) by writing $(1 - \Psi(x))/n = 1 - W(d_n + xc_n)$. Moreover, from (1.3.11) and the first equivalence we conclude that $[1 - G(d_n + xc_n)]/[1 - W(d_n + xc_n)] \to 1$, $n \to \infty$, and hence, obviously, the second equivalence is also valid. $\qquad\square$

Notice that, necessarily, $b_n + xa_n \to \omega(F)$, $n \to \infty$, if $F^n(b_n + xa_n) \to G(x)$, $n \to \infty$, and $\alpha(G) < x < \omega(G)$. Thus, Theorem 5.1.1 reveals that F belongs to the weak domain of attraction of G if, and only if, the upper tail of F can asymptotically be made equivalent to $G(d_n + xc_n)$. Below we shall prefer to work with the generalized Pareto d.f.'s W instead of the extreme value d.f.'s G because of technical advantages and other reasons which will become apparent when treating joint distributions of extremes.

Strong Domain of Attraction

Recall that the symbol G is used for the d.f. as well as for the corresponding probability measure. In analogy to the notion of the weak domain of attraction, F is said to belong to the strong domain of attraction of G if

$$\sup_{B} |P\{a_n^{-1}(X_{n:n} - b_n) \in B\} - G(B)| \to 0, \qquad n \to \infty, \qquad (5.1.16)$$

where the sup is taken over all Borel sets B.

Notice that condition (5.1.16) implies that F belongs to the weak domain of attraction of G. Thus, necessarily the normalizing constants are again those of the weak convergence. Moreover, it can easily be verified that (5.1.11) carries over to the strong convergence.

The following result was already indicated in (1.3.14).

Lemma 5.1.2.

$$\sup_{B} |P\{n(U_{n:n} - 1) \in B\} - G_{2,1}(B)| \to 0, \qquad n \to \infty. \qquad (5.1.17)$$

PROOF. From Theorem 1.3.2 we deduce that $n(U_{n:n} - 1)$ has the density f_n given by $f_n(x) = (1 + x/n)^{n-1}$, $-n < x < 0$, and $= 0$, otherwise. Thus, $f_n(x) \to e^x = g_{2,1}(x)$, $n \to \infty$, $x < 0$, and hence the Scheffé lemma implies the assertion. $\qquad\qquad\square$

Next, we study conditions under which F belongs to the strong domain of attraction of an extreme value d.f.

Tail Equivalence of Densities

In this sequel let us assume that F has a density f.

Denote by w the density of the generalized Pareto d.f. W. Notice that

$$w = g/G$$

on appropriate intervals where G is the corresponding extreme value d.f. and $g = G'$. Explicitly, we have

$$w_{1,\alpha}(x) = \begin{cases} 0 \\ \alpha x^{-(1+\alpha)} \end{cases} \quad \text{if} \quad \begin{array}{l} x < 1 \\ x \geq 1 \end{array} \qquad \text{``Pareto''} \qquad (5.1.18)$$

$$w_{2,\alpha}(x) = \begin{cases} 0 & x < -1 \\ \alpha(-x)^{\alpha-1} & \text{if} \quad -1 \leq x \leq 0 \qquad \text{``Type II''} \qquad (5.1.19) \\ 0 & x > 0 \end{cases}$$

$$w_3(x) = \begin{cases} 0 \\ e^{-x} \end{cases} \quad \text{if} \quad \begin{array}{l} x < 0 \\ x \geq 0. \end{array} \qquad \text{``Exponential''} \qquad (5.1.20)$$

The generalized Pareto densities as well as the extreme value densities are unimodal. The particular feature of the generalized Pareto densities is the tail equivalence to the corresponding extreme value densities at the right end-point of the support.

The counterpart to Theorem 5.1.1—with respect to the strong convergence—is the following.

Lemma 5.1.3. *Assume that the constants $a_n > 0$ and b_n are chosen so that the weak convergence holds, that is, $F^n(b_n + xa_n) \to G(x)$, $n \to \infty$, for every x, where $G \in \{G_{1,\alpha}, G_{2,\alpha}, G_3 \colon \alpha < 0\}$. Then,*

$$\sup_{B} |P\{a_n^{-1}(X_{n:n} - b_n) \in B\} - G(B)| \to 0, \qquad n \to \infty, \qquad (5.1.21)$$

if, and only if, for every subsequence $i(n)$ there exists a subsequence $k(n) = i(j(n))$ such that

$$\frac{a_{k(n)} f(b_{k(n)} + xa_{k(n)})}{c_{k(n)} w(d_{k(n)} + xc_{k(n)})} \to 1, \qquad n \to \infty, \qquad (5.1.22)$$

for Lebesgue almost all $x \in (\alpha(G), \omega(G))$ where w is the corresponding generalized Pareto density and c_n and d_n are the constants of Theorem 5.1.1.

Condition (5.1.22) is equivalent to the condition that for every subsequence $i(n)$ there exists a subsequence $k(n) = i(j(n))$ such that

$$k(n)a_{k(n)} f(b_{k(n)} + xa_{k(n)}) \to \psi(x) = g/G, \qquad n \to \infty, \qquad (5.1.22')$$

for almost all $x \in (\alpha(G), \omega(G))$. The equivalence of (5.1.22) and (5.1.22') becomes obvious by noting that $d_n + xc_n \in (\alpha(W), \omega(W))$ for every $x \in (\alpha(G), \omega(G))$, and $\psi(x)/n = c_n w(d_n + xc_n)$, eventually.

Without the condition $F^n(b_n + xa_n) \to G(x)$, $n \to \infty$, (5.1.22) does not necessarily imply (5.1.21) as can be shown by examples. If the weak convergence holds then a sufficient condition for the convergence w.r.t. the variational distance is

$$\frac{a_n f(b_n + xa_n)}{c_n w(d_n + xc_n)} \to 1, \qquad n \to \infty, \quad x \in (\alpha(G), \omega(G)). \qquad (5.1.23)$$

Note that the rate of convergence in (5.1.23) will also determine the rate at which the strong convergence of the distributions holds. We remark that the generalized Pareto density w can be replaced by the density g of G in condition (5.1.23).

Notice that (5.1.23) is equivalent to

$$na_n f(b_n + xa_n) \to \psi(x), \qquad n \to \infty, \quad x \in (\alpha(G), \omega(G)). \qquad (5.1.23')$$

PROOF OF LEMMA 5.1.3. Since

$$x \to na_n f(b_n + xa_n) F^{n-1}(b_n + xa_n)$$

is the density of $a_n^{-1}(X_{n:n} - b_n)$ it is immediate from the Scheffé lemma 3.3.4 that (5.1.21) is equivalent to (5.1.22′).   □

Lemma 5.1.3 will be the decisive tool to prove the following equivalence: F belongs to the strong domain of attraction of an extreme value distribution if, and only if, the corresponding result holds for the joint distribution of the k largest extremes for every positive integer k. For details we refer to Section 5.3.

From the mathematical point of view, condition (5.1.22) is more satisfactory than the sufficient condition (5.1.23). However, for practical purposes condition (5.1.23) can be useful; e.g. to verify that a given d.f. belongs to the strong domain of attraction of a particular extreme value distribution $G \in \{G_{1,\alpha}, G_{2,\alpha}, G_3: \alpha > 0\}$. It was proved by Falk (1985a) that the von Mises conditions (5.1.24) imply (5.1.23) and that (5.1.23) implies the convergence in the strong sense. Sweeting (1985) was able to show that the von Mises conditions (5.1.24) are equivalent to the uniform convergence of the densities in (5.1.23′) on finite intervals if the density f is positive on a left neighborhood of $\omega(F)$.

Von Mises-Type Conditions

Hereafter, we assume that F has a positive derivative f on $(x_0, \omega(F))$ where $x_0 < \omega(F)$. The following conditions $(1, \alpha)$, $(2, \alpha)$, and (3) are sufficient for F to belong to the strong domain of attraction of $G_{1,\alpha}$, $G_{2,\alpha}$, and G_3, respectively.

$(1, \alpha)$: $\omega(F) = \infty$, and $\lim\limits_{t \to \infty} tf(t)/[1 - F(t)] = \alpha$;

$(2, \alpha)$: $\omega(F) < \infty$, and $\lim\limits_{t \uparrow \omega(F)} [\omega(F) - t]f(t)/[1 - F(t)] = \alpha$;

$$(3): \quad \int_{-\infty}^{\omega(F)} (1 - F(u))\,du < \infty, \quad \text{and}$$

        (5.1.24)

$$\lim_{t \uparrow \omega(F)} f(t) \int_t^{\omega(F)} (1 - F(u))\,du / [1 - F(t)]^2 = 1.$$

Another set of sufficient conditions can be formulated if, in addition, F has a second derivative on $(x_0, \omega(F))$ where $x_0 < \omega(F)$:

$$\lim_{t \uparrow \omega(F)} [(1 - F)/f]'(t) = \begin{cases} \dfrac{1}{\alpha} & i = 1 \\[2mm] -\dfrac{1}{\alpha} & \text{if } i = 2 \\[2mm] 0 & i = 3. \end{cases} \qquad (5.1.25)$$

If $i = 3$ then the normalizing constant $a_n^* = g(b_n^*)$ as given in (5.1.10) can be replaced by

$$a_n = 1/(nf(b_n^*)) \tag{5.1.26}$$

where again $b_n^* = F^{-1}(1 - 1/n)$.

Notice that

$$[(1 - F)/f]' = -(1 - F)f'/f^2 - 1. \tag{5.1.27}$$

Thus, (5.1.25), $i = 3$, is equivalent to $\lim_{t \uparrow \omega(F)} (1 - F(t))f'(t)/f^2(t) = -1$.

(5.1.25) can be formulated in the following way: If the limit in (5.1.25) exists then F belongs to the strong domain of attraction of the von Mises d.f. H_β with parameter

$$\beta = \lim_{t \uparrow \omega(F)} [(1 - F)/f]'(t).$$

Since the conditions (5.1.5)–(5.1.7) and (5.1.24)–(5.1.26) are deduced from (5.1.4) it is not very amazing that these conditions are trivially fulfilled for the generalized Pareto d.f.'s W, that is, the equalities $tf(t)/[1 - F(t)] = \alpha$ etc. hold for every t in the support of W.

The von Mises-type conditions are sufficient for a d.f. to belong to the strong domain of attraction of an extreme value d.f. However, as examples show these conditions are not necessary. This is intuitively clear since for every density f which fulfills a von Mises-type condition we can find—by slightly varying f in the tail of the distribution—a density g which violates the von Mises-type condition whereas the stochastical properties of the sample maximum remain to hold asymptotically.

The main purpose of the following example is to clarify the connection between the different normalizing constants used in literature for the maximum of normal r.v.'s.

EXAMPLE 5.1.4. Let $X_{n:n}$ be the maximum of standard normal r.v.'s. Write again $\varphi = \Phi'$. Since $\varphi'(x) = -x\varphi(x)$ we get

$$\left(\frac{1 - \Phi}{\varphi}\right)'(x) = \frac{(1 - \Phi(x))x}{\varphi(x)} - 1.$$

It is immediate from (3.2.3) that this expression tends to zero as $x \to \infty$. Thus, condition (5.1.25), $i = 3$, implies that Φ belongs to the domain of attraction of the Gumbel d.f. G_3. Hence, according to (5.1.26), with $b_n = \Phi^{-1}(1 - 1/n)$,

$$\sup_B |P\{n\varphi(b_n)(X_{n:n} - b_n) \in B\} - G_3(B)| \to 0, \qquad n \to \infty. \tag{1}$$

Direct calculations or an application of Example 5.2.4 shows that (1) holds with a remainder term of order $O(1/\log n)$.

Next $a_n = 1/n\varphi(b_n)$ and b_n will be replaced by other normalizing constants that satisfy (5.1.11). Obviously, b_n is the solution of the equation

$$1 - \Phi(b) = 1/n. \tag{2}$$

Since $1 - \Phi(x) \sim \varphi(x)/x$ as $x \to \infty$ it is immediate that $(2 \log n)^{1/2}$ may be taken as a first approximate solution of (2). Moreover, (2) may be written

$$(1 - \Phi(b))/\varphi(b) = 1/(n\varphi(b))$$

and hence a solution of the equation

$$n\varphi(b) = b, \tag{3}$$

say, b_n' will be an approximate solution of (2). It can be shown (compare also with Example 5.2.4) that (1) still holds with a remainder term of order $O(1/\log n)$ if a_n and b_n are replaced by a_n' and b_n' where $a_n' = (b_n')^{-1}$.

(3) is equivalent to the equation

$$b = (2\log n - \log 2\pi - 2\log b)^{1/2}. \tag{4}$$

A Taylor expansion of length two about $2\log n$ leads to the equation

$$b = (2\log n)^{1/2} - \frac{\log 2\pi + 2\log b}{2(2\log n)^{1/2}}.$$

Replacing b on the right-hand side by $(2\log n)^{1/2}$ we get

$$b_n^* = (2\log n)^{1/2} - \frac{\log 4\pi + \log\log n}{2(2\log n)^{1/2}}. \tag{5}$$

Use P.5.7 to prove that

$$\sup_B |P\{(2\log n)^{1/2}(X_{n:n} - b_n^*) \in B\} - G_3(B)| = O\left(\frac{(\log\log n)^2}{\log n}\right). \tag{6}$$

We remark that the bound in (6) is sharp. Moreover, the same rates are obtained if d.f.'s are considered.

The kth Largest Order Statistic

The results given above can easily be extended to the case of the kth largest order statistic. It is well known that

$$P\{a(n)^{-1}(X_{n:n} - b(n)) \le x\} \to G_{i,\alpha}(x), \qquad n \to \infty,$$

implies for every fixed k that

$$P\{a(n)^{-1}(X_{n-k+1:n} - b(n)) \le x\} \to G_{i,\alpha,k}(x), \qquad n \to \infty, \tag{5.1.28}$$

where the d.f.'s $G_{i,\alpha,k}$ are given by

$$G_{1,\alpha,k}(x) = \exp(-x^{-\alpha}) \sum_{j=0}^{k-1} \frac{x^{-j\alpha}}{j!}, \qquad x > 0,$$

$$G_{2,\alpha,k}(x) = \exp(-(-x)^{\alpha}) \sum_{j=0}^{k-1} \frac{(-x)^{j\alpha}}{j!}, \qquad x < 0, \tag{5.1.29}$$

$$G_{3,k}(x) = \exp(-e^{-x}) \sum_{j=0}^{k-1} \frac{e^{-jx}}{j!}, \qquad -\infty < x < \infty.$$

With the convention $G_{3,\alpha,k} \equiv G_{3,k}$ we have

$$G_{i,\alpha,k} = G_{i,\alpha} \sum_{j=0}^{k-1} (-\log G_{i,\alpha})^j/j! \qquad (5.1.30)$$

on the support of $G_{i,\alpha}$.

To prove (5.1.28) recall that, necessarily, for $G \in \{G_{1,\alpha}, G_{2,\alpha}, G_3: \alpha > 0\}$ and $x \in (\alpha(G), \omega(G))$,

$$n(1 - F(u_n)) \to -\log G(x), \qquad n \to \infty,$$

with $u_n = b_n + a_n x$. According to (1.1.8), as $n \to \infty$,

$$P\{X_{n-k+1:n} \leq u_n\} = P\left\{\sum_{i=1}^n 1_{(u_n,\infty)}(\xi_i) \leq k - 1\right\}$$

$$= B_{(n,\, 1-F(u_n))}(\{0, 1, \ldots, k - 1\}) \qquad (5.1.31)$$

$$\to P_{-\log G(x)}(\{0, 1, \ldots, k - 1\})$$

where P_t denotes the Poisson distribution with parameter $t > 0$. Thus, (5.1.28) holds.

Moreover, it is well known that every nondegenerate limiting d.f. of the kth largest order statistic $X_{n-k+1:n}$ has to be one of the d.f.'s in (5.1.29) (see e.g. Galambos (1987), Theorem 2.8.1) where it is always understood that we have to include a location and scale parameter if the d.f. of $X_{n-k+1:n}$ is not properly standardized.

Note that in analogy to (1.3.15) the nondegenerate limiting d.f.'s $F_{i,\alpha,k}$ of the kth smallest order statistics $X_{k:n}$ are given by

$$F_{1,\alpha,k}(x) = 1 - G_{1,\alpha,k}(-x), \qquad x < 0,$$

$$F_{2,\alpha,k}(x) = 1 - G_{2,\alpha,k}(-x), \qquad x > 0, \qquad (5.1.32)$$

$$F_{3,k}(x) = 1 - G_{3,k}(-x)$$

where again $\alpha > 0$.

Obviously, $G_{2,1,k}$ is the "negative" gamma d.f. with parameter k; thus, the density $g_{2,1,k}$ of $G_{2,1,k}$ is given by

$$g_{2,1,k}(x) = e^x(-x)^{k-1}/(k - 1)!, \qquad x < 0, \qquad (5.1.33)$$

and $= 0$, otherwise.

We also note the explicit form of the densities $g_{i,\alpha,k}$ of $G_{i,\alpha,k}$. Since

$$G_{i,\alpha,k} = G_{2,1,k}(\log G_{i,\alpha}) \quad \text{on} \quad (\alpha(G_{i,\alpha}), \omega(G_{i,\alpha}))$$

we know that

$$g_{i,\alpha,k}(x) = g_{2,1,k}(\log G_{i,\alpha}(x))\frac{g_{i,\alpha}(x)}{G_{i,\alpha}(x)}, \qquad x \in (\alpha(G_{i,\alpha}), \omega(G_{i,\alpha})),$$

and $= 0$, otherwise. Explicitly, we have

$$g_{1,\alpha,k}(x) = \alpha\exp(-x^{-\alpha})\frac{x^{-(\alpha k+1)}}{(k-1)!}, \qquad x > 0,$$

$$g_{2,\alpha,k}(x) = \alpha\exp(-(-x)^{\alpha})\frac{(-x)^{\alpha k-1}}{(k-1)!}, \qquad x < 0, \qquad (5.1.34)$$

$$g_{3,k}(x) = \exp(-e^{-x})\frac{e^{-kx}}{(k-1)!}, \qquad -\infty < x < \infty.$$

Notice that

$$g_{i,\alpha,k} = g_{i,\alpha}(-\log G_{i,\alpha})^{k-1}/(k-1)!.$$

Lemma 1.6.6 yields that $G_{2,1,k}$ is the d.f. of the partial sum $S_k = \sum_{i=1}^{k}\xi_i$ where $\xi_1,\ldots,\xi_k$ are i.i.d. random variables with common d.f. $F(x) = e^x$, $x < 0$. Next it will be proved that $n(U_{n-k+1:n} - 1)$ is asymptotically distributed according to $G_{2,1,k}$ (in other words, can asymptotically be represented by S_k).

As an extension of Lemma 5.1.2 we obtain

Lemma 5.1.5. *For every positive integer k,*

$$\sup_{B} |P\{n(U_{n-k+1:n} - 1) \in B\} - G_{2,1,k}(B)| \to 0, \qquad n \to \infty. \quad (5.1.35)$$

PROOF. Obvious by noting that $n(U_{n-k+1:n} - 1)$ has the density f_n given by

$$f_n(x) = \left(\prod_{i=1}^{k-1}\left(1 - \frac{i}{n}\right)\right)\left(1 + \frac{x}{n}\right)^{n-k}\frac{(-x)^{k-1}}{(k-1)!}, \qquad -n < x < 0,$$

and $= 0$, otherwise. $\qquad\qquad\square$

Obviously, Lemma 5.1.5 can be written

$$\sup_{B} |P\{nV_{n-k+1:n} \in B\} - G_{2,1,k}(B)| \to 0, \qquad n \to \infty, \quad (5.1.36)$$

where $V_{n-k+1:n}$ is the kth largest order statistic of n i.i.d. random variables that are uniformly distributed on $(-1,0)$.

Recall that the uniform distribution on $(-1,0)$ is the generalized Pareto distribution $W_{2,1}$. (5.1.36) can easily be extended to the other generalized Pareto distributions $W_{i,\alpha}$ by using the transformation technique.

Let again $T_{i,\alpha}$ be defined as in (1.6.10). For $x < 0$, we have $T_{1,\alpha}(x) = (-x)^{-1/\alpha}$, $T_{2,\alpha}(x) = -(-x)^{1/\alpha}$ and $T_{3,1}(x) = -\log(-x)$.

Since $T_{i,\alpha}(nV_{r:n}) = c_n^{-1}(X_{r:n} - d_n)$ where c_n, d_n are the constants of Theorem 5.1.1 and since $G_{i,\alpha,k}$ is induced by $G_{2,1,k}$ and $T_{i,\alpha}$ [recall that $T_{i,\alpha}^{-1} = G_{2,1}^{-1} \circ G_{i,\alpha} = \log G_{i,\alpha}$] the following result is immediate from Lemma 5.1.5.

Corollary 5.1.6. *Let $X_{n-k+1:n}$ be the kth largest order statistic of n i.i.d. random variables with common generalized Pareto d.f. $W \in \{W_{1,\alpha}, W_{2,\alpha}, W_3 : \alpha > 0\}$.*

Then, for every fixed k, as $n \to \infty$,

$$\sup_B |P\{n^{-1/\alpha} X_{n-k+1:n} \in B\} - G_{1,\alpha,k}(B)| \to 0 \qquad \text{if } W = W_{1,\alpha} \qquad (5.1.37)$$

$$\sup_B |P\{n^{1/\alpha} X_{n-k+1:n} \in B\} - G_{2,\alpha,k}(B)| \to 0 \qquad \text{if } W = W_{2,\alpha} \qquad (5.1.38)$$

$$\sup_B |P\{(X_{n-k+1:n} - \log n) \in B\} - G_{3,k}(B)| \to 0 \qquad \text{if } W = W_3. \qquad (5.1.39)$$

In Section 5.4 it will be shown that Lemma 5.1.5 (and thus also Corollary 5.1.6) is valid with a remainder term of order $O(k/n)$.

Intermediate Order Statistics

From Chapter 4 we already know that intermediate order statistics are asymptotically normal under weak regularity conditions. For example, according to Theorem 4.2.1,

$$\sup_B |P\{a_{r,n}^{-1}(U_{r:n} - b_{r,n}) \in B\} - N_{(0,1)}(B)| \le C(n/r(n-r))^{1/2} \qquad (5.1.40)$$

where $C > 0$ is a universal constant, and $a_{r,n} > 0$ and $b_{r,n}$ are normalizing constants. In Section 5.4 it will be proved that

$$\sup_B |P\{n(U_{n-k+1:n} - 1) \in B\} - G_{2,1,k}(B)| \le Ck/n, \qquad (5.1.41)$$

where $G_{2,1,k}$ is the "negative" gamma distribution. We also refer to P.5.18 where a rate of order $O(k^{1/2}/n)$ is achieved in (5.1.41) by using other normalizing constants. Approximations of joint distributions of intermediate order statistics are established in Sections 4.5, 5.4, and 5.5.

The following theorem is taken from Falk (1989b).

Theorem 5.1.7. *Assume that one of the von Mises conditions (5.1.24) holds. Let $k(n) \in \{1, \dots, n\}$ be such that $k(n) \to \infty$ and $k(n)/n \to 0$ as $n \to \infty$.*

Then, with $b_n = F^{-1}(1 - k(n)/n)$, we have

$$\sup_B \left| P\left\{ \frac{nf(b_n)}{k(n)^{1/2}} (X_{n-k(n)+1:n} - b_n) \in B \right\} - N_{(0,1)}(B) \right| \to 0, \qquad n \to \infty. \qquad (5.1.42)$$

The proof of (5.1.42) is based on (5.1.40) and the transformation technique.

5.2. Hellinger Distance between Exact and Approximate Distributions of Sample Maxima

Given n i.i.d. random variables $\xi_1, \dots, \xi_n$ with common d.f. F we know that the d.f. of the sample maximum $M_n = X_{n:n}$ is given by F^n. In Section 5.1 we gave a short outline of classical results concerning the weak convergence of F^n (if appropriately normalized) to a limiting d.f. G. Moreover, we know that under von Mises-type conditions the weak convergence is equivalent to the convergence w.r.t. the variational distance. In the present section we study the accuracy of such approximations. Again we use the same symbol for a d.f. and the pertaining probability measure to simplify the notation.

The Hellinger Distance

In statistical applications it is desirable to use the Hellinger distance instead of the variational distance. To highlight this point consider the sample maxima

$$M_{n,i} = \max(\xi_{i,1}, \ldots, \xi_{i,n}), \qquad i = 1, \ldots, N,$$

where the random variables $\xi_{1,1}, \ldots, \xi_{1,n}, \xi_{2,1}, \ldots, \xi_{2,n}, \ldots, \xi_{N,1}, \ldots, \xi_{N,n}$ are i.i.d. with common d.f. F. Thus, $M_{n,1}, \ldots, M_{n,N}$ are i.i.d. random variables with common d.f. F^n. If $\eta_1, \ldots, \eta_N$ are i.i.d. random variables with d.f. G then we know from Corollary 3.3.11 that for every Borel set B,

$$|P\{(M_{n,1}, \ldots, M_{n,N}) \in B\} - P\{(\eta_1, \ldots, \eta_N) \in B\}| \le N^{1/2} H(F^n, G) \quad (5.2.1)$$

where $H(F^n, G)$ is the Hellinger distance between F^n and G.

Given d.f.'s F and G with Lebesgue densities f and g, the Hellinger distance of F and G is defined by

$$H(F, G) = \left[\int (f^{1/2}(x) - g^{1/2}(x))^2 \, dx \right]^{1/2}. \qquad (5.2.2)$$

In general, if F and G have densities f and g with respect to some σ-finite measure μ then

$$H(F, G) = \left[\int (f^{1/2} - g^{1/2})^2 \, d\mu \right]^{1/2} \qquad (5.2.3)$$

and the distance is independent of a particular representation. Thus (5.2.2) and (5.2.3) lead to the same distance (if Lebesgue densities exist). We refer to Section 3.3 for further details.

(5.2.1) also holds with $N^{1/2} H(F^n, G)$ replaced by $N \|F^n - G\|$ where $\|F^n - G\|$ is the variational distance between F^n and G. However, the use of $N \|F^n - G\|$ yields an inaccurate inequality in those cases where $\|F^n - G\|$ and $H(F^n, G)$ are of the same magnitude.

An Auxiliary Approximation

According to (5.1.4)

$$F^n(b_n + xa_n) \to G(x) =: \exp(-h(x)), \qquad n \to \infty,$$

if, and only if, $\qquad\qquad\qquad\qquad\qquad\qquad\qquad\qquad\qquad\qquad\qquad\qquad$ (5.2.4)

$$n(1 - F(b_n + xa_n)) \to h(x), \qquad n \to \infty.$$

Since F is a d.f. it is obvious that also D_n defined by

$$D_n = [\exp[-n(1 - F_n)] - e^{-n}]/(1 - e^{-n})$$

is a d.f. where $F_n(x) = F(b_n + xa_n)$. Now (5.2.4) may be written

$$F_n^n \to G \quad \text{iff} \quad D_n \to G \tag{5.2.4'}$$

w.r.t. the pointwise convergence.

According to Lemma 5.2.1, $D_n \to G$ implies $F_n^n \to G$ where the convergence is taken w.r.t. the Hellinger distance H. Notice that the d.f. G in Lemma 5.2.1 is not necessarily an extreme value d.f. In particular, G may also depend on n.

Lemma 5.2.1. *There exists a universal constant $C > 0$ such that for every n and all d.f.'s F and G the following inequality holds:*

$$H(F^n, G) \le H(D_n, G) + C/n$$

where $D_n = [\exp[-n(1 - F)] - e^{-n}]/(1 - e^{-n})$.

PROOF. Since $H(F^n, G) \le H(F^n, D_n) + H(D_n, G)$ we know that the assertion holds if

$$H(F^n, D_n) \le C/n. \tag{1}$$

First, (1) will be verified in the special case of $F_0(x) = 1 + x/n$, $-n < x < 0$. Notice that F_0^n is the d.f. of $n(U_{n:n} - 1)$.

In this case we have $D_n(x) \equiv D_{0,n}(x) = (e^x - e^{-n})/(1 - e^{-n})$, $-n < x < 0$, and, therefore, $D_{0,n}$ is the normalized restriction of the extreme value d.f. $G_{2,1}$ to the interval $(-n, 0)$.

Denote by f_0 and $d_{0,n}$ the densities of F_0 and $D_{0,n}$. Since

$$H(F_0^n, D_{0,n}) \le \left[\int (nf_0 F_0^{n-1}/d_{0,n} - 1)^2 \, dD_{0,n} \right]^{1/2}$$

(see Lemma 3.3.9(ii)) it is immediate that (1) holds for F_0 and $D_{0,n}$ if

$$\int_{-n}^{0} \left[e^{-x}\left(1 + \frac{x}{n}\right)^{n-1}(1 - e^{-n}) - 1 \right]^2 e^x/(1 - e^{-n}) \, dx \le (C/n)^2. \tag{2}$$

This inequality can be verified by means of some straightforward calculations.

The extension to arbitrary d.f.'s is obtained by means of the transformation technique. If ξ and η are r.v.'s with d.f.'s F_0 and $D_{0,n}$ then $F^{-1}(1 + \xi/n)$ and $F^{-1}(1 + \eta/n)$ are r.v.'s with d.f.'s F^n and $D_n = [\exp[-n(1 - F)] - e^{-n}]/(1 - e^{-n})$. Now, Lemma 3.3.13, which concerns the Hellinger distance between induced probability measures, implies (1) in the general case.   □

Let F_0 be defined as in the proof to Lemma 5.2.1, that is, F_0^n is the d.f. of $n(U_{n:n} - 1)$. It is easy to see that also

$$H(F_0^n, G_{2,1}) \le C/n. \tag{5.2.5}$$

According to our considerations in Section 5.1, this inequality can easily be extended to sample maxima under arbitrary generalized Pareto d.f.'s.

The Main Results

Notice that Lemma 5.2.1 holds for arbitrary d.f.'s F and G. Hereafter, we shall assume that F and G possess densities f and g.

In the next step we establish an upper bound for $H(D_n, G)$—and thus for $H(F^n, G)$—which depends on F through the density f only.

Lemma 5.2.2. *Let F and G be d.f.'s with densities f and g. Define $\psi = g/G$ on the support of G. Then, for every $x_0 \ge -\infty$,*

$$H(F^n, G) \le \left[2G(B^c) + \int_B [nf/\psi - 1 - \log(nf/\psi)]\, dG \right.$$
$$\left. + \int_B (1 + \log G)\, dG + \int_{\{g=0\}} nG\, dF \right]^{1/2} + C/n \tag{5.2.6}$$

where $B = \{x: x > x_0, f(x) > 0\}$ and $C > 0$ is a universal constant.

PROOF. Let the d.f. D_n be defined as in Lemma 5.2.1. Notice that D_n has the density $x \to nf \exp[-n(1-F)]/(1-e^{-n})$. To prove this apply e.g. Remark 1.5.3. Now, by Lemma 5.2.1 and Lemma A.3.5, applied to $H(D_n, G)$, we obtain

$$H(F^n, G) \le \left[2G(B^c) + \int_B [n(1-F) - \log(nf) + \log(G\psi)]\, dG \right]^{1/2} + \frac{C}{n}. \tag{1}$$

Recall that $G = g/\psi$ on the set $\{g > 0\}$. Hence, by Fubini's theorem

$$\int_B (1-F)\, dG = \int_{x_0}^\infty \left(\int_x^\infty f(y)\, dy \right) dG(x)$$

$$= \iint 1_{[x_0, \infty)}(x) 1_{(-\infty, y]}(x) f(y) g(x)\, dx\, dy$$

$$= \int f(y) 1_{[x_0, \infty)}(y) \left(\int_{x_0}^y g(x)\, dx \right) dy \tag{2}$$

$$\le \int_{x_0}^\infty f(y) G(y)\, dy$$

$$\le \int_B (f/\psi)\, dG + \int_{\{g=0\}} G(y)\, dF(y).$$

Combining (1) and (2) we obtain inequality (5.2.6).   □

In special cases the term on the right-hand side of (5.2.6) simplifies considerably.

Corollary 5.2.3. *Assume in addition to the conditions of Lemma 5.2.2 that F and G are mutually absolutely continuous (that is, $G\{f > 0\} = F\{g > 0\} = 1$). Then,*

$$H(F^n, G) \le \left[\int (nf/\psi - 1 - \log(nf/\psi)) \, dG \right]^{1/2} + C/n. \qquad (5.2.7)$$

PROOF. Lemma 5.2.2 will be applied to $x_0 = -\infty$. It suffices to prove that

$$\int \log G \, dG = -1. \qquad (1)$$

Notice that according to Lemma 1.2.4,

$$\int (1 + \log G) \, dG = \int_0^1 (1 + \log(G \circ G^{-1})(x)) \, dx$$

$$= \int_0^1 (1 + \log x) \, dx = x \log x|_0^1 = 0$$

since $x \log x \to 0$ as $x \to 0$.   □

The proof of (1) shows that $\int \log G \, dG = -1$ for continuous d.f.'s G. If G has a density g then $\int g(x)(-\log G(x)) \, dx = 1$ so that $g(x)(-\log G(x))$ is a probability density. In Section 5.1, we already obtained a special case, namely, that $g_{i,\alpha,2} = g_{i,\alpha}(-\log G_{i,\alpha})$ where $g_{i,\alpha}$ is the limiting density of the second largest order statistic.

Thus, if g is an approximation to the density of the standardized sample maximum then $g(-\log G)$ will be the proper candidate as an approximate density of the second largest order statistic. The extension of this argument to $k > 2$ is straightforward and can be left to the reader.

Since $x - 1 - \log x \le x - 1 + 1/x - 1 = (x - 1)^2/x$ we obtain from Corollary 5.2.3 that

$$H(F^n, G) \le \left[\int \frac{(nf/\psi - 1)^2}{nf/\psi} \, dG \right]^{1/2} + C/n \qquad (5.2.8)$$

where again $\psi = g/G$. This inequality shows once more (see also Section 5.1) that the approximating d.f. G should be chosen in such a way that nf/ψ is close to one.

EXAMPLE 5.2.4. Let $F(x) = \Phi(b_n + b_n^{-1}x)$ where Φ is the standard normal d.f. and b_n is the solution of $b_n = n\varphi(b_n)$ with $\varphi = \Phi'$. Then,

$$H(F^n, G_3) = O(1/\log n). \qquad (5.2.9)$$

To prove this, we apply (5.2.7), with $\psi(x) = e^{-x}$. We have

$$H(F^n, G_3)$$

$$\leq \left[\int (n\varphi(b_n + b_n^{-1}x)b_n^{-1}e^x - 1 - \log(n\varphi(b_n + b_n^{-1}x)b_n^{-1}e^x))\,dG_3(x) \right]^{1/2} + C/n$$

$$= \left[\int (\exp(-x^2/2b_n^2) - 1 + x^2/2b_n^2)\,dG_3(x) \right]^{1/2} + C/n$$

$$\leq \left[\int (x^4/8b_n^4)\,dG_3(x) \right]^{1/2} + C/n \leq C(b_n^{-2} + n^{-1}).$$

Thus, (5.2.9) holds since $b_n^{-2} = O(1/\log n)$.

Next, Lemma 5.2.2 will be applied to extreme value d.f.'s $G \in \{G_{1,\alpha},\ G_{2,\alpha}, G_3 : \alpha > 0\}$. Note that the function $\psi = g/G$ is given by

$$\psi_{1,\alpha}(x) = \alpha x^{-(1+\alpha)}, \qquad x > 0$$

$$\psi_{2,\alpha}(x) = \alpha(-x)^{-(1-\alpha)}, \qquad x < 0 \qquad (5.2.10)$$

$$\psi_3(x) = e^{-x}, \qquad -\infty < x < \infty.$$

Recall that $\omega(F) < \infty$ if F belongs to the domain of attraction of $G_{2,\alpha}$. In this case, the usual choice of the constant b_n is $\omega(F)$ so that we may assume w.l.g. that $\omega(F) = \omega(G_{2,\alpha}) = 0$. If F belongs to the domain of attraction of $G_{1,\alpha}$ then $\omega(F) = \omega(G_{1,\alpha}) = \infty$. Let us also assume that $\omega(F) = \omega(G_3) = \infty$ if $i = 3$ to make the inequality in Theorem 5.2.5 as simple as possible without losing too much generality.

Theorem 5.2.5. *Let* $G \in \{G_{1,\alpha}, G_{2,\alpha}, G_3 : \alpha > 0\}$. *Let* F *be a d.f. with density* f *such that* $f(x) > 0$ *for* $x_0 < x < \omega(F)$. *Assume that* $\omega(F) = \omega(G)$. *Then,*

$$H(F^n, G)$$

$$\leq \left[\int_{x_0}^{\omega(G)} [nf/\psi - 1 - \log(nf/\psi)]\,dG + 2G(x_0) - G(x_0)\log G(x_0) \right]^{1/2} + C/n$$

where $C > 0$ *is a universal constant.*

PROOF. Immediate from Lemma 5.2.2 since $\int_{\{g=0\}} nG\,dF = 0$, and

$$2G(B^c) + \int_B (1 + \log G)\,dG = G_{i,\alpha}(B^c) + G_{i,\alpha,2}(B^c)$$

$$= G_{i,\alpha}(x_0) + G_{i,\alpha,2}(x_0)$$

$$= 2G(x_0) - G(x_0)\log(G(x_0)). \qquad \square$$

Limit Distributions

The results above provide us with useful auxiliary inequalities which, in a next step, have to be applied to special examples or certain classes of underlying d.f.'s to obtain a more explicit form of the error bound.

Our first example again reveals the exceptional role of the generalized Pareto d.f.'s $W_{i,\alpha}$ (at least, from a technical point of view).

EXAMPLE 5.2.6. (i) Let $W \in \{W_{1,\alpha}, W_{2,\alpha}, W_3 : \alpha > 0\}$ and c_n, d_n be the constants of Theorem 5.1.1. Put

$$F_n(x) = W(d_n + xc_n).$$

The density f_n of F_n is given by

$$f_n(x) = c_n w(d_n + xc_n) = \psi(x)/n$$

for every x with $f_n(x) > 0$. Thus, we have

$$\int_{\{f_n > 0\}} (nf_n/\psi - 1 - \log(nf_n/\psi)) \, dG = 0.$$

Applying Theorem 5.2.5 to $x_0 = (\alpha(W) - d_n)/c_n$ we obtain again

$$H(F_n^n, G) \le C/n.$$

(ii) Let in (i) the generalized Pareto d.f. W be replaced by a d.f. F which has the same tail as W. More precisely,

$$f(x) = w(x), \qquad T(x_0) < x < w(G),$$

where $-1 < x_0 < 0$ and T is the corresponding transformation as defined in (1.6.10). Then,

$$H(F_n^n, G) \le C_0/n$$

where C_0 is a constant which only depends on x_0.

Notice that the condition $T(x_0) < x$ in Example 5.2.6(ii) makes the accuracy of the approximation independent of the special underlying d.f. F.

Example 5.2.6 will be generalized to classes of d.f.'s which include the generalized Pareto d.f.'s as well as the extreme value d.f.'s. Since our calculations are always carried out within an error bound of order $O(n^{-1})$ it is clear that the estimates will be inaccurate for extreme value d.f.'s.

Assume that the underlying density f is of the form

$$f = \psi e^h$$

where $h(x) \to 0$, $x \to \omega(G)$. Equivalently, one may use the representation $f = \psi(1 + \tilde{h})$ by writing $f = \psi e^h = \psi(1 + (e^h - 1))$.

Corollary 5.2.7. *Assume that* $G \in \{G_{1,\alpha}, G_{2,\alpha}, G_3 : \alpha > 0\}$ *and* ψ, T *are the corresponding auxiliary functions with* $\psi = g/G$ *and* $T = G^{-1} \circ G_{2,1}$.

Assume that the density f of the d.f. F has the representation

$$f(x) = \psi(x)e^{h(x)}, \qquad T(x_0) < x < \omega(G), \qquad (5.2.11)$$

and $= 0$, if $x > \omega(G)$, where $x_0 < 0$ and h satisfies the condition

$$|h(x)| \leq \begin{matrix} Lx^{-\alpha\delta} & i = 1 \\ L(-x)^{\alpha\delta} & \text{if} & i = 2 \\ Le^{-\delta x} & i = 3 \end{matrix} \qquad (5.2.12)$$

and L, δ are positive constants. Write

$$F_n(x) = F(d_n + xc_n)$$

where c_n, d_n are the constants of Theorem 5.1.1. We have $d_n = 0$ if $i = 1, 2$, and $d_n = \log n$ if $i = 3$; moreover, $c_n = n^{1/\alpha}$ if $i = 1$, $c_n = n^{-1/\alpha}$ if $i = 2$, and $c_n = 1$ if $i = 3$.

Then, the following inequality holds:

$$H(F_n^n, G) \leq \begin{matrix} Dn^{-\delta} \\ Dn^{-1} \end{matrix} \quad \text{if} \quad \begin{matrix} 0 < \delta \leq 1 \\ \delta > 1 \end{matrix} \qquad (5.2.13)$$

where D is a constant which only depends on x_0, L, and δ.

PROOF. W.l.g. we may assume that $G = G_{2,1}$. The other cases can easily be deduced by using the transformations $T \equiv T_{i,\alpha}$.

Theorem 5.2.5 will be applied to $x_{0,n} = nx_0$. It is straightforward that the term $2G_{2,1}(nx_0) - G_{2,1}(nx_0)\log G_{2,1}(nx_0)$ can be neglected. Put $f_n(x) = f(x/n)/n$. Since h is bounded on $(x_0, 0)$ we have

$$\int_{nx_0}^0 (nf_n/\psi_{2,1} - 1 - \log(nf_n/\psi_{2,1}))\,dG_{2,1}$$

$$= \int_{nx_0}^0 (e^{h(x/n)} - 1 - h(x/n))\,dG_{2,1}(x)$$

$$\leq \tilde{D} \int_{nx_0}^0 (h(x/n))^2\,dG_{2,1}(x) \leq \tilde{D}L^2 n^{-2\delta} \int_{-\infty}^0 |x|^{2\delta}\,dG_{2,1}(x)$$

where $\tilde{D}$ only depends on x_0, L and δ. Now the assertion is immediate from Theorem 5.2.5. $\qquad\qquad\qquad\square$

Extreme value d.f.'s have representations as given in (5.2.11) with $\delta = 1$ and $h(x) = -x^{-\alpha}$ if $i = 1$, $h(x) = -(-x)^{\alpha}$ if $i = 2$, and $h(x) = -e^{-x}$ if $i = 3$. Moreover, the special case of $h = 0$ concerns the generalized Pareto densities.

Remark 5.2.8. Corollary 5.2.7 can as well be formulated for densities having the representation

$$f(x) = \psi(x)(1 + h(x)), \qquad T(x_0) < x < w(G), \qquad (5.2.14)$$

and $= 0$, if $x > \omega(G)$, where h satisfies the condition (5.2.12).

Maximum of Normal R.V.'s: Penultimate Distributions

Inequality (5.2.6) is also applicable to problems where approximate distributions which are different from the limiting ones are taken. The first example will show that Weibull distributions $G_{2,\alpha(n)}$ with $\alpha(n) \to \infty$ as $n \to \infty$ provide more accurate approximations to distributions of sample maxima of normal r.v.'s than the limiting distribution G_3.

The use of a "penultimate" distribution was already suggested by Tippett in 1925. For a numerical comparison of the "ultimate" and "penultimate" approximation we also refer to Fisher and Tippett (1928).

EXAMPLE 5.2.9. Let $F(x) = \Phi(b - b^{-1} + b^{-1}x)$ where b is the solution of the equation

$$n\varphi(b - b^{-1}) = b.$$

Notice that b and thus also F depends on n; we have $b^{-2} = O(1/\log n)$. Below we shall use the von Mises parametrization (see (1.3.17)) of Weibull d.f.'s, namely,

$$H_{-b^{-2}}(x) = G_{2,b^2}(-1 + x/b^2).$$

Applying Lemma 5.2.2 to $G = H_{-b^{-2}}$ we obtain after some straightforward but tedious calculations that

$$H(F^n, H_{-b^{-2}}) = O((\log n)^{-2}). \tag{5.2.15}$$

We indicate some details of the proof of (5.2.15). Check that

$$\frac{nf(x)}{\psi(x)} = \frac{\exp(-x + x/b^2 - x^2/2b^2)}{(1 - x/b^2)^{b^2-1}}, \qquad x < b^2.$$

To establish a sharp estimate of $\int (nf/\psi - 1 - \log nf/\psi)\, dG$ proceed in the following way: (a) Apply Lemma A.2.1 to the integrand evaluated over the interval $[-cb, cb]$ with c being sufficiently small. (b) Use the crude inequalities $e^{\beta x/\alpha} \geq (1 + x/\alpha)^\beta$ for $x > -\alpha$ and $(1 + x/\alpha)^\alpha \geq 1 + x + (\alpha - 1)x^2/2\alpha$ for $x > 0$ and $\alpha \geq 2$ to obtain estimates of the integral over $(-\infty, -cb)$ and (cb, b^2).

Maximum of Normal R.V.'s: Expansions of Length Two

From Lemma 5.2.10 it will become obvious that

$$x \to G_3(x)(1 + e^{-x}x^2/4\log n) \tag{5.2.16}$$

provides an expansion of length two of $\Phi^n(b - b^{-1} + b^{-1}x)$.

However, since this expansion is not monotone increasing it is evident that (5.2.15) cannot be formulated with H_{-b^2} replaced by this expansion since the Hellinger distance is only defined for d.f.'s. One might overcome this problem by extending the definition of the Hellinger distance to signed measures. Another possibility is to redefine the expansion in such a way that one obtains a probability measure; this was e.g. achieved in Example 5.2.9. To reformulate (5.2.15) we need the following lemma which concerns an expansion of length two of von Mises d.f.'s H_β.

Lemma 5.2.10. *For every real β denote by μ_β the signed measure which corresponds to the measure generating function*

$$x \to G_3(x)(1 - \beta e^{-x} x^2/2).$$

Let again H_β denote the von Mises distribution with parameter β. Then,

$$\sup_B |H_\beta(B) - \mu_\beta(B)| = O(\beta^{-2}).$$

PROOF. Apply Lemma A.2.1 and Lemma A.3.2.   □

Thus as an analogue to (5.2.15) we get

$$\sup_B |P\{b_n(X_{n:n} - (b_n - b_n^{-1})) \in B\} - \mu_{-b_n^{-2}}(B)| = O((\log n)^{-2}) \quad (5.2.17)$$

where $X_{n:n}$ is the maximum of n i.i.d. standard normal r.v.'s, and b_n is the solution of the equation $n\varphi(b - b^{-1}) = b$.

Figures 5.2.1–5.2.3 concern the density f_n of $\Phi^n(b_n + a_n \cdot)$, with $b_n = \Phi^{-1}(1 - 1/n)$ and $a_n = 1/(n\varphi(b_n))$ (compare with P.5.8), the Gumbel density g_3 and the derivative $g_3(1 + h_n)$ of the expansion in (5.2.16).

Observe that f_n and $g_3(1 + h_n)$ have modes larger than zero; moreover, $g_3(1 + h_n)$ provides a better approximation to f_n than g_3.

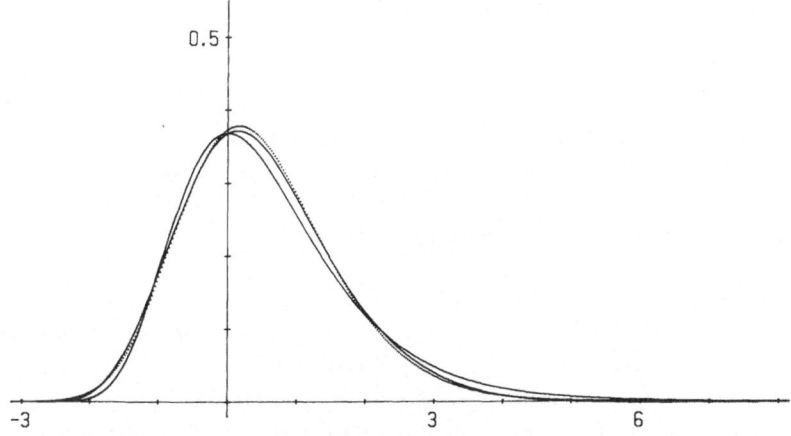

Figure 5.2.1. Normalized density f_n (dotted line) of maximum of normal r.v.'s, Gumbel density g_3, and expansion $g_3(1 + h_n)$ for $n = 40$.

In order to get a better insight into the approximation, indicated by Figure 5.2.1, we also give illustrations concerning the error of the approximation.

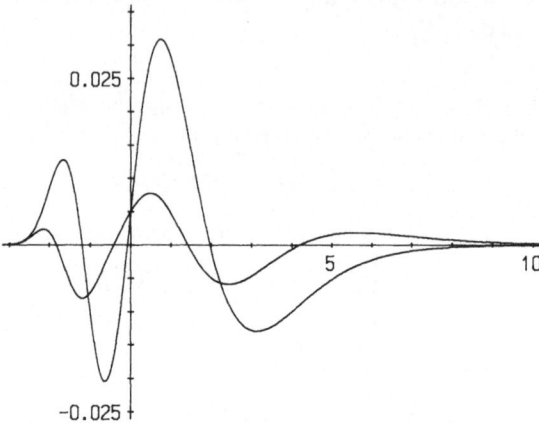

Figure 5.2.2. $f_n - g_3, f_n - g_3(1 + h_n)$ for $n = 40$.

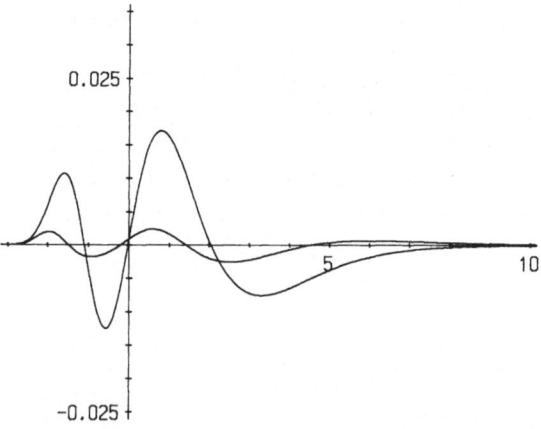

Figure 5.2.3. $f_n - g_3, f_n - g_3(1 + h_n)$ for $n = 400$.

We are well aware that some statisticians take the slow convergence rate of order $O(1/\log n)$ as an argument against the asymptotic theory of extremes, perhaps, believing that a rate of order $O(n^{-1/2})$ ensures a much better accuracy of an approximation for small sample sizes. However, one may argue that from the historical and mathematical point of view it is always challenging to tackle this and related problems. Moreover, one should know that typical statistical problems in extreme value theory do not concern normal r.v.'s.

The illustrations above and further numerical computations show that the Gumbel approximation to the normalized d.f. and density of the maximum of normal r.v.'s is of a reasonable accuracy for small sample sizes. This may serve as an example that the applicability of an approximation not only depends on the rate of convergence but also on the constant involved in the error bound.

If a more accurate approximation is needed then, instead of increasing the sample size, it is advisable to use an expansion of length two or a penultimate distribution. Comparing Figures 5.2.2 and 5.2.3 we see that the expansion of length two for $n = 40$ is of a higher accuracy than the Gumbel approximation for $n = 400$.

The limit theorem and the expansion give some insight into the asymptotic behavior of the sample maximum. Keep in mind that the d.f. Φ^n of the sample maximum itself may serve as an approximate d.f. in certain applications (see Reiss, 1978a).

Expansions of Length Two

Another example of an expansion of length two is obtained by treating a refinement of Corollary 5.2.7 and Remark 5.2.8. In Remark 5.2.8 we studied distributions of sample maxima under densities of the form $f = \psi(1 + h)$ where h varies over a certain class of functions. Next, we consider densities of the form

$$f = \psi(1 + p + h)$$

with p being fixed. Moreover, ψ is given as in (5.2.10).

Below, an expansion of length 2 of distributions of sample maxima is established where the leading term of the expansion is an extreme value distribution G and the second term depends on G and p. Let

$$p(x) = \begin{array}{ll} -Kx^{-\alpha\rho} & i = 1 \\ -K(-x)^{\alpha\rho} & \text{if} \quad i = 2 \\ -Ke^{-\rho x} & i = 3 \end{array} \qquad (5.2.18)$$

for some fixed $K \geq 0$ and $\rho > 0$, and

$$|h(x)| \leq \begin{array}{ll} Lx^{-\alpha\delta} & i = 1 \\ L(-x)^{\alpha\delta} & \text{if} \quad i = 2 \\ Le^{-\delta x} & i = 3 \end{array} \qquad (5.2.19)$$

where $L > 0$ and $0 < \rho \leq \delta \leq 1$. The expansion of length two is given by

$$G_{p,n}(x) = G(x)\left[1 - n^{-\rho} \int_x^{\omega(G)} p(y)\psi(y)\,dy \right] \qquad (5.2.20)$$

for $\alpha(G) < x < \omega(G)$. This may be written

$$G_{p,n}(x) = G(x)\left[1 + n^{-\rho}\frac{K}{1 + \rho} \cdot \begin{array}{ll} x^{-(1+\rho)\alpha} & i = 1 \\ (-x)^{(1+\rho)\alpha} & \text{if} \quad i = 2 \\ e^{-(1+\rho)x} & i = 3. \end{array} \right] \qquad (5.2.21)$$

Notice that $f = \psi(1 + p + h)$ and $G_{p,n}$ arise from the special case with $i = 2$ and $\alpha = 1$ via the transformation $T_{i,\alpha} = G_{i,\alpha}^{-1} \circ G_{2,1}$.

It is easy to check that $G_{p,n}$ is a d.f. if n is sufficiently large; more precisely, this holds if, and only if,

$$K\rho^p/(1 + \rho) \le n^p. \tag{5.2.22}$$

Theorem 5.2.11. *Let G, ψ and T be as in Corollary 5.2.7. Assume that the underlying density f has the representation*

$$f(x) = \psi(x)(1 + p(x) + h(x)), \qquad T(x_0) < x < \omega(G), \tag{5.2.23}$$

and $= 0$, if $x > \omega(G)$, where $x_0 < 0$ and p, h satisfy (5.2.18) and (5.2.19).
    Put

$$F_n(x) = F(d_n + xc_n)$$

where c_n, d_n are the constants of Theorem 5.1.1. Then,

$$H(F_n^n, G_{p,n}) = O(n^{-\min(\delta, 2\rho)}).$$

PROOF. Apply Lemma 5.2.2.   □

It was observed by Radtke (1988) (compare with P.5.16) that for a special case the expansion $G_{p,n}(x)$ can be replaced by $G(b_n + a_n x)$ where G is the leading term of the expansion and $b_n \to 0$ and $a_n \to 1$ as $n \to \infty$. Notice that $G(b_n + a_n x)$ can be written—up to terms of higher order—as

$$G(x)[1 + \psi(x)(b_n + (a_n - 1)x)]$$

where again $\psi = G'/G$. One can easily check that such a representation holds in (5.2.21) if, and only if, $i = 1$ and $\rho = 1/\alpha$.

5.3. The Structure of Asymptotic Joint Distributions of Extremes

Let us reconsider the stochastical model which was studied in Section 5.2. The sample maxima $M_{n,i} := \max(\xi_{n(i-1)+1}, \ldots, \xi_{ni})$ are the observed r.v.'s, and it is assumed that

(a) $M_{n,1}, \ldots, M_{n,N}$ are i.i.d. random variables,
(b) the (possibly, non-observable) r.v.'s $\xi_{n(i-1)+1}, \ldots, \xi_{ni}$ are i.i.d. for every $i = 1, \ldots, N$.

The r.v.'s $\xi_{n(i-1)+1}, \ldots, \xi_{ni}$ may correspond to data which are collected within the ith period (as e.g. the amount of daily rainfall within a year). Then, the sample $M_{n,1}, \ldots, M_{n,N}$ of the annual maxima can be used to estimate the unknown distribution of the maximum daily rainfall within a year. Condition

(a) seems to be justified in this example, however, the second condition is severely violated. It would be desirable to get some insight (within a mathematical model) into the influence of a deviation from condition (b), however, this problem is beyond the scope of this book. With the present state-of-the-art one can take some comfort from experience and from statements as e.g. made in Pickands (1975, page 120) that "the method has been shown to be very robust against dependence" of the r.v.'s $\xi_{n(i-1)+1}, \ldots, \xi_{ni}$.

It may happen that a certain amount of information is lost if the statistical influence is only based on maxima. Thus, a different method was proposed by Pickands (1975), namely, to consider the k largest observations of the original data. This method is only applicable if these data can be observed. For the mathematical treatment of this problem it is assumed (by combining the conditions (a) and (b)) that $\xi_1, \ldots, \xi_{nN}$ are i.i.d. random variables. The statistical inference will be based on the k largest order statistics $X_{nN-k+1:nN} \leq \cdots \leq X_{nN:nN}$ of $\xi_1, \ldots, \xi_{nN}$. In this sequel, the sample size will again be denoted by n instead of nN.

In special cases, a comparison of the two different methods will be made in Section 9.6. The information which is lost or gained by one or the other method can be indicated by the relative efficiency between statistical procedures which are constructed according to the respective methods.

One should keep in mind that such a comparison heavily depends on the conditions stated above. For example one can argue that the dependence of the rainfall on consecutive days has less influence on the stochastic properties of the annual maxima compared to the influence on the k largest observations within the whole period. Thus, the second method may be less robust against the departure from the condition of independence.

The main purpose of this section is to introduce the asymptotic distributions of the k largest order statistics. Moreover, it will be of great importance to find appropriate representations for these distributions. For the aims of this section it suffices to consider order statistics from generalized Pareto r.v.'s as introduced in (1.6.11). Notice again that the same symbol will be used for the d.f. and the pertaining probability measure.

Upper Extremes of Uniform R.V.'s

Let $V_{n-k+1:n}$ be the kth largest order statistic of n i.i.d. random variables with common d.f. $W_{2,1}$ (the uniform distribution on $(-1, 0)$). In Section 5.1 it was proved that $nV_{n-k+1:n}$ is asymptotically equal (in distribution) to a "negative" gamma r.v.

$$S_k = \sum_{i=1}^{k} \xi_i$$

where $\xi_1, \ldots, \xi_k$ are i.i.d. random variables with common "negative" exponential d.f. $F(x) = e^x$ for $x < 0$. An extension of the result for a single order statistic to joint distributions of upper extremes can easily be established by utilizing the following lemma.

Lemma 5.3.1. *For every* $k = 1, \ldots, n$ *we have*

$$\sup_B |P\{(nV_{n:n}, nV_{n-1:n}, \ldots, nV_{n-k+1:n}) \in B\} - P\{(S_1, S_2, \ldots, S_k) \in B\}|$$

$$= \sup_B |P\{nV_{n-k+1:n} \in B\} - P\{S_k \in B\}|.$$

It is obvious that "$\geq$" holds. At first sight the equality looks surprising, however, the miracle will have a simple explanation when the distributions are represented in an appropriate way.

From Corollary 1.6.11 it is immediate that

$$\left(\frac{V_{n:n}}{V_{n-1:n}}, \ldots, \frac{V_{n-k+2:n}}{V_{n-k+1:n}}, V_{n-k+1:n}\right) \overset{d}{=} \left(\frac{S_1}{S_2}, \ldots, \frac{S_{k-1}}{S_k}, \frac{S_k}{-S_{n+1}}\right). \qquad (5.3.1)$$

Thus we easily get

$$\sup_B |P\{(nV_{n:n}, nV_{n-1:n}, \ldots, nV_{n-k+1:n}) \in B\} - P\{(S_1, S_2, \ldots, S_k) \in B\}|$$

$$= \sup_B \left| P\left\{\left(\frac{S_1}{S_2}, \ldots, \frac{S_{k-1}}{S_k}, \frac{S_k}{-S_{n+1}/n}\right) \in B\right\} \right.$$

$$\left. - P\left\{\left(\frac{S_1}{S_2}, \ldots, \frac{S_{k-1}}{S_k}, S_k\right) \in B\right\} \right| =: A.$$

Notice that the first $k - 1$ components in the random vectors above are equal. Moreover, it is straightforward to verify that the components in each vector are independent since according to Corollary 1.6.11(iii) the r.v.'s $S_1/S_2, \ldots, S_n/S_{n+1}, S_{n+1}$ are independent. An application of inequality (3.3.4) (which concerns an upper bound for the variational distance of product measures via the variational distances of the single components) yields

$$A \leq \sup_B \left| P\left\{\frac{S_k}{-S_{n+1}/n} \in B\right\} - P\{S_k \in B\} \right|$$

$$= \sup_B |P\{nV_{n-k+1:n} \in B\} - P\{S_k \in B\}|.$$

Thus, Lemma 5.3.1 is proved.

Combining Lemma 5.1.5 and Lemma 5.3.1 we get

Lemma 5.3.2. *For every fixed* $k \geq 1$ *as* $n \to \infty$,

$$\sup_B |P\{(nV_{n:n}, nV_{n-1:n}, \ldots, nV_{n-k+1:n}) \in B\} - P\{(S_1, S_2, \ldots, S_k) \in B\}| \to 0.$$

The limiting distribution in Lemma 5.3.2 will be denoted by $G_{2,1,k}$. It is apparent that $G_{2,1,j}$—the limiting distribution of the jth largest order statistic—is the jth marginal distribution of $G_{2,1,k}$. From Lemma 1.6.6(iii) we know that the density, say, $g_{2,1,k}$ of $G_{2,1,k}$ is given by

$$g_{2,1,k}(\mathbf{x}) = \exp(x_k), \qquad x_k < x_{k-1} < \cdots < x_1 < 0, \tag{5.3.2}$$

and $= 0$, otherwise.

Upper Extremes of Generalized Pareto R.V.'s

The extension of Lemma 5.3.2 to other generalized Pareto d.f.'s $W_{i,\alpha}$ is straightforward.

Let again $T_{i,\alpha}$ denote the transformation in (1.6.10). We have $T_{1,\alpha}(x) = (-x)^{-1/\alpha}$, $T_{2,\alpha}(x) = -(-x)^{1/\alpha}$, and $T_3(x) = -\log(-x)$ for $-\infty < x < 0$.

Denote by $G_{i,\alpha,k}$ the distribution of the random vector

$$(T_{i,\alpha}(S_1), \ldots, T_{i,\alpha}(S_k)). \tag{5.3.3}$$

The transformation theorem for densities (see (1.4.4)) enables us to compute the density, say, $g_{i,\alpha,k}$ of $G_{i,\alpha,k}$. We have

$$g_{1,\alpha,k}(\mathbf{x}) = \alpha^k \exp(-x_k^{-\alpha}) \prod_{j=1}^{k} x_j^{-(\alpha+1)}, \qquad 0 < x_k < x_{k-1} < \cdots < x_1;$$

$$g_{2,\alpha,k}(\mathbf{x}) = \alpha^k \exp(-(-x_k)^{\alpha}) \prod_{j=1}^{k} (-x_j)^{\alpha-1}, \qquad x_k < x_{k-1} < \cdots < x_1 < 0,$$

$$g_{3,k}(\mathbf{x}) = \exp(-e^{-x_k}) \exp\left(-\sum_{j=1}^{k} x_j\right), \qquad x_k < x_{k-1} < \cdots < x_1, \tag{5.3.4}$$

and the densities are zero, otherwise.

Notice that the following representation of the density $g_{i,\alpha,k}$ holds:

$$g_{i,\alpha,k}(\mathbf{x}) = G_{i,\alpha}(x_k) \prod_{j=1}^{k} \psi_{i,\alpha}(x_j) = g_{i,\alpha}(x_k) \prod_{j=1}^{k-1} \psi_{i,\alpha}(x_j) \tag{5.3.5}$$

if $\alpha(G_{i,\alpha}) < x_k < \cdots < x_1 < \omega(G_{i,\alpha})$ where again $\psi_{i,\alpha} = g_{i,\alpha}/G_{i,\alpha}$.

Corollary 5.3.3. *Let $X_{r:n}$ be the rth order statistic of n i.i.d. random variables with common generalized Pareto d.f. $W_{i,\alpha}$. Then,*

$$\sup_{B} |P\{(c_n^{-1}(X_{n-j+1:n} - d_n))_{j=1}^{k} \in B\} - G_{i,\alpha,k}(B)| \to 0, \qquad n \to \infty,$$

where c_n and d_n are the constants of Theorem 5.1.1.

PROOF. Straightforward from Lemma 5.3.2, the definition of $G_{i,\alpha,k}$ and the fact that

$$(c_n^{-1}(X_{n-j+1:n} - d_n))_{j=1}^{k} \overset{\mathrm{d}}{=} (T_{i,\alpha}(nV_{n-j+1:n}))_{j=1}^{k}. \qquad \square$$

Domains of Attraction

This section concludes with a characterization of the domains of attractions of joint distributions of a fixed number of upper extremes by means of the corresponding result for sample maxima.

First, we refer to the well-known result (see e.g. Galambos (1987), Theorem 2.8.2) that a d.f. belongs to the weak domain of attraction of an extreme value d.f. $G_{i,\alpha}$ if, and only if, the corresponding result holds for the kth largest order statistic with $G_{i,\alpha,k}$ as the limiting d.f.

Our interest is focused on the convergence w.r.t. the variational distance.

Theorem 5.3.4. *Let F be a d.f. with density f. Then, the following two statements are equivalent*:
(i) *F belongs to the strong domain of attraction of an extreme value distribution* $G \in \{G_{1,\alpha}, G_{2,\alpha}, G_3 : \alpha > 0\}$.
(ii) *There exist constants* $a_n > 0$ *and* b_n *such that for every positive integer* k *there is a nondegenerate distribution* $G^{(k)}$ *such that*

$$\sup_B |P\{(a_n^{-1}(X_{n-j+1:n} - b_n))_{j=1}^k \in B\} - G^{(k)}(B)| \to 0, \qquad n \to \infty.$$

In addition, if (i) *holds for* $G = G_{i,\alpha}$ *then* (ii) *is valid for* $G^{(k)} = G_{i,\alpha,\mathbf{k}}$.

PROOF. (ii) $\Rightarrow$ (i): Obvious.
(i) $\Rightarrow$ (ii): Let $a_n > 0$ and b_n be such that for every x

$$F^n(b_n + xa_n) \to G(x), \qquad n \to \infty, \tag{1}$$

where $G \in \{G_{1,\alpha}, G_{2,\alpha}, G_3 : \alpha > 0\}$. According to Lemma 5.1.3, (i) is equivalent to the condition that for every subsequence $i(n)$ there exists a subsequence $m(n) := i(j(n))$ such that

$$m(n)a_{m(n)}f(b_{m(n)} + xa_{m(n)}) \to \psi(x), \qquad n \to \infty,$$

for Lebesgue almost all $x \in (\alpha(G), \omega(G))$ where again $\psi = G'/G$. Thus, also

$$\prod_{j=1}^k m(n)a_{m(n)}f(b_{m(n)} + x_j a_{m(n)}) \to \prod_{j=1}^k \psi(x_j), \qquad n \to \infty, \tag{2}$$

for Lebesgue almost all $\mathbf{x} = (x_1, \dots, x_k) \in (\alpha(G), \omega(G))^k$. Furthermore, deduce with the help of (1.4.4) that the density of $(a_n^{-1}(X_{n-j+1:n} - b_n))_{j=1}^k$, say, $f_{n,k}$ is given by

$$f_{n,k}(\mathbf{x}) = F^{n-k}(b_n + x_k a_n) \prod_{j=1}^k [(n - j + 1)a_n f(b_n + x_j a_n)], \qquad x_k < \cdots < x_1, \tag{3}$$

and $= 0$, otherwise. Combining (1)–(3) with (5.3.5) we obtain for $G = G_{i,\alpha}$ that

$$f_{m(n),k}(\mathbf{x}) \to g_{i,\alpha,\mathbf{k}}(\mathbf{x}), \qquad n \to \infty,$$

for Lebesgue almost all $\mathbf{x}$ with $\alpha(G) < x_k < \cdots < x_1 < \omega(G)$. Thus the Schéffe Lemma 3.3.2 implies (ii) with $G^{(k)} = G_{i,\alpha,\mathbf{k}}$. $\qquad\square$

5.4. Expansions of Distributions of Extremes of Generalized Pareto Random Variables

In this section we establish higher order approximations to the distribution of upper extremes of generalized Pareto r.v.'s. First, we prove an expansion of the distribution of the kth largest order statistic of uniform r.v.'s. The leading term of the expansion is a "negative" gamma distribution $G_{2,1,k}$. By using the transformation technique the result is extended to generalized Pareto r.v.'s. Finally, the results of Section 5.3 enable us to examine joint distributions of upper extremes.

Let $V_{n-k+1:n}$ again be the kth largest order statistic of n i.i.d. $(-1,0)$-uniformly distributed r.v.'s. From (5.1.35) we already know that

$$\sup_B |P\{nV_{n-k+1:n} \in B\} - G_{2,1,k}(B)| \to 0, \qquad n \to \infty.$$

We shall prove that the remainder term is bounded by Ck/n where C is a universal constant. The expansion of length 2 will show that this bound is sharp. The extension from $W_{2,1}$ to a generalized Pareto d.f. $W \in \{W_{1,\alpha}, W_{2,\alpha}, W_3: \alpha > 0\}$ is straightforward. We have

$$\sup_B |P\{c_n^{-1}(X_{n-k+1:n} - d_n) \in B\} - G_{i,\alpha,k}(B)| \le Ck/n \qquad (5.4.1)$$

where c_n and d_n are the usual normalizing constants.

In Section 5.5 we shall see that if the generalized Pareto d.f. W is replaced by an extreme value d.f. $G \in \{G_{1,\alpha}, G_{2,\alpha}, G_3: \alpha > 0\}$ then the bound in (5.4.1) is of order $O(k^{3/2}/n)$.

Moreover, as it will be indicated at the end of this section, F has the tail of a generalized Pareto d.f. if an inequality of the form (5.4.1) holds. Therefore, in a certain sense, the generalized Pareto d.f.'s occupy the place of the max-stable extreme value d.f.'s as far as joint distributions of extremes are concerned.

Extremes of Uniform R.V.'s

Let us begin with a simple result concerning central moments of the gamma distribution $G_{2,1,k}$.

Lemma 5.4.1. *The ith central moment*

$$u(i,k) = \int (x+k)^i \, dG_{2,1,k}(x)$$

of $G_{2,1,k}$ *fulfills the recurrence relation*

$$u(i + 2, k) = (i + 1)[ku(i, k) - u(i + 1, k)]. \tag{5.4.2}$$

Moreover,

$$\int |x + k|^i \, dG_{2,1,k}(x) \leq i! k^{i/2}. \tag{5.4.3}$$

As special cases we note $u(1, k) = 0$, $u(2, k) = k$, $u(3, k) = -2k$, $u(4, k) = 6k + 3k^2$.

PROOF. Recall that the density of $G_{2,1,k}$ is given by $g_{2,1,k}(x) = e^x(-x)^{k-1}/(k-1)!$, $x < 0$. By partial integration we get

$$-\int (i + 1)(x + k)^i x \, dG_{2,1,k}(x) = \int (x + k)^{i+1} x \, dG_{2,1,k}(x) + ku(i + 1, k).$$

Now, (5.4.2) is straightforward since

$$u(i + 2, k) = \int (x + k)^{i+1} x \, dG_{2,1,k}(x) + ku(i + 1, k)$$

$$= -\int (i + 1)(x + k)^i x \, dG_{2,1,k}(x) = (i + 1)[ku(i, k) - u(i + 1, k)].$$

Moreover, because of $(i + 1)[(i + 1)! + i!] = (i + 2)!$ we obtain by induction over i that $|u(i, k)| \leq i! k^{i/2}/2$. This implies (5.4.3) for every even i. Finally, the Schwarz inequality yields

$$\int |x + k|^{2i+1} x \, dG_{2,1,k}(x) \leq (2i + 1)! k^{(2i+1)/2}.$$

The proof is complete. $\qquad\qquad\qquad\qquad\qquad\qquad\qquad\qquad\qquad\qquad$ $\square$

A preliminary higher order approximation is obtained in Lemma 5.4.2.

Lemma 5.4.2. *For every positive integer m there exists a constant $C_m > 0$ such that for n and $k \in \{1, \ldots, n\}$ with k/n sufficiently small (so that the denominators below are bounded away from zero) the following inequality holds:*

$$\sup_B \left| P\{nV_{n-k+1:n} \in B\} - \frac{G_{2,1,k}(B) + \sum_{i=2}^{2(m-1)} \beta(i, n - k) \int_B (x + k)^i \, dG_{2,1,k}(x)}{1 + \sum_{i=2}^{2(m-1)} \beta(i, n - k)u(i, k)} \right|$$

$$\leq C_m(k/n)^m.$$

Moreover,

$$\beta(i, n) = \sum_{j=0}^{i} (-1)^j \binom{n}{i - j} n^{-(i-j)}/j!$$

and $u(i, k)$ is the ith central moment of $G_{2,1,k}$.

As special cases we note $\beta(2, n) = -1/2n$, $\beta(3, n) = 1/3n^2$, $\beta(4, n) = 1/8n^2 - 1/4n^3$. Moreover, $|\beta(2i - 1, n)|$, $|\beta(2i, n)| \le C_m n^{-i}$, $i = 1, \ldots, m - 1$.

PROOF. Put

$$g_n(x) = e^{-k}\left(1 + \frac{x + k}{n - k}\right)^{n-k} \frac{(-x)^{k-1}}{(k - 1)!} 1_{(-n, 0)}(x).$$

From Theorem 1.3.2 we conclude that $g_n/\int g_n(x)\,dx$ is the density of $nV_{n-k+1:n}$. Moreover, we write

$$f_n(x) = \left[1 + \sum_{i=2}^{2(m-1)} \beta(i, n - k)(x + k)^i\right] g_{2,1,k}(x).$$

Lemma A.2.1 yields

$$|g_n(x) - f_n(x)| \le C(n - k)^{-m}[|x + k|^{2m-1} + (x + k)^{2m}] g_{2,1,k}(x) \qquad (1)$$

for every $x \in A_n := \{x < 0: |x + k| \le (n - k)^{1/2}\}$ where, throughout, C will be used as a generic constant that only depends on m.

From (5.4.3) and from the upper bound of $\beta(i, n - k)$ as given in Lemma A.2.1 we conclude that $\int f_n(x)\,dx \ge 1/2$ if k/n is sufficiently small. Thus, by (1), Lemma A.3.2, and (5.4.3) we obtain

$$\sup_B \left| P\{nV_{n-k+1:n} \in B\} - \int_B f_n(x)\,dx \Big/ \int f_n(x)\,dx \right|$$

$$\le C \int_{A_n} |g_n(x) - f_n(x)|\,dx + \int_{A_n^c} |g_n(x) - f_n(x)|\,dx \qquad (2)$$

$$\le C(k/n)^m + \int_{A_n^c} |g_n(x) - f_n(x)|\,dx.$$

Moreover, because of $(1 + x/n)^n \le \exp(x)$ we have $g_n \le g_{2,1,k}$. Thus, the Schwarz inequality yields

$$\int_{A_n^c} |g_n(x) - f_n(x)|\,dx$$

$$\le 2G_{2,1,k}(A_n^c) + \sum_{i=2}^{2(m-1)} |\beta(i, n - k)| \int_{A_n^c} |x + k|^i\,dG_{2,1,k}(x)$$

$$\le 2G_{2,1,k}(A_n^c) + \sum_{i=2}^{2(m-1)} |\beta(i, n - k)| [G_{2,1,k}(A_n^c)]^{1/2} \left[\int |x + k|^{2i}\,dG_{2,1,k}(x)\right]^{1/2}$$

$$\le C(k/n)^m.$$

Combining this and (2) the proof is completed. $\qquad \square$

The following theorem is an immediate consequence of Lemma 5.4.2 and Lemma 3.2.5. Moreover, we remark that the polynomials $p_{j,k,n}$ can easily be constructed by means of formula (3.2.9).

Theorem 5.4.3. *For every positive integer m there exists a constant $C_m > 0$ such that for every n and $k \in \{1, \dots, n\}$ the following inequality holds:*

$$\sup_B \left| P\{nV_{n-k+1:n} \in B\} - \left[G_{2,1,k}(B) + \sum_{j=1}^{m-1} \int_B p_{j,k,n} \, dG_{2,1,k} \right] \right| \le C_m (k/n)^m$$

where $p_{j,k,n}$ are polynomials of degree 2j.

We note the explicit form of $p_{1,k,n}$ and $p_{2,k,n}$. We have

$$p_{1,k,n}(x) = -[(x+k)^2 - k]/2(n-k)$$

and   (5.4.4)

$$p_{2,k,n}(x) = \beta(4, n-k)[(x+k)^4 - u(4,k)] + \beta(3, n-k)[(x+k)^3 - u(3,k)]$$
$$- \beta(2, n-k)u(2,k)[(x+k)^2 - u(2,k)].$$

Lemma 5.4.2 as well as Theorem 5.4.3, applied to $m = 1$, yield (5.4.1) in the particular case of $W = W_{2,1}$.

Extremes of Generalized Pareto R.V.'s

The extension of the results above to the kth largest order statistics $X_{n-k+1:n}$ under a generalized Pareto d.f. $W \in \{W_{1,\alpha}, W_{2,\alpha}, W_3 : \alpha > 0\}$ is immediate. By using the transformation technique we easily obtain (5.4.1) and the following expansion

$$\sup_B \left| P\{c_n^{-1}(X_{n-k+1:n} - d_n) \in B\} - \left[G_{i,\alpha,k}(B) \right. \right.$$

$$\left. \left. + \sum_{j=1}^{m-1} \int_B p_{j,k,n}(\log G_{i,\alpha}) \, dG_{i,\alpha,k} \right] \right| \le C_m (k/n)^m$$ (5.4.5)

where c_n and d_n are the constants of Theorem 5.1.1 and $p_{j,k,n}$ are the polynomials of Theorem 5.4.3.

Next, we prove the corresponding result for joint distributions of upper extremes.

Theorem 5.4.4. *Let $X_{n:n}, \dots, X_{n-k+1:n}$ be the k largest order statistics under the generalized Pareto d.f. $W \in \{W_{1,\alpha}, W_{2,\alpha}, W_3 : \alpha > 0\}$. Let c_n, d_n, C_m, and $p_{j,k,n}$ be as above. Then,*

$$\sup_B \left| P\{(c_n^{-1}(X_{n:n} - d_n), \dots, c_n^{-1}(X_{n-k+1:n} - d_n)) \in B\} - \left[G_{i,\alpha,\mathbf{k}}(B) \right. \right.$$

$$\left. \left. + \sum_{j=1}^{m-1} \int_B p_{j,k,n}(\log G_{i,\alpha}(x_k)) \, dG_{i,\alpha,\mathbf{k}}(\mathbf{x}) \right] \right| \le C_m (k/n)^m.$$ (5.4.6)

PROOF. It suffices to prove the assertion in the special case of $i = 2$ and $\alpha = 1$. The general case can easily be deduced by means of the transformation technique. Thus, we have to prove that

$$
\sup_B \left| P\{(nV_{n:n}, \ldots, nV_{n-k+1:n}) \in B\} - \left[G_{2,1,\mathbf{k}}(B) \right. \right.
$$

$$
\left. \left. + \sum_{j=1}^{m-1} \int_B p_{j,k,n}(x_k) \, dG_{2,1,\mathbf{k}}(\mathbf{x}) \right] \right| \le C_m (k/n)^m. \tag{5.4.7}
$$

If $m = 1$ then the proof of Lemma 5.3.2 carries over if Lemma 5.1.5 is replaced by Theorem 5.4.3.

If $m > 1$ then one has to deal with signed measures, however, the method of the proof to Lemma 5.3.2 is still applicable. Notice that the approximating signed measure in (5.4.7) has the density

$$
\mathbf{x} \to \left(1 + \sum_{j=1}^{m-1} p_{j,k,n}(x_k) \right) g_{2,1,\mathbf{k}}(\mathbf{x}).
$$

By inducing with $\mathbf{x} \to (x_1/x_2, \ldots, x_{k-1}/x_k, x_k)$ one obtains a product measure where the kth component has the density

$$
\left(1 + \sum_{j=1}^{m-1} p_{j,k,n} \right) g_{2,1,k}.
$$

Now inequality (A.3.3), which holds for signed measures, and Theorem 5.4.3 imply the assertion.   □

Next, Theorem 5.4.4 will be stated once more in the particular case of $m = 1$. In an earlier version of this book we conjectured that a d.f. F has the tail of a generalized Pareto d.f. if an inequality of the form (5.4.1) (formulated for d.f.'s) holds. This was confirmed in Falk (1989a).

Theorem 5.4.5. (i) *If* $X_{n:n}, \ldots, X_{n-k+1:n}$ *are the* k *largest order statistics under a generalized Pareto d.f.* $W \in \{W_{1,\alpha}, W_{2,\alpha}, W_3 : \alpha > 0\}$ *then there exists a constant* $C > 0$ *such that for every* $k \in \{1, \ldots, n\}$,

$$
\sup_B |P\{(c_n^{-1}(X_{n:n} - d_n), \ldots, c_n^{-1}(X_{n-k+1:n} - d_n)) \in B\} - G_{i,\alpha,\mathbf{k}}(B)| \tag{5.4.8}
$$
$$
\le Ck/n
$$

with c_n *and* d_n *as in Theorem 5.1.1.*

(ii) *Let* F *be a d.f. which is strictly increasing and continuous on a left neighborhood of* $\omega(F)$. *If* (5.4.8) *holds with* W, c_n, *and* d_n *replaced by* F *and any normalizing constants* $a_n > 0$ *and* b_n *then there exist* $c > 0$ *and* d *such that*

$$
F((x - d)/c) = W_{i,\alpha}(x)
$$

for x *in a neighborhood of* $\omega(W_{i,\alpha})$.

For a slightly stronger formulation of (ii) and for the proof we refer to Falk (1989a).

5.5. Variational Distance between Exact and Approximate Joint Distributions of Extremes

In this section we prove a version of Theorem 5.2.5 valid for the joint distribution of the upper extremes. In view of our applications and to avoid technical complications the results will be proved w.r.t the variational distance.

The Main Results

In a preparatory step we prove the following technical lemma. Notice that the upper bound in (5.5.1) still depends on the underlying distribution through the d.f. F. The main purpose of the subsequent considerations will be to cancel the d.f. F in the upper bound to facilitate further computations. We remark that the results below are useful modifications of results of Falk (1986a).

Lemma 5.5.1. *Given* $G_{\mathbf{k}} \in \{G_{1,\alpha,\mathbf{k}}, G_{2,\alpha,\mathbf{k}}, G_{3,\mathbf{k}}: \alpha > 0\}$ *let* G *denote the first marginal d.f. Let* $X_{n:n} \geq \cdots \geq X_{n-k+1:n}$ *be the* k *largest order statistics of* n *i.i.d. random variables with d.f.* F *and density* f. *Define again* $\psi = g/G$ *on the support of* G *where* g *is the density of* G. *Moreover, fix* $x_0 \geq -\infty$.
   Then,

$$\sup_{B} |P\{(X_{n:n}, \ldots, X_{n-k+1:n}) \in B\} - G_{\mathbf{k}}(B)|$$

(5.5.1)

$$\leq \left[2G_{\mathbf{k}}(M^c) + \int_M \left[n(1 - F(x_k)) + \log G(x_k) - \sum_{j=1}^k \log(nf/\psi)(x_j) \right] dG_{\mathbf{k}}(\mathbf{x}) \right]^{1/2}$$
$$+ Ck/n$$

where $M = \{\mathbf{x}: x_j > x_0, f(x_j) > 0, j = 1, \ldots, k\}$ *and* C *is a universal constant.*

PROOF. The quantile transformation and inequality (5.4.9) yield

$$P\{(X_{n:n}, \ldots, X_{n-k+1:n}) \in B\} = P\{[F^{-1}(1 + (nV_{n-j+1:n})/n]_{j=1}^k \in B\}$$
$$= \mu_n(B) + O(k/n)$$

(1)

uniformly over n, k, and Borel sets B where the measure μ_n is defined by

$$\mu_n(B) = G_{2,1,\mathbf{k}}\{\mathbf{x}: -n < x_k < \cdots < x_1, [F^{-1}(1 + x_j/n)]_{j=1}^k \in B\}.$$

In analogy to the proof of Theorem 1.4.5, part III (see also Remark 1.5.3) deduce that μ_n has the density h_n defined by

$$h_n(\mathbf{x}) = \exp[-n(1 - F(x_k))] \prod_{j=1}^{k} (nf(x_j)), \qquad x_k < \cdots < x_1,$$

and $= 0$, otherwise.

In (1), the measure μ_n can be replaced by the probability measure $Q_n = \mu_n/b_n$ where

$$b_n = G_{2,1,\mathbf{k}}\{\mathbf{x}: -n < x_k < \cdots < x_1\} = G_{2,1,k}(-n, 0]$$

$$= 1 - \exp(-n) \sum_{j=0}^{k-1} n^j/j! = 1 + O(k/n).$$

Denote by $g_\mathbf{k}$ the density of $G_\mathbf{k}$. Recall that $g_\mathbf{k}(\mathbf{x}) = G(x_k) \prod_{j=1}^{k} \psi(x_j)$ for $\alpha(G) < x_k < \cdots < x_1 < \omega(G)$. Now, Lemma A.3.5, applied to Q_n and $G_\mathbf{k}$, implies the asserted inequality (5.5.1).   □

Next we formulate a simple version of Theorem 5.5.4 as an analogue to Corollary 5.2.3. The proof can be left to the reader.

Corollary 5.5.2. *Denote by G_j the jth marginal d.f. of $G_\mathbf{k} \in \{G_{1,\alpha,\mathbf{k}}, G_{2,\alpha,\mathbf{k}}, G_{3,\mathbf{k}}:$ $\alpha > 0\}$, and write $G = G_1$. If, in addition to the conditions of Lemma 5.5.1, $G\{f > 0\} = 1$ and $\omega(F) = 0$ for $i = 2$, then*

$$\sup_B |P\{(X_{n:n}, \ldots, X_{n-k+1:n}) \in B\} - G_\mathbf{k}(B)|$$

$$\leq \left[\sum_{j=1}^{k} \int [nf/\psi - 1 - \log(nf/\psi)] \, dG_j \right]^{1/2} + Ck/n$$

$$\leq \left[\sum_{j=1}^{k} \int [(nf/\psi - 1)^2/(nf/\psi)] \, dG_j \right]^{1/2} + Ck/n.$$

As a consequence of Corollary 5.5.2 one gets the following example taken from Falk (1986a).

EXAMPLE 5.5.3. Let φ denote the standard normal density. Define b_n by the equation $b_n = \varphi(b_n)$. Let $X_{n:n} \geq \cdots \geq X_{n-k+1:n}$ be the k largest order statistics of n i.i.d. standard normal r.v.'s. Then,

$$\sup_B |P\{[b_n(X_{n-j+1:n} - b_n)]_{j=1}^{k} \in B\} - G_{3,\mathbf{k}}(B)| \leq Ck^{1/2} \frac{(\log(k+1))^2}{\log n}.$$

The following theorem can be regarded as the main result of this section. Notice that the integrals in the upper bound have only to be computed on $(x_0, \omega(F))$. Moreover, the condition $G\{f > 0\} = 1$ as used in Corollary 5.5.2 is omitted.

Theorem 5.5.4. *Denote by G_j the jth marginal d.f. of $G_\mathbf{k} \in \{G_{1,\alpha,\mathbf{k}}, G_{2,\alpha,\mathbf{k}}, G_{3,\mathbf{k}}: \alpha > 0\}$, and put $G = G_1$. Let F be a d.f. with density f such that $f(x) > 0$*

for $x_0 < x < \omega(F)$. *Assume that* $\omega(F) = \omega(G)$. *Define again* $\psi = g/G$ *on the support of* G *where* g *is the density of* G. *Then,*

$$\sup_B |P\{(X_{n:n}, \ldots, X_{n-k+1:n}) \in B\} - G_{\mathbf{k}}(B)|$$

$$\leq \left[\sum_{j=1}^{k} \int_{x_0}^{\omega(G)} [nf/\psi - 1 - \log(nf/\psi)] \, dG_j + G_k(x_0) + kG_{k+1}(x_0) \right. \qquad (5.5.2)$$

$$\left. + \sum_{j=1}^{k-1} \int_{\{x_j > x_0, x_k \leq x_0\}} [\log(nf/\psi)(x_j)] \, dG_{\mathbf{k}}(x) \right]^{1/2} + C(k/n).$$

PROOF. To prove (5.5.2) one has to establish an upper bound of the right-hand side of (5.5.1).

Note that under the present conditions

$$M = \{\mathbf{x}: x_j > x_0, f(x_j) > 0, j = 1, \ldots, k\}$$

$$= \{\mathbf{x}: x_0 < x_j < \omega(G), j = 1, \ldots, k\}.$$

Moreover, recall that $x_1 \geq \cdots \geq x_k$ for every $\mathbf{x}$ in the support of $G_{\mathbf{k}}$.

Obviously,

$$G_{\mathbf{k}}(M^c) = G_{\mathbf{k}}\{\mathbf{x}: x_k \leq x_0\} = G_k(x_0). \qquad (1)$$

Denote by g_k the density of G_k. Recall that $G_k = G \sum_{j=0}^{k-1} (-\log G)^j/j!$ and $g_k = g(-\log G)^{k-1}/(k-1)!$. In analogy to inequality (2) in the proof to Lemma 5.5.2. we obtain

$$\int_M [1 - F(x_k)] \, dG_{\mathbf{k}}(\mathbf{x}) = \int_{x_0}^{\omega(G)} (1 - F) \, dG_k \leq \int_{x_0}^{\omega(G)} f(y) G_k(y) \, dy$$

$$= \int_{x_0}^{\omega(G)} \left[(f/\psi) \left(\sum_{j=0}^{k-1} (-\log G)^j/j! \right) \right] dG \qquad (2)$$

$$= \sum_{j=1}^{k} \int_{x_0}^{\omega(G)} (f/\psi) \, dG_j.$$

Moreover,

$$\int_M (\log G(x_k)) \, dG_{\mathbf{k}}(\mathbf{x}) = \int_{x_0}^{\omega(G)} (\log G(x)) \, dG_k(x)$$

$$= -k \int_{x_0}^{\omega(G)} g(x)(-\log G(x))^k/k! \, dx \qquad (3)$$

$$= -k(1 - G_{k+1}(x_0)).$$

Now the proof can easily be completed by combining (5.5.1) with (1)–(3). $\qquad \square$

Notice that Theorem 5.2.5 is a special case of Theorem 5.5.4.

Special Classes of Densities

Finally, Theorem 5.5.4 will be applied to the particular densities as dealt with in Corollary 5.2.7.

Corollary 5.5.5. *Assume that* $G \in \{G_{1,\alpha}, G_{2,\alpha}, G_3 : \alpha > 0\}$ *and* ψ, T *are the corresponding auxiliary functions with* $\psi = g/G$ *and* $T = G^{-1} \circ G_{2,1}$.
  Assume that the density f of the d.f. F has the representation

$$f(x) = \psi(x)e^{h(x)}, \qquad T(x_0) < x < \omega(G), \tag{5.5.3}$$

and $= 0$ *if* $x > \omega(G)$, *where* $x_0 < 0$ *and h satisfies the condition*

$$|h(x)| \leq \begin{array}{ll} Lx^{-\alpha\delta} & i = 1 \\ L(-x)^{\alpha\delta} & \text{if} \quad i = 2 \\ Le^{-\delta x} & i = 3 \end{array} \tag{5.5.4}$$

and L, δ *are positive constants. Then,*

$$\sup_B |P\{[c_n^{-1}(X_{n-j+1:n} - d_n)]_{j=1}^k \in B\} - G_k(B)| \leq D[(k/n)^\delta k^{1/2} + k/n]$$

where c_n, d_n *are the constants of Theorem 5.1.1 and* $D > 0$ *is a constant which only depends on* x_0, δ, *and* L.
  We have $d_n = 0$ *if* $i = 1, 2$, *and* $d_n = \log n$ *if* $i = 3$; *moreover,* $c_n = n^{1/\alpha}$ *if* $i = 1$, $c_n = n^{-1/\alpha}$ *if* $i = 2$, *and* $c_n = 1$ *if* $i = 3$.

PROOF. Again it suffices to prove the result for the particular case $G = G_{2,1}$. Theorem 5.5.4 will be applied to $x_{0,n} = nx_0$ and $f_n(x) = f(x/n)/n$. We obtain

$$\sup_B |P\{(nX_{n:n}, \ldots, nX_{n-k+1:n}) \in B\} - G_k(B)|$$

$$\leq \left[\sum_{j=1}^k \int_{-nx_0}^0 [e^{h(x/n)} - 1 - h(x/n)] \, dG_j(x) \right. \tag{1}$$

$$\left. + (1 + (k-1)(-x_0^\delta))G_k(nx_0) + kG_{k+1}(nx_0) \right]^{1/2} + Ck/n.$$

Check that $G_k(x) = O((k/|x|)^m)$ uniformly in k and $x < 0$ for every positive integer m. Moreover, since h is bounded on $(x_0, 0)$ we have

$$\sum_{j=1}^k \int_{-nx_0}^0 [e^{h(x/n)} - 1 - h(x/n)] \, dG_j(x)$$

$$\leq Dn^{-2\delta} \sum_{j=1}^k \int_{-\infty}^0 |x|^{2\delta} \, dG_j(x)$$

$$\leq Dn^{-2\delta} \sum_{j=1}^k \int_{-\infty}^0 |x|^{2\delta+j-1} \exp(x)/(j-1)! \, dx \tag{2}$$

$$\leq Dn^{-2} \sum_{j=1}^k \Gamma(2\delta + j)/\Gamma(j)$$

where $\Gamma(t) = \int_0^\infty x^{t-1} \exp(-x) \, dx$ denotes the Γ-function.

Finally, observe that (compare with Erdélyi et al. (1953), formula (5), page 47)

$$\sum_{j=1}^{k} \Gamma(2\delta + j)/\Gamma(j) \le D \sum_{j=1}^{k} j^{2\delta}. \tag{3}$$

Now by choosing $m \ge 2\delta$ the asserted inequality is immediate from (1)–(3).

$\square$

EXAMPLE 5.5.6. If $f \in \{g_{1,\alpha}, g_{2,\alpha}, g_3 : \alpha > 0\}$—that is the case of extreme value densities—then Corollary 5.5.5 is applicable with $\delta = 1$. Thus, the error bound is of order $O(k^{3/2}/n)$ which is a rate worse than that in the case of generalized Pareto densities. Direct calculations show that the bound $O(k^{3/2}/n)$ is sharp for $k > 1$.

5.6. Variational Distance between Empirical and Poisson Processes

In this section we shall study the asymptotic behavior of extremes according to their multitude in Borel sets. This topic does not directly concern order statistics. It is the purpose of this section to show that the results for order statistics can be applied to obtain approximations for empirical point processes.

Preliminaries

Let $\xi_1, \ldots, \xi_n$ be i.i.d. random variables with common d.f. F which belongs to the weak domain of attraction of $G \in \{G_{1,\alpha}, G_{2,\alpha}, G_3 : \alpha > 0\}$. Hence according to (5.1.4) there exist $a_n > 0$ and b_n such that

$$n(1 - F(b_n + a_n x)) \to -\log G(x), \qquad n \to \infty, \tag{5.6.1}$$

for $x \in (\alpha(G), \omega(G))$. According to the Poisson approximation to binomial r.v.'s we know that

$$\sum_{j=1}^{n} 1_{(x, \infty)}(a_n^{-1}(\xi_j - b_n)) \tag{5.6.2}$$

is asymptotically a Poisson r.v. with parameter $\lambda = -\log G(x)$.

Our investigations will be carried out within the framework of point processes and in this context the expression in (5.6.2) is usually written in the form

$$\sum_{j=1}^{n} \varepsilon_{(\xi_j - b_n)/a_n}(B) \tag{5.6.3}$$

where $\varepsilon_z(B) = 1_B(z)$ and $B = (x, \infty)$. With B varying over all Borel sets we obtain the empirical (point) process

$$N_n = \sum_{j=1}^{n} \varepsilon_{(\xi_j - b_n)/a_n} \tag{5.6.4}$$

of a sample of size n with values in the set of point measures.

Recall that μ is a point measure if there exists a denumerable set of points $x_j, j \in J$, such that

$$\mu = \sum_{j \in J} \varepsilon_{x_j}$$

and $\mu(K) < \infty$ for every relatively compact set K. The set of all point measures $\mathbf{M}$ is endowed with the smallest σ-field $\mathcal{M}$ such that the "projections" $\mu \to \mu(B)$ are measurable. It is apparent that $N: \Omega \to \mathbf{M}$ is measurable if $N(B): \Omega \to [0, \infty]$ is measurable for every Borel set B. If N is measurable then N is called a point process. Hence, the empirical process is a point process. Certain Poisson processes will be the limiting processes of empirical processes.

Homogeneous Poisson Process

Let $\xi_1, \ldots, \xi_n$ be i.i.d. random variables with common d.f. $W_{2,1}$ the uniform d.f. on $(-1, 0)$. In this case, the empirical process is given by

$$N_n = \sum_{j=1}^{n} \varepsilon_{n\xi_j}. \tag{5.6.5}$$

In the limit this point process will be the homogeneous Poisson process N_0 with unit rate. The Poisson process N_0 is defined by

$$N_0 = \sum_{j=1}^{\infty} \varepsilon_{S_j} \tag{5.6.6}$$

where S_j is the sum of j i.i.d. standard "negative" exponential r.v.'s. Moreover, $\mathbf{M}$ is the set of all point measures on the Borel sets in $(-\infty, 0)$.

For every $s > 0$ and $n = 0, 1, 2, \ldots$ define the truncation $N_n^{(s)}$ by

$$N_n^{(s)}(B) = N_n(B \cap [-s, 0)). \tag{5.6.7}$$

Theorem 5.6.1. *There exists a universal constant $C > 0$ such that for every positive integer n and $s \geq \log(n)$ the following inequality holds:*

$$\sup_{M \in \mathcal{M}} |P\{N_n^{(s)} \in M\} - P\{N_0^{(s)} \in M\}| \leq Cs/n. \tag{5.6.8}$$

PROOF. Let $V_{n:n} \geq \cdots \geq V_{1:n}$ be the order statistics of n i.i.d. random variables with uniform distribution on $(-1, 0)$. Let $k \equiv k(n)$ be the smallest integer such that

$$P\{nV_{n-k+1:n} \geq -s\} \leq n^{-1}. \tag{1}$$

In this sequel, C will denote a constant which is independent of n and $s \geq \log(n)$. It follows from the exponential bound theorem for order statistics (see Lemma 3.1.1) that $k \leq Cs$. Write

$$N_{0,k}^{(s)} = \sum_{i=1}^{k} \varepsilon_{S_i}(\cdot \cap [-s, 0))$$

and   (2)

$$N_{n,k}^{(s)} = \sum_{i=1}^{k} \varepsilon_{nV_{n-i+1:n}}(\cdot \cap [-s, 0)).$$

It is immediate from (1) that for $n \geq 1$,

$$\sup_{M \in \mathcal{M}} |P\{N_n^{(s)} \in M\} - P\{N_{n,k}^{(s)} \in M\}| \leq n^{-1}.$$ (3)

From Theorem 5.4.4 we know that

$$\sup_{B} |P\{(nV_{n:n}, \ldots, nV_{n-k+1:n}) \in B\} - P\{(S_1, \ldots, S_k) \in B\}| \leq Ck/n.$$ (4)

Note that $N_{n,k}^{(s)}$, $n \geq 1$, and $N_{0,k}^{(s)}$ may be written as the composition of the random vectors $(nV_{n:n}, \ldots, nV_{n-k+1:n})$, $n \geq 1$, and $(S_1, \ldots, S_k)$, respectively, and the measurable map

$$(x_1, \ldots, x_k) \to \sum_{i=1}^{k} \varepsilon_{x_i}$$

having its values in the set of point measures.

Therefore, (4) yields

$$\sup_{M \in \mathcal{M}} |P\{N_{n,k}^{(s)} \in M\} - P\{N_{0,k}^{(s)} \in M\}| \leq Ck/n.$$ (5)

Moreover, (1) and (4) yield

$$P\{S_k \geq -s\} \leq Ck/n$$ (6)

and hence, in analogy to (3),

$$\sup_{M \in \mathcal{M}} |P\{N_{0,k}^{(s)} \in M\} - P\{N_0^{(s)} \in M\}| \leq Ck/n.$$ (7)

Now (3), (5), (7), and the triangle inequality imply the asserted inequality.   □

The bound in Theorem 5.6.1 is sharp. Notice that for every $k \in \{1, \ldots, n\}$

$$\sup_{-s \leq -t} |P\{N_n(-t, 0) < k - 1\} - P\{N_0(-t, 0) < k - 1\}|$$
                                                                                       (5.6.9)
$$= \sup_{-s \leq -t} |P\{nV_{n-k+1:n} \leq -t\} - G_{2,1,k}(-t)|.$$

Hence a remainder term of a smaller order than that in (5.6.8) would yield a result for order statistics which does not hold according to the expansion of length 2 in Theorem 5.4.3.

Extensions

Denote by v_0 the Lebesgue measure restricted to $(-\infty, 0)$. Recall that v_0 is the intensity measure of the homogeneous Poisson process N_0. We have

$$v_0(B) = EN_0(B). \tag{5.6.10}$$

Write again $T_{i,\alpha} = G_{i,\alpha}^{-1} \circ G_{2,1}$ (see (1.6.10)). Denote by $\mathbf{M}_i$ the set of point measures on $(\alpha(G_{i,\alpha}), \omega(G_{i,\alpha}))$ and by $\mathcal{M}_i$ the pertaining σ-field. Denote by $T_{i,\alpha}$ also the map from $\mathbf{M}_1$ to $\mathbf{M}_i$ where $T_{i,\alpha}\mu$ is the measure induced by μ and $T_{i,\alpha}$. Notice that if $\mu = \sum_{i \in J} \varepsilon_{x_j}$ then

$$T_{i,\alpha}\mu = \sum_{j \in J} \varepsilon_{T_{i,\alpha}(x_j)}.$$

Define

$$N_{i,\alpha,n} = T_{i,\alpha}(N_n) \tag{5.6.11}$$

for N_n as in (5.6.5) and (5.6.6). It is obvious that for $n = 1, 2, \ldots$

$$N_{i,\alpha,n} \overset{\mathrm{d}}{=} \sum_{k=1}^{n} \varepsilon_{(\xi_k - d_n)/c_n} \tag{5.6.12}$$

where $\xi_1, \ldots, \xi_n$ are i.i.d. random variables with common generalized Pareto d.f. $W_{i,\alpha}$; moreover, $c_n > 0$ and d_n are the usual normalizing constants as defined in (1.3.13).

It is well known that $N_{i,\alpha} \equiv N_{i,\alpha,0}$ is a Poisson process with intensity measure $v_{i,\alpha} = T_{i,\alpha}v_0$ (having the mean value function $\log(G_{i,\alpha})$). Recall that the distribution of $N_{i,\alpha}$ is uniquely characterized by the following two properties:

(a) $N_{i,\alpha}(B)$ is a Poisson r.v. with parameter $v_{i,\alpha}(B)$ if $v_{i,\alpha}(B) < \infty$, and
(b) $N_{i,\alpha}(B_1), \ldots, N_{i,\alpha}(B_m)$ are independent r.v.'s for mutually disjoint Borel sets $B_1, \ldots, B_m$.

Define the truncated point processes $N_{i,\alpha,n}^{(s)}$ by

$$N_{i,\alpha,n}^{(s)}(B) = N_{i,\alpha}(B \cap [T_{i,\alpha}(-s), \omega(G_{i,\alpha}))). \tag{5.6.13}$$

From Theorem 5.6.1 and (5.6.11) it is obvious that the following result holds.

Corollary 5.6.2. *There exists a universal constant $C > 0$ such that for every positive integer n and $s \geq \log(n)$ the following inequality holds:*

$$\sup_{M \in \mathcal{M}_i} |P\{N_{i,\alpha,n}^{(s)} \in M\} - P\{N_{i,\alpha,0}^{(s)} \in M\}| \leq Cs/n. \tag{5.6.14}$$

Notice that Corollary 5.6.2 specialized to $i = 2$ and $\alpha = 1$ yields Theorem 5.6.1.

Final Remarks

Theorem 5.6.1 and Corollary 5.6.2 can easily be extended to a large class of d.f.'s F belonging to a neighborhood of a generalized Pareto d.f. $W_{i,\alpha}$ with $N_{i,\alpha,0}^{(s)}$ again being the approximating Poisson process. This can be proved just by replacing Theorem 5.4.3 in the proof of Theorem 5.6.1 (for appropriate inequalities we refer to Section 5.5). Moreover, in view of (5.6.9) and Theorem 5.4.5(ii) it is apparent that a bound of order $O(s/n)$ can only be achieved if F has the upper tail of a generalized Pareto d.f. The details will be omitted since this topic will not be pursued further in this book.

In statistical applications one gets in the most simple case a model of independent Poisson r.v.'s by choosing mutually disjoint sets. The value of s has to be large to gain efficiency; on the other hand, the Poisson model provides an accurate approximation only if s is sufficiently small compared to n. The limiting model is represented by the unrestricted Poisson processes $N_{i,\alpha}$. One has to consider Poisson processes with intensity measures depending on location and scale parameters if the original model includes such parameters. This family of Poisson processes can again be studied within a 3-parameter representation.

P.5. Problems and Supplements

1. Check that the max-stability $G^n(d_n + xc_n) = G(x)$ of extreme value d.f.'s has its counterpart in the equation

$$n(1 - W(d_n + xc_n)) = 1 - W(x)$$

for the generalized Pareto d.f.'s $W \in \{W_{1,\alpha}, W_{2,\alpha}, W_3 : \alpha > 0\}$.

2. Check that the necessary and sufficient conditions (5.1.5)–(5.1.7) are trivially satisfied by the generalized Pareto d.f.'s in the following sense:
   (i) For $x > 0$ and t such that $tx > 1$:

$$(1 - W_{1,\alpha}(tx))/(1 - W_{1,\alpha}(t)) = x^{-\alpha}.$$

   (ii) For $x < 0$ and $t > 0$ such that $tx > -1$:

$$(1 - W_{2,\alpha}(tx))/(1 - W_{2,\alpha}(-t)) = (-x)^{\alpha}.$$

   (iii) For $t, x > 0$:

$$g(t) = \int_t^\infty (1 - W_3(y))\, dy/(1 - W_3(t)) = 1$$

   and

$$(1 - W_3(t + x))/(1 - W_3(t)) = e^{-x}.$$

3. Let $F_1, F_2, F_3, \dots$ be d.f.'s. Define $G_n^*(x) = F_n(b_n^* + a_n^* x)$ and $G_n(x) = F_n(b_n + a_n x)$ where $a_n^*, a_n > 0$. Assume that for some nondegenerate d.f. G^*,

$$G_n^* \to G^* \quad \text{weakly.}$$

(i) The following two assertions are equivalent:
   (a) For some nondegenerate d.f. G,

$$G_n \to G \quad \text{weakly.}$$

   (b) For some constants $a > 0$ and b,

$$a_n/a_n^* \to a \quad \text{and} \quad (b_n - b_n^*)/a_n^* \to b \quad \text{as } n \to \infty.$$

(ii) Moreover, if (a) or (b) holds then

$$G(x) = G^*(b + ax) \quad \text{for all real } x.$$

[Hint: Use Lemma 1.2.9; see also de Haan, 1976.]

4. (i) Let c be the unique solution of the equation

$$x^2 \sin(1/x) + 4x + 1 = 0$$

on the interval $(-1, 0)$. Define the d.f. F by

$$F(x) = x^2 \sin(1/x) + 4x + 1, \qquad x \in (c, 0).$$

Then, for every x,

$$F^n(x/4n) \to G_{2,1}(x) \quad \text{as } n \to \infty.$$

However, F does not belong to the strong domain of attraction of $G_{2,1}$.

(Falk, 1985b)

(ii) The Cauchy d.f. F and density f are given by

$$F(x) = 1/2 + (1/\pi)\arctan x$$

and

$$f(x) = (\pi(1 + x^2))^{-1}.$$

Verify the von Mises-condition (5.1.24) with $i = 1$ and $\alpha = 1$.

[Hint: Use the de l'Hospital rule.]

5. (Asymptotic d.f.'s of intermediate order statistics)
   Let $k(n) \in \{1, \dots, n\}$ be such that $k(n) \uparrow \infty$ and $k(n)/n \to 0$ as $n \to \infty$.
   (i) The nondegenerate limiting d.f.'s of the $k(n)$th order statistic are given by

$$\Phi(G_3^{-1}(G)) \quad \text{on } (\alpha(G), \omega(G))$$

where $G \in \{G_{1,\alpha}, G_{2,\alpha}, G_3 : \alpha > 0\}$.

(Chibisov, 1964; Wu, 1966)

(ii) The weak convergence of the distribution of $a_n^{-1}(X_{k(n):n} - b_n)$ to the limiting d.f. defined by G holds if, and only if,

$$[nF(b_n + a_n x) - k(n)]/k(n)^{1/2} \to G_3^{-1}(G(x)), \qquad n \to \infty, x \in (\alpha(G), \omega(G)).$$

(Chibisov, 1964)

6. Let $\xi_1, \xi_2, \xi_3, \dots$ be i.i.d. symmetric random variables (that is, $\xi_i \overset{d}{=} -\xi_i$). Prove that

$$\sup_B |P\{\max(|\xi_1|,\ldots,|\xi_n|) \in B\} - P\{\max(\xi_1,\ldots,\xi_{2n}) \in B\}| = O(n^{-1}).$$

[Hint: Apply (4.2.10).]

7. Let b_n^* be defined as in Example 5.1.4(5) and $b_n = \Phi^{-1}(1 - 1/n)$. Show that

$$|b_n - b_n^*| = O((\log\log n)^2/(\log n)^{3/2}).$$

Let $a_n = 1/n\varphi(b_n)$ and $a_n^* = (2\log n)^{-1/2}$. Show that

$$|a_n - a_n^*| = O((\log\log n)/(\log n)^{3/2}).$$

Show that

$$|a_n b_n - 1| = O((\log n)^{-1}).$$

(Reiss, 1977a, Lemma 15.11)

8. For $\gamma > 0$ and a real number x_0 let F_γ be a d.f. with

$$F_\gamma(x) = \gamma\Phi(x) + (1 - \gamma) \quad \text{for } x \geq x_0.$$

Put $b_n = \Phi^{-1}(1 - 1/n\gamma)$ and $a_n = 1/n\varphi(b_n)$. Show that

$$\sup_x |F_\gamma^n(b_n + a_n x) - G_3(x)(1 + x^2 e^{-x}/(4\log n)| = O((\log n)^{-2})$$

and, thus,

$$\sup_x |F_\gamma^n(b_n + x/b_n) - G_3(x)| = O((\log n)^{-1}).$$

(Reiss, 1977a, Theorem 15.17 and Remark 15.18)

9. (Graphical representation of generalized Pareto densities)
Recall that for Pareto densities $w_{1,\alpha}(x) = \alpha x^{-(1+\alpha)}$, $x \geq 1$, we have $w_{1,\alpha}(1) = \alpha$ (Fig. P.5.1). For the generalized Pareto type II densities $w_{2,\alpha}$ we have $w_{2,\alpha}(x) = \alpha(-x)^{\alpha-1} \sim g_{2,\alpha}(x)$ as $x \uparrow 0$ (Fig. P.5.2).

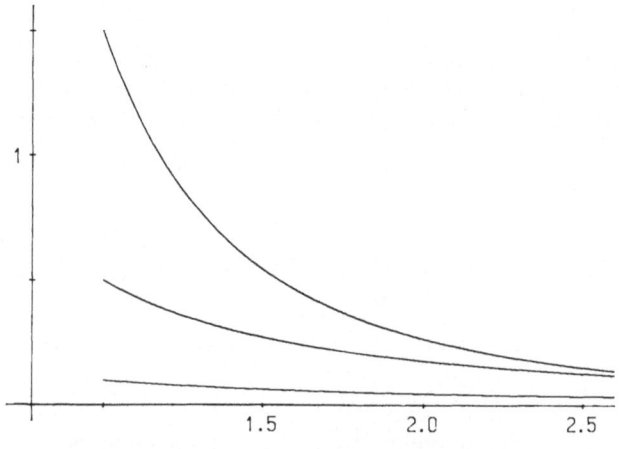

Figure P.5.1. Pareto densities $w_{1,\alpha}$ with $\alpha = 0.1, 0.5, 1.5$.

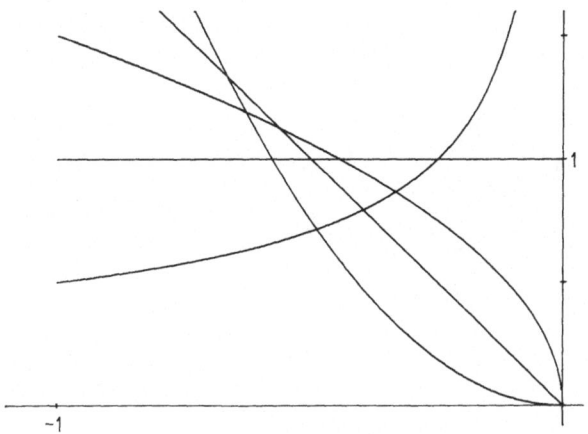

Figure P.5.2. Generalized Pareto densities $w_{2,\alpha}$ with $\alpha = 0.5, 1, 1.5, 2, 3$.

10. (Von Mises parametrization of generalized Pareto d.f.'s)
    For $\beta > 0$, define

$$V_\beta(x) = 1 - (1 + \beta x)^{-1/\beta} \quad \text{if } 0 < x.$$

For $\beta < 0$, define

$$V_\beta(x) = \begin{cases} 1 - (1 + \beta x)^{-1/\beta} & 0 < x < -\dfrac{1}{\beta} \\ 1 & x \geq -\dfrac{1}{\beta}. \end{cases} \quad \text{if}$$

For $\beta = 0$, define

$$V_0(x) = 1 - e^{-x} \quad \text{for } x > 0.$$

Show that

$$W_{1,1/\beta}(x) = V_\beta\left(\frac{x-1}{\beta}\right) \quad \text{if } \beta > 0,$$

$$W_{2,1/|\beta|}(x) = V_\beta\left(\frac{x+1}{|\beta|}\right) \quad \text{if } \beta < 0,$$

$$W_3(x) = V_0(x).$$

The density v_β of V_β is equal to zero for $x < 0$. Moreover, if $\beta > 0$ then

$$v_\beta(x) = (1 + \beta x)^{-(1+1/\beta)}.$$

If $\beta < 0$ then

$$v_\beta(x) = \begin{cases} (1 + \beta x)^{-(1+1/\beta)} & 0 < x < -1/\beta \\ 0 & x \geq -1/\beta. \end{cases}$$

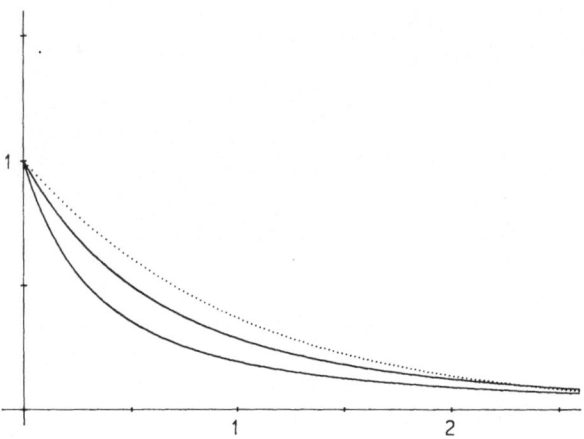

Figure P.5.3. Standard exponential and Pareto densities v_β with $\beta = 0, 0.6, 2$.

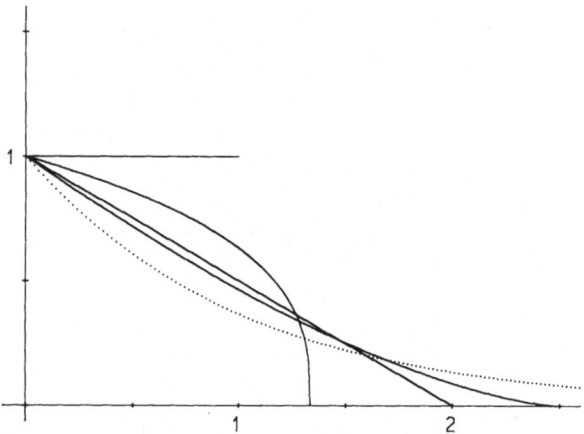

Figure P.5.4. Standard exponential and generalized Pareto type II densities v_β with $\beta = -1, -0.75, -0.5, -0.4, 0$.

The Pareto densities v_β with $\beta \downarrow 0$ (Figure P.5.3) and the generalized Pareto type II densities v_β with $\beta \uparrow 0$ (Figure P.5.4) approach the standard exponential density v_0 (dotted curve).

11. (Maxima with random indices)
   Let $\xi_i, i = 1, 2, \ldots$ be i.i.d. random variables and let $N(i), i = 0, 1, \ldots$ be positive integer-valued r.v.'s.
   (i) If G is an extreme value d.f., and
      (a) $P\{a_n^{-1}(X_{n:n} - b_n) \le x\} \to G(x), n \to \infty$,
      (b) $N(i)/i \to N(0), i \to \infty$, in probability,
         then
$$P\{a_i^{-1}(X_{N(i):N(i)} - b_i) \le x\} \to EG(x)^{N(0)}, \qquad i \to \infty.$$

(Barndorff–Nielsen, 1964)

(ii) If the sequence (ξ_i) and $N(j)$ are independent for every j then the condition (i)(b) can be replaced by
  (b′) $N(i)/i \overset{d}{\to} N(0), \qquad i \to \infty.$
(iii) Show that the independence condition in (ii) cannot be omitted without compensation. [Hint: Define $N(i) = \min\{j: \xi_j > \log i\}$ for standard exponential r.v.'s.]

(M. Falk)

12. Show that the Cauchy d.f. with scale parameter $\sigma = \pi$ satisfies condition (5.2.14) for $i = 1$ and $\alpha = 1$ with $\delta = 1$. As a consequence one gets for the maximum $X_{n:n}$ of standard Cauchy r.v.'s that

$$\sup_B |P\{(\pi/n)X_{n:n} \in B\} - G_{1,1}(B)| = O(n^{-1}).$$

13. Under the Weibull d.f. F_α on the positive half-line defined by

$$F_\alpha(x) = 1 - \exp(-x^\alpha), \qquad x > 0,$$

one gets for every $\alpha \neq 1$,

$$\sup_B |P\{\alpha(\log n)^{1-1/\alpha}(X_{n:n} - (\log n)^{1/\alpha}) \in B\} - G_3(B)| = O(1/\log n).$$

14. (Bounds for remainder terms involving von Mises conditions)
Assume that the d.f. F has three continuous derivatives and that $f = F' > 0$ on the interval $(x_0, \omega(F))$. Put

$$H = (1 - F)/f.$$

Assume that the von Mises condition (5.1.25) holds for some $i \in \{1, 2, 3\}$ and $\alpha > 0$ (with $\alpha = 1$ if $i = 3$). Thus, we have

$$h_{i,\alpha}(x) := \alpha H'(x) - T_{i,\alpha}(-1) \to 0 \quad \text{as } x \uparrow \omega(F)$$

where again $T_{i,\alpha} = G_{i,\alpha}^{-1} \circ G_{2,1}$. Notice that $T_{1,\alpha}(-1) = 1$, $T_{2,\alpha}(-1) = -1$ and $T_3(-1) = 0$. Then for $\alpha(G_{i,\alpha}) < x < \omega(G_{i,\alpha})$,

$$|F^n(b_n + a_n x) - G_{i,\alpha}(x)| = O(|h_{i,\alpha}(x_n)| + n^{-1})$$

with $x_n = F^{-1}(1 - 1/n)$ and the normalizing constants are given by

$$a_n = \alpha/(nf(x_n))$$

and

$$b_n = x_n - T_{i,\alpha}(-1)a_n.$$

(Radtke, 1988)

15. (Expansions involving von Mises conditions)
Assume, in addition to the conditions of P.5.14, that $f'' > 0$ on the interval $(x_0, \omega(F))$. Then for $\alpha(G_{i,\alpha}) < x < \omega(G_{i,\alpha})$,

$$|F^n(b_n + a_n x) - G_{i,\alpha}(x)(1 - h_{i,\alpha}(x_n)\psi_{i,\alpha}(x)[x - T_{i,\alpha}(-1)]^2/2)|$$

$$= O(h_{i,\alpha}(x_n)^2 + |h_{i,\alpha}(x_n)||g_{i,\alpha}(x_n)| + n^{-1})$$

where $g_{i,\alpha}$ is another auxiliary function. We have

$$g_{i,\alpha} = h'_{i,\alpha}H/h_{i,\alpha} + T_{i,\alpha}(-1)/\alpha$$

implicitly assuming that $h_{i,\alpha} \neq 0$. Moreover, assume that $\lim_{x \uparrow \omega(F)} g_{i,\alpha}(x)$ exists in $(-\infty, \infty)$, and there exist real numbers K_t such that

$$g_{1,\alpha}(tx) = g_{1,\alpha}(x)(K_t + o(x^0)) \quad \text{as } x \to \omega(F) \text{ for all } t > 0 \text{ if } i = 1,$$

$$g_{2,\alpha}(\omega(F) - tx) = g_{2,\alpha}(\omega(F) - x)(K_t + o(x^0)) \quad \text{as } x \downarrow 0 \text{ for all } t > 0 \text{ if } i = 2,$$

$$g_3(x + tH(x)) = g_3(x)(K_t + o(x^0)) \quad \text{as } x \uparrow \omega(F) \text{ for all reals } t \text{ if } i = 3.$$

(Radtke, 1988)

16. (Special cases)
    (i) Let

$$F(x) = 1 - x^{-\alpha}(1 + x^{-\alpha\rho}), \qquad x \geq 1,$$

for some $\alpha > 0$ and $0 < \rho \leq 1$. Then

$$|F^n(b_n + a_n x) - G_{1,\alpha}(x)| = \begin{array}{ll} O(n^{-2\rho} + n^{-1}) & \quad \rho\alpha = 1 \\ O(n^{-\rho}) & \quad \rho\alpha \neq 1 \end{array} \quad \text{if}$$

with a_n and b_n as above. Moreover, $g_{i,\alpha}(x_n)$ does not converge to zero as $n \to \infty$ (compare with P.5.15).

    (ii) Let

$$F(x) = 1 - x^{-\alpha}\exp\left[1 - \frac{1 + \log x}{x}\right], \qquad x \geq 1,$$

for $\alpha > 0$. Then

$$|F^n(b_n + a_n x) - G_{1,\alpha}(x)(1 - h_{1,\alpha}(x_n)\psi_{1,\alpha}(x)[x - 1]^2/2)|$$
$$= O((\log n)^2/n^{2\alpha} + n^{-1})$$

and

$$h_{1,\alpha}(x_n) = O(n^{-\alpha}).$$

17. (i) Prove that for a d.f. F and a positive integer k the following two statements are equivalent:
    (a) F belongs to the weak domain of attraction of an extreme value d.f. $G \in \{G_{1,\alpha}, G_{2,\alpha}, G_3 : \alpha > 0\}$.
    (b) There are constants $a_n > 0$ and b_n such that the d.f.'s $F_{n,k}$ defined by

$$F_{n,k}(x) = P\{a_n^{-1}(X_{n:n} - b_n) \leq x_1, \ldots, a_n^{-1}(X_{n-k+1:n} - b_n) \leq x_k\}$$

converge weakly to a nondegenerate d.f. $G^{(k)}$.
    (ii) In addition, if (a) holds for $G = G_{i,\alpha}$ then (b) is valid for $G^{(k)} = G_{i,\alpha,k}$.

18. (i) There exists a constant $C > 0$ such that for every positive integer n and $k \in \{1, 2, \ldots, [n/2]\}$ the following inequality holds:

$$\sup_B \left| P\left\{ \left(\frac{n^{3/2}}{(n-k)^{1/2}} \left(U_{n-k+1:n} - \frac{n-k}{n} \right) - k \right) \in B \right\} - G_{2,1,k}(B) \right| \leq Ck^{1/2}/n.$$

(Kohne and Reiss, 1983)

    (ii) It is unknown whether the standardized distribution of $U_{n-k+1:n}$ admits an expansion of length m arranged in powers of $k^{1/2}/n$ where again $G_{2,1,k}$ is the leading term of the expansion.
    (iii) Reformulate (i) by using $N_{(0,1)}$ in place of $G_{2,1,k}$.

19. (Asymptotic independence of spacings)
    There exists a constant $C > 0$ such that for every positive integer n and $k \in \{1, 2, \ldots, n\}$ the following inequality holds:

    $$\sup_B |P\{(nU_{1:n}, n(U_{2:n} - U_{1:n}), \ldots, n(U_{k:n} - U_{k-1:n})) \in B\}$$
    $$- P\{(\xi_1, \ldots, \xi_k) \in B\}| \le Ck/n$$

    where $\xi_1, \ldots, \xi_k$ are i.i.d. random variables with standard exponential d.f.

20. Show that under the triangular density

    $$f(x) = 1 - |x|, \qquad x \le 1,$$

    one gets

    $$\sup_B |P\{(n/2)^{1/2}(X_{n-i+1:n} - 1)_{i=1}^k \in B\} - G_{2,2,k}(B)| \le Ck/n$$

    where $C > 0$ is a universal constant.

21. (Problem) Prove inequalities w.r.t. the Hellinger distance corresponding to those in Lemma 5.5.1 and Theorem 5.5.5.

22. For the k largest order statistics of standard Cauchy r.v.'s one gets

    $$\sup_B \left| P\left\{ \left(\frac{\pi}{n} X_{n:n}, \ldots, \frac{\pi}{n} X_{n-k+1:n} \right) \in B \right\} - G_{1,1,k}(B) \right| \le Ck^{3/2}/n$$

    where $C > 0$ is a universal constant.

23. Extend Corollary 5.6.2 to d.f.'s that satisfy condition (5.2.11).

Bibliographical Notes

An excellent survey of the literature concerning classical extreme value theory can be found in the book of Galambos (1987). Therefore it suffices here to repeat only some of the basic facts of the classical part and, in addition, to give a more detailed account of the recent developments concerning approximations w.r.t. the variational distance etc. and higher order approximations.

Out of the long history, of the meanwhile classical part of the extreme value theory, we have already mentioned the pioneering work of Fisher and Tippett (1928), who provided a complete list of all possible limiting d.f.'s of sample maxima. Gnedenko (1943) found necessary and sufficient conditions for a d.f. to belong to the weak domain of attraction of an extreme value d.f. De Haan (1970) achieved a specification of the auxiliary function in Gnedenko's characterization of F to belong to the domain of attraction of the Gumbel d.f. G_3.

The conditions $(1, \alpha)$ and $(2, \alpha)$ in (5.1.24) which are sufficient for a d.f. to belong to the weak domain of attraction of the extreme value d.f.'s $G_{1,\alpha}$ and $G_{2,\alpha}$ are due to von Mises (1936). The corresponding condition (5.1.24)(3) for the Gumbel d.f. G_3 was found by de Haan (1970). Another set of "von Mises conditions" is given in (5.1.25) for d.f.'s having two derivatives. For $i = 3$ this condition is due to von Mises (1936). Its extension to the cases $i = 1, 2$ appeared in Pickands (1986).

In conjunction with strong domain of attraction, the von Mises conditions have gained new interest. The pointwise convergence of the densities of sample maxima under the von Mises condition (5.1.25), $i = 3$, was proved in Pickands (1967) and independently in Reiss (1977a, 1981d). A thorough study of this subject was carried out by de Haan and Resnick (1982), Falk (1985b), and Sweeting (1985).

Sweeting, in his brilliant work, was able to show that the von Mises conditions (5.1.24) are equivalent to the uniform convergence of densities of normalized maxima on finite intervals. We also mention the article of Pickands (1986) where a result closely related to that of Sweeting is proved under certain differentability conditions imposed on F.

In (5.1.31) the number of exceedances of n i.i.d. random variables over a threshold u_n was studied to establish the limit law of the kth largest order statistic. The key argument was that the number of exceedances is asymptotically a Poisson r.v. This result also holds under weaker conditions. We mention Leadbetter's conditions $D(u_n)$ and $D'(u_n)$ for a stationary sequence (for details see Leadbetter et al. (1983)).

A necessary and sufficient condition (see P.5.5(ii)) for the weak convergence of normalized distributions of intermediate order statistics is due to Chibisov (1964). The possible limiting d.f.'s were characterized by Chibisov (1964) and Wu (1966) (see P.5.5(i)). Theorem 5.1.7, formulated for $G_{3,k}$ instead of $N_{(0,1)}$, is given in Reiss (1981d) under the stronger condition that the von Mises condition (5.1.25), $i = 3$, holds; by the way, this result was proved via the normal approximation. The weak convergence of intermediate order statistics was extensively dealt with by Cooil (1985, 1988). Cooil proved the asymptotic joint normality of a fixed number of suitably normalized intermediate order statistics under conditions that correspond to that in Theorem 5.1.7. For the treatment of intermediate order statistics under dependence conditions we refer to Watts et al. (1982).

Bounds for the remainder terms of limit laws concerning maxima were established by various authors. We refer to W.J. Hall and J.A. Wellner (1979), P. Hall (1979), R.A. Davis (1982), and the book of Galambos (1987) for bounds with explicit constants.

As pointed out by Fisher and Tippett (1928), extreme value d.f.'s different from the limiting ones (penultimate d.f.'s) may provide a more accurate approximation to d.f.'s of sample maxima. This line of research was taken up by Gomes (1978, 1984) and Cohen (1982a, b). Cohen (1982b), Smith (1982), and Anderson (1984) found conditions that allow the computation of the rate of convergence w.r.t. the Kolmogorov–Smirnov distance. Another notable article pertaining to this is Zolotarev and Rachev (1985) who applied the method of metric distances.

It can easily be deduced from a result of Matsunawa and Ikeda (1976) that the variational distance between the normalized distribution of the $k(n)$th largest order statistic of n independent, identically $(0, 1)$-uniformly distributed r.v.'s and the gamma distribution with parameter $k(n)$ tends to zero as $n \to \infty$ if $k(n)/n$ tends to zero as $n \to \infty$. In Reiss (1981d) it was proved that the accuracy of this approximation is $\leq Ck/n$ for some universal constant C. This result was taken up by Falk (1986a) to prove an inequality related to (5.2.6) w.r.t. the variational distance. A further improvement was achieved in Reiss (1984): By proving the result in Reiss (1981d) w.r.t. the Hellinger distance and by using an inequality for induced probability measures (compare with Lemma 3.3.13) it was shown that Falk's result still holds if the variational distance is replaced by the Hellinger distance. The present result is a further improvement since the upper bound only depends on the upper tail of the underlying distribution.

The investigation of extremes under densities of the form (5.2.14) was initiated by L. Weiss (1971) who studied the particular case of a neighborhood of Weibull densities. The class of densities defined by (5.2.18) and (5.2.19) corresponds to the class of d.f.'s introduced by Hall (1982a).

It is evident that if the underlying d.f. only slightly deviates from an extreme value d.f. then the rate of convergence of the d.f. of the normalized maximum to the limit d.f. can be of order $o(n^{-1})$. The rate is of exponential order if F has the same upper tail as an extreme value d.f. It was shown by Rootzén (1984) that this is the best order achievable under a d.f. unequal to an extreme value d.f. It would be of interest to explore, in detail, the rates for the second largest order statistic.

Because of historical reasons we note the explicit form of the interesting expansion in Uzgören (1954), which could have served as a guide to the mathematical research of expansions in extreme value theory:

$$\log(-\log F^n(b_n + xg(b_n)))$$

$$= -x + \frac{x^2}{2!}g'(b_n) + \frac{x^3}{3!}[g(b_n)g''(b_n) - 2g'^2(b_n)] + \cdots + \cdots$$

$$+ \frac{e^{-x+\cdots}}{2n} + \frac{5}{24n^2}e^{-2x+\cdots} - \frac{1}{8n^3}e^{-3x+\cdots} + \cdots$$

where $b_n = F^{-1}(1 - 1/n)$ and $g = (1 - F)/f$. The first two terms of the expansion formally agree to that in (5.2.16) in the Gumbel case. However, as reported by T.J. Sweeting (talk at the Oberwolfach meeting on "Extreme Value Theory," 1987) the expansion is not valid as far as the third term is concerned. Other references pertaining to this are Dronskers (1958), who established an approximate density of the $k(n)$th largest order statistic and Haldane and Jayakar (1963), who studied the particular case of extremes of normal r.v.'s.

Expansions of length 2 related to that in (5.2.16) are well known in literature (e.g. Anderson (1971) and Smith (1982)). These expansions were established in a particularly appealing form by Radtke (1988) (see P.5.15). From P.5.15 we see that the rate of convergence, at which the von Mises condition holds, also determines the rate at which the convergence to the limiting extreme value d.f. holds. The available results do not fit to our present program since only expansions of d.f.'s are treated. In spite of the importance of these results, details are given in the Supplements. It is an open problem under which conditions the expansions in P.5.15 lead to higher order approximations that are valid w.r.t. the variational or the Hellinger distance. (5.2.15) and (5.2.16) only provide a particular example. A certain characterization of possible types of expansions of distributions of maxima was given by Goldie and Smith (1987).

Weinstein (1973) and Pantcheva (1985) adopted a nonlinear normalization in order to derive a more accurate approximation of the d.f. of sample maxima by means of the limiting extreme value d.f. From our point of view, a systematic treatment of this approach would be the following: First, find an expansion of finite length; second, construct a nonlinear normalization by using the "inverse" of the expansion as it was done in Section 4.6 (see also Theorem 6.1.2).

The method to base the statistical inference on the k largest order statistics may be regarded as Type II censoring. Censoring plays an important role in applications like reliability and life-testing. This subject is extensively studied in books by N.R. Mann et al. (1974), A.J. Gross (1975), L.J. Bain (1978), W. Nelson (1982), and J.F. Lawless (1982).

Upper bounds for the variational distance between the counting processes $N_n[-t, 0)$, $0 \le t \le s$, and $N_{2,1}[-t, 0)$, $0 \le t \le s$, may also be found in Kabanov and Lipster (1983) and Jacod and Shiryaev (1987). The bounds given there are of order s^2/n and $s/n^{1/2}$ and therefore not sharp. Another reference is Karr (1986) who proved an upper bound of order n^{-1} for fixed s.

In Chapter 4 of the book by Resnick (1987), the weak convergence of certain point processes connected to extreme value theory is studied. For this purpose one has to verify that the σ-field $\mathcal{M}$ on the set of point measures is the Borel-σ-field generated by the topology of vague convergence. The weak convergence of empirical processes can be formulated in such a way that it is equivalent to the condition that the underlying d.f. belongs to the domain of attraction of an extreme value d.f. Note that the "empirical point processes" studied by Resnick (1987, Corollary 4.19) are of the form

$$\sum_{k=1}^{\infty} \varepsilon_{(k/n, (\xi_k - d_n)/c_n)},$$

thus allowing a simultaneous treatment of the time scale and the sequence of observations.

From the statistical point of view the weak convergence is not satisfactory. The condition that F belongs to the domain of attraction of an extreme value d.f. is not strong enough to yield e.g. the existence of a consistent estimator of the tail index (that is, the index α of the domain of attraction). Thus, the weak convergence cannot be of any help either if F satisfies stronger regularity conditions.

We briefly mention a recent article by Deheuvels and Pfeifer (1988) who independently proved a result related to Theorem 5.6.1 by using the coupling method. We do not know whether their method is also applicable to prove the extension, as indicated at the end of Section 5.6, where F belongs to a neighborhood of a generalized Pareto d.f.

CHAPTER 6

Other Important Approximations

In Chapters 4 and 5 we studied approximations to distributions of central and extreme order statistics uniformly over Borel sets.

The approximation over Borel sets is equivalent to the approximation of integrals over bounded measurable functions. In Section 6.1 we shall indicate the extension of such approximations to unbounded functions, thus, getting approximations to moments of order statistics.

From approximations of joint distributions of order statistics one can easily deduce limit theorems for certain functions of order statistics. Results of this type will be studied in Section 6.2. We also mention other important results concerning linear combinations of order statistics which, however, have to be proved by means of a different approach.

Sections 6.3 and 6.4 deal with approximations of a completely different type. In Section 6.3 we give an outline of the well-known stochastic approximation of the sample d.f. to the sample q.f., connected with the name of R.R. Bahadur.

Section 6.4 deals with the bootstrap, a resampling method introduced by B. Efron in 1979. We indicate the stochastic behavior of the bootstrap d.f. of the sample q-quantile.

6.1. Approximations of Moments and Quantiles

This section provides approximations to functional parameters of distributions of order statistics by means of the corresponding functional parameters of the limiting distributions and of finite expansions. We shall only consider central and intermediate order statistics.

Moments of Central and Intermediate Order Statistics

We shall utilize the result of Section 4.7, that concerns Edgeworth type expansions of densities of central and intermediate order statistics.

Theorem 6.1.1. *Let $q \in (0, 1)$ be fixed. Assume that the d.f. F has $m + 1$ bounded derivatives on a neighborhood of $F^{-1}(q)$, and that $f(F^{-1}(q)) > 0$ where $f = F'$. Assume that $(r(n)/n - q) = O(n^{-1})$. Put $\sigma^2 = q(1 - q)$.*
Moreover, assume that

$$E|X_{s:j}| < \infty \quad \text{for some positive integer } j \text{ and } s \in \{1, \dots, j\}.$$

Then, for every measurable function h with $|h(x)| \le |x|^k$ the following relation holds:

$$\left| Eh\left(\frac{n^{1/2} f(F^{-1}(q))}{\sigma} (X_{r(n):n} - F^{-1}(q)) \right) - \int h \, dG_{r(n),n} \right| = O(n^{-m/2}) \quad (6.1.1)$$

where

$$G_{r(n),n} = \Phi + \varphi \sum_{i=1}^{m-1} n^{-i/2} S_{i,n} \quad (6.1.2)$$

and $S_{i,n}$ is a polynomial of degree $\le 3i - 1$ with coefficients uniformly bounded over n.
In particular, we have

$$S_{1,n}(t) = \left[\frac{2q - 1}{3\sigma} + \frac{\sigma f'(F^{-1}(q))}{2f(F^{-1}(q))^2} \right] t^2 + \left[\frac{-q + nq - r(n) + 1}{\sigma} + \frac{2(2q - 1)}{3\sigma} \right]. \quad (6.1.3)$$

PROOF. Denote by $f_{r(n),n}$ the density of the normalized distribution of $X_{r(n):n}$ and by $g_{r(n),n}$ the density (that is, the derivative) of $G_{r(n),n}$. Put $B_n = [-\log n, \log n]$. By P.4.5,

$$\left| \int h(x) f_{r(n),n}(x) \, dx - \int h(x) g_{r(n),n}(x) \, dx \right| \quad (1)$$

$$\le \int_{B_n} |h(x)| |f_{r(n),n}(x) - g_{r(n),n}(x)| \, dx + \int_{B_n^c} |h(x)| (f_{r(n),n}(x) + |g_{r(n),n}(x)|) \, dx$$

$$= O\left(n^{-m/2} \int_{B_n} |x|^k \varphi(x)(1 + |x|^{3m}) \, dx + \int_{B_n^c} |x|^k (f_{r(n),n}(x) + |g_{r(n),n}(x)|) \, dx \right).$$

It remains to prove an upper bound for the second term on the right-hand side of (1). Straightforward calculations yield

$$\int_{B_n^c} |x|^k |g_{r(n),n}(x)| \, dx = O(n^{-m/2}).$$

The decisive step is to prove that

$$\alpha_n := \int_{B_n^c} |x|^k f_{r(n),n}(x)\, dx = O(n^{-m/2}).$$

Apparently,

$$\alpha_n = \int_{\{X_{r(n):n} < F^{-1}(q)-t_n\}} |X_{r(n):n} - F^{-1}(q)|^k\, dP$$

$$+ \int_{\{X_{r(n):n} > F^{-1}(q)+t_n\}} |X_{r(n):n} - F^{-1}(q)|^k\, dP =: \alpha_{n,1} + \alpha_{n,2}$$

where $t_n = (\log n)\sigma/[n^{1/2} f(F^{-1}(q))]$. Applying Lemma 3.1.4 and Corollary 1.2.7 we get

$$\alpha_{n,1} = O\left(P\{X_{r(n):n} < F^{-1}(q) - t_n\} + \int_{\{X_{r(n):n} < F^{-1}(q)-t_n\}} |X_{r(n):n}|^k\, dP \right)$$

$$= O\left(P\{U_{r(n):n} \le F(F^{-1}(q) - t_n)\} \right.$$

$$+ \frac{b(r(n) - ks, n - (j+1)k - r(n) + ks + 1)}{b(r(n), n - r(n) + 1)} P\{U_{r(n)-ks:n-(j+1)k}$$

$$\left. \le F(F^{-1}(q) - t_n)\} \right)$$

where b denotes the beta function. Applying Lemma 3.1.1 one obtains that $\alpha_{n,1} = O(n^{-m/2})$. We may also prove $\alpha_{n,2} = O(n^{-m/2})$ which completes the proof.   □

As a special case of (6.1.1) we obtain

$$E|F_n^{-1}(q) - F^{-1}(q)|^k = \frac{(q(1-q))^{k/2}}{n^{k/2} f(F^{-1}(q))^k} \int |x|^k\, d\Phi(x) + O(n^{-(k+1)/2}) \qquad (6.1.4)$$

Expansions of Quantiles of Distributions of Order Statistics

Recall the result of Section 4.6 where we obtained an expansion concerning the "inverse" of an Edgeworth expansion. A corresponding result holds for expansions of d.f.'s of order statistics.

Theorem 6.1.2. *Let $q \in (0, 1)$ be fixed. Suppose that the d.f. F has $m + 1$ derivatives on a neighborhood of $F^{-1}(q)$, and that $f(F^{-1}(q)) > 0$ where $f = F'$. Suppose that $(r(n)/n - q) = O(n^{-1})$.*

Then there exist polynomials $R_{i,n}, i = 1, \ldots, m - 1$, such that uniformly over $|x| \le \log n$,

$$P\left\{\frac{n^{1/2}f(F^{-1}(q))}{(q(1-q))^{1/2}}(X_{r(n):n}-F^{-1}(q))\leq x+\sum_{i=1}^{m-1}n^{-i/2}R_{i,n}(x)\right\}=\Phi(x)+O(n^{-m/2}).$$

$$(6.1.5)$$

With $S_{i,n}$ denoting the polynomials in (6.1.2) we have

$$R_{1,n}=-S_{1,n}$$

and $$(6.1.6)$$

$$R_{2,n}(x)=S_{1,n}(x)S'_{1,n}(x)-\frac{x}{2}(S_{1,n}(x))^2-S_{2,n}(x).$$

PROOF. Apply P.4.5 and use the arguments of Section 4.6.   □

(6.1.5), applied to $x=\Phi^{-1}(\alpha)$, yields

$$P\left\{X_{r(n):n}\leq F^{-1}(q)+\left(\Phi^{-1}(\alpha)+\sum_{i=1}^{m-1}n^{-i/2}R_{i,n}(\Phi^{-1}(\alpha))\right)\frac{(q(1-q))^{1/2}}{n^{1/2}f(F^{-1}(q))}\right\}$$

$$(6.1.7)$$

$$=\alpha+O(n^{-m/2}).$$

This result may be adopted to justify a formal expansion given by F.N. David and N.L. Johnson (1954) in case of sample medians (see P.6.2).

6.2. Functions of Order Statistics

With regard to functions of order statistics a predominant role is played by linear combinations of order statistics. A comprehensive presentation of this subject is given in the book by Helmers (1982) so that it suffices to make some introductory remarks. In special cases we are able to prove supplementary results by using the tools developed in this book.

Chapters 4 and 5 provide approximations of joint distributions of central and extreme order statistics by means of normal and extreme value distributions. Thus, asymptotic distributions of certain functions of order statistics can easily be established. In this context, we also refer to Sections 9.5 and 10.4 where we shall study Hill's estimator and a certain χ^2-test.

Asymptotic Normality of a Linear Combination of Uniform R.V.'s

From a certain technical point of view the existing results for linear combinations of order statistics are very satisfactory. However, the question is still open whether one can find a condition which guarantees the asymptotic normality of a linear combination of order statistics related to the Lindeberg-condition for sums of independent r.v.'s or martingales.

Such a condition (see (6.2.4)) was found by Hecker (1976) in the special case of order statistics of i.i.d. uniform r.v.'s. This theorem is a simple application of the central limit theorem.

Theorem 6.2.1. *Given a triangular array of constants* $a_{i,n}$, $i = 1, \ldots, n$, *define*

$$b_{j,n} = \sum_{i=j}^{n} \left(1 - \frac{i}{n+1}\right) a_{i,n} - \sum_{i=1}^{j-1} \frac{i}{n+1} a_{i,n}, \qquad j = 1, \ldots, n+1, \quad (6.2.1)$$

and

$$\tau_n^2 = \sum_{j=1}^{n+1} b_{j,n}^2. \tag{6.2.2}$$

Then,

$$P\left\{\tau_n^{-1} \sum_{i=1}^{n} (n+1)a_{i,n}\left(U_{i:n} - \frac{i}{n+1}\right) \le t\right\} \to \Phi(t), \qquad n \to \infty, \quad (6.2.3)$$

for every t if, and only if,

$$\max_{j=1}^{n+1} \tau_n^{-1} |b_{j,n}| \to 0, \qquad n \to \infty. \tag{6.2.4}$$

PROOF. Let $\eta_1, \eta_2, \eta_3, \ldots$ be i.i.d. standard exponential r.v.'s. Put $S_i = \sum_{j=1}^{i} \eta_j$. From Corollary 1.6.9 it is immediate that

$$\sum_{i=1}^{n} (n+1)a_{i,n}\left(U_{i:n} - \frac{i}{n+1}\right) \overset{d}{=} \sum_{i=1}^{n} a_{i,n}[S_i - iS_{n+1}/(n+1)]/[S_{n+1}/(n+1)]. \quad (1)$$

Check that

$$\sum_{i=1}^{n} a_{i,n}[S_i - iS_{n+1}/(n+1)] = \sum_{j=1}^{n+1} b_{j,n}\eta_j. \tag{2}$$

From (2) and the fact that $E\eta_j = 1$ it is clear that

$$E \sum_{j=1}^{n+1} b_{j,n}\eta_j = \sum_{j=1}^{n+1} b_{j,n} = 0. \tag{3}$$

Consequently, τ_n^2 is the variance of $\sum_{j=1}^{n+1} b_{j,n}\eta_j$. Moreover, since $S_{n+1}/(n+1) \to 1$ in probability as $n \to \infty$ we deduce from (1)–(3) that (6.2.3) holds if, and only if,

$$P\left\{\tau_n^{-1} \sum_{j=1}^{n+1} b_{j,n}\eta_j \le t\right\} \to \Phi(t), \qquad n \to \infty. \tag{4}$$

The equivalence of (4) and (6.2.4) is a particular case of the Lindeberg–Lévy–Feller theorem as proved in Chernoff et al. (1967), Lemma 1. $\qquad\square$

EXAMPLE 6.2.2. If $a_{r,n} = 1$ and $a_{i,n} = 0$ for $i \neq r$ (that is, we consider the order statistic $U_{r:n}$) then $\tau_n^2 = r(n - r + 1)/(n + 1)^3$ in (6.2.2). Furthermore, (6.2.4) is equivalent to $r(n) \to \infty$ or $n - r(n) \to \infty$ as $n \to \infty$ with $r(n)$ in place of r.

As an immediate consequence of Theorem 6.2.1 and Theorem 4.3.1 we obtain the following result of preliminary character: Assume that the density f is strictly larger than zero and has three derivatives on the interval $(F^{-1}(q) - \varepsilon, F^{-1}(q) + \varepsilon)$ for some $q \in (0, 1)$. Define $I_n = \{r(n) + 4i: i = 1, \ldots, k(n)\}$ where $r(n)/n \to q$ and $k(n)/n \to 0$ as $n \to \infty$. Assume that $a_{i,n} = 0$ for $i \notin I_n$.

Then, with τ_n as in (6.2.2), as $n \to \infty$,

$$P\left\{\tau_n^{-1} \sum_{i=1}^{n} (n + 1)a_{i,n}f\left(F^{-1}\left(\frac{i}{n+1}\right)\right)\left(X_{i:n} - F^{-1}\left(\frac{i}{n+1}\right)\right) \leq t\right\}$$
$$\to \Phi(t), \qquad (6.2.5)$$

for every t if, and only if, (6.2.4) holds.

Of course, this result is very artificial. It would be interesting to know whether the index set I_n can be replaced by $\{r(n) + i: i = 1, \ldots, k(n)\}$ etc. It is left to the reader to formulate other theorems of this type by using Theorem 4.3.1 or Theorem 4.5.3.

The Trimmed Mean

The trimmed mean $\sum_{i=r}^{s} X_{i:n}$ is another exceptional case of a linear combination of order statistics which can easily be treated by conditioning on the order statistics $X_{r:n}$ and $X_{s:n}$.

Denote by $Y_{i:s-r-1}$ the ith order statistic of $s - r - 1$ r.v.'s with common d.f. $F_{x,y}$ [the truncation of F on the left of x and on the right of y]. Moreover, denote by $Q_{r,s,n}$ the joint distribution of $X_{r:n}$ and $X_{s:n}$. Then, according to Example 1.8.3(iii),

$$P\left\{\sum_{i=r}^{s} X_{i:n} \leq t\right\} = \int P\left\{x + \sum_{i=r}^{s-r-1} Y_{i:s-r-1} + y \leq t\right\} dQ_{r,s,n}(x, y)$$
$$= \int P\left\{\sum_{i=r}^{s-r-1} Y_i \leq t - (x + y)\right\} dQ_{r,s,n}(x, y)$$

where $Y_1, \ldots, Y_{s-r-1}$ are i.i.d. random variables with common d.f. $F_{x,y}$.

Now we are able to apply the classical results for sums of i.i.d. random variables to the integrand. Moreover, Section 4.5 provides a normal approximation to $Q_{r,s,n}$. Concerning more details we refer again to Helmers (1982).

Systematic Statistics

The notion of systematic statistics goes back to Mosteller (1946); we mention this expression for historical reasons only because nowadays one would speak of a linear combination of order statistics when treating this type of statistics. Based on the asymptotic normality of a fixed number of sample q-quantiles one can easily verify the asymptotic normality of a linear combination of these order statistics. Given a location and scale parameter family of distributions one can e.g. try to find the optimum estimator based on k order statistics. Below we shall only touch on the most simple case, namely, that of $k = 2$.

Lemma 6.2.3. (i) *Let* $0 < q_0 < 1$. *Assume that the d.f.* F *has two bounded derivatives on a neighborhood of* $F^{-1}(q_0)$ *and that* $f_0 := F'(F^{-1}(q_0)) > 0$.
Then,

$$\sup_t \left| P\left\{ \frac{n^{1/2}}{\sigma_0}(X_{[nq_0]:n} - F^{-1}(q_0)) \le t \right\} - \Phi(t) \right| = O(n^{-1/2})$$

where $\sigma_0^2 = q_0(1 - q_0)/f_0^2$.
(ii) *Let* $0 < q_1 < q_2 < 1$. *Assume that the d.f.* F *has two bounded derivatives on a neighborhood of* $F^{-1}(q_i)$ *and that* $f_i := F'(F^{-1}(q_i)) > 0$ *for* $i = 1, 2$.
Then,

$$\sup_t \left| P\left\{ \frac{n^{1/2}}{\sigma_1}(X_{[nq_2]:n} - X_{[nq_1]:n} - (F^{-1}(q_2) - F^{-1}(q_1))) \le t \right\} - \Phi(t) \right|$$
$$= O(n^{-1/2})$$

where

$$\sigma_1^2 = q_1(1 - q_1)/f_1^2 - 2q_1(1 - q_2)/(f_1 f_2) + q_2(1 - q_2)/f_2^2.$$

PROOF. Immediate from Theorem 4.5.3 by routine calculations. □

Sample quantiles and spacings ($\equiv$ difference of order statistics) provide quick estimators of the location and scale parameter. Recall that a d.f.

$$F_{\mu,\sigma}(x) := F((x - \mu)/\sigma)$$

has the q.f.

$$F_{\mu,\sigma}^{-1}(q) = \mu + \sigma F^{-1}(q).$$

Under the conditions of Lemma 6.2.3 we obtain for the sample quantiles $X_{[nq_i]:n}$ of n i.i.d. random variables with common d.f. $F_{\mu,\sigma}$ that with $\sigma_i = \sigma_i(F)$ as in Lemma 6.2.3:

$$\sup_t \left| P\left\{ \frac{n^{1/2}}{\sigma\sigma_0(F)}(X_{[nq_0]:n} - \mu) \le t \right\} - \Phi(t) \right| = O(n^{-1/2}) \qquad (6.2.6)$$

if w.l.g. $F^{-1}(q_0) = 0$; moreover,

$$\sup_t \left| P\left\{ \frac{n^{1/2}}{\sigma\sigma_2(F)}(\hat{\sigma}_n - \sigma) \le t \right\} - \Phi(t) \right| = O(n^{-1/2}) \qquad (6.2.7)$$

where the estimator $\hat{\sigma}_n$ is given by

$$\hat{\sigma}_n = \frac{X_{[nq_2]:n} - X_{[nq_1]:n}}{F^{-1}(q_2) - F^{-1}(q_1)}$$

and

$$\sigma_2(F) = \sigma_1(F)/(F^{-1}(q_2) - F^{-1}(q_1)).$$

An Expansion of Length Two for the Convex Combination of Consecutive Order Statistics

Let $X_{r:n}$ be the rth order statistic of n i.i.d. random variables with common continuous d.f.

We shall study statistics of the form

$$(1 - \gamma)X_{r:n} + \gamma X_{r+1:n}, \qquad \gamma \in [0, 1],$$

which may be used as estimators of the q-quantile. The most important case is the sample median for even sample sizes.

It is apparent that this statistic has the same asymptotic behavior for every $\gamma \in [0, 1]$ as far as the first order performance is concerned. The different performance of the statistics for varying γ can be detected if the second order term is studied. For this purpose we shall establish an expansion of length 2.

Denote by $F_{r,n}$ the d.f. of $c^{-1}(X_{r:n} - d)$. From Corollary 1.8.5 it is immediate that for γ and t,

$$P\{(1 - \gamma)X_{r:n} + \gamma X_{r+1:n} \le d + ct\}$$

$$= F_{r,n}(t) - \int_{-\infty}^{t} P\{\gamma[Y_{1:n-r} - (d + cx)] > c(t - x)\} \, dF_{r,n}(x) \tag{6.2.8}$$

where $Y_{1:n-r}$ is the sample minimum of $n - r$ i.i.d. random variables with common d.f. F_{d+cx} [the truncation of F on the left of $d + cx$].

Let $G_{n,r}$ be an approximation to $F_{n,r}$ such that

$$\sup_B \left| P\{c^{-1}(X_{r:n} - d) \in B\} - \int_B dG_{n,r} \right| = O(n^{-1}). \tag{6.2.9}$$

From Corollary 1.2.7 and Theorem 5.4.3 we get uniformly in t and x,

$$P\{\gamma[Y_{1:n-r} - (d + cx)] > c(t - x)\}$$

$$= P\{(n - r)U_{1:n-r} > (n - r)F_{d+cx}[d + cx + c(t - x)/\gamma]\} \tag{6.2.10}$$

$$= \exp[-(n - r)F_{d+cx}[d + cx + c(t - x)/\gamma]] + O(n^{-1}).$$

Combining (6.2.8)–(6.2.10) and applying P.3.5 we get

$$\sup_t \left| P\{(1 - \gamma)X_{r:n} + \gamma X_{r+1:n} \le d + ct\} - \left[G_{r,n}(t) \right. \right.$$

$$\left. \left. - \int_{-\infty}^{t} \exp[-(n - r)F_{d+cx}[d + cx + c(t - x)/\gamma]] \, dG_{r,n}(x) \right] \right| = O(n^{-1}). \tag{6.2.11}$$

Notice that if $\gamma = 0$ then, in view of (6.2.10), the integral in (6.2.11) can be replaced by zero.

Specifying normalizing constants and an expansion $G_{r,n}$ of $F_{r,n}$ we obtain the following theorem.

Theorem 6.2.4. *Let $q \in (0, 1)$ be fixed. Assume that F has three bounded derivatives on a neighborhood of $F^{-1}(q)$ and that $f(F^{-1}(q)) > 0$ where $f = F'$. Moreover, assume that $(r(n)/n - q) = o(n^{-1/2})$. Put $\sigma^2 = q(1 - q))$.*

Then, uniformly in $\gamma \in [0, 1]$,

$$\sup_t \left| P\left\{ \frac{n^{1/2} f(F^{-1}(q))}{\sigma} [(1 - \gamma)X_{r(n):n} + \gamma X_{r(n)+1:n} - F^{-1}(q)] \le t \right\} \right.$$

$$\left. - (\Phi(t) + n^{-1/2}\varphi(t)R_n(t)) \right| = o(n^{-1/2})$$

where

$$R_n(t) = -\left[\frac{1 - 2q}{3\sigma} - \frac{\sigma f'(F^{-1}(q))}{2f(F^{-1}(q))^2} \right] t^2$$

$$- \left[\frac{q - nq + r(n) + \gamma - 1}{\sigma} + \frac{2(1 - 2q)}{3\sigma} \right].$$

PROOF. The basic formula (6.2.11) will be applied to $d = F^{-1}(q)$ and $c = \sigma/(n^{1/2}f(d))$. In view of P.4.5 which supplies us with an expansion $G_{r,n} = \Phi + n^{1/2}\varphi S_{r,n}$ of $F_{r,n}$ it suffices to prove that

$$\int_{-\infty}^t \exp[-(n - r)F_{d+cx}[d + cx + c(t - x)/\gamma]]\varphi(x)\,dx \tag{1}$$

$$= n^{-1/2}\varphi(t)\gamma/\sigma + o(n^{-1/2})$$

and

$$\int_{-\infty}^t \exp[-(n - r)F_{d+cx}[d + cx + c(t - x)/\gamma]] |(\varphi S_{r,n})'(x)|\,dx = o(n^0) \tag{2}$$

uniformly in γ and t. The proof of (1) will be carried out in detail. Similar arguments lead to (2).

Since $\Phi(-(\log n)/2) = O(n^{-1})$ it is obvious that $\int_{-\infty}^t$ can be replaced by $\int_{-\log n}^t$ where $t \le (\log n)/2$. Then, the integrand is of order $O(n^{-1})$ for those x with $c(t - x)/\gamma > s(\log n)/n$ for some sufficiently large $s > 0$. Thus, $\int_{-\infty}^t$ can be replaced by $\int_{u(n)}^t$ where $u(n) = \max(-\log n, t - \gamma s(\log n)/cn)$.

Under the condition that F has three bounded derivatives it is not difficult to check that for $u(n) \le x \le t$,

$$F_{d+cx}[d + cx + c(t - x)/\gamma]$$

$$= \frac{f(d)c(t - x)}{(1 - q)\gamma} + O[c|x|(c(t - x)/\gamma + (c(t - x)/\gamma)^2)] \tag{3}$$

$$= \frac{f(d)c(t - x)}{(1 - q)\gamma} + O[(\log n)^2/n^{3/2}].$$

Thus, (1) has to be verified with the left-hand side replaced by the term

$$\int_{u(n)}^{t} \exp\left[-(n-r)n^{-1/2}\frac{\sigma(t-x)}{(1-q)\gamma}\right]\varphi(x)\,dx \tag{4}$$

which, by substituting $y = n^{1/2}\sigma(t-x)/\gamma$, can easily be verified to be equal to

$$n^{-1/2}(\gamma/\sigma)\int_{0}^{v(n)} \exp\left[-\frac{1-r/n}{1-q}y\right]\varphi(t-n^{-1/2}\gamma y/\sigma)\,dy \tag{5}$$

where $v(n) = s(\log n)/f(d)$. Since

$$\exp\left[-\frac{1-r/n}{1-q}y\right] = \exp(-y)[1 + o(n^0)]$$

and   (6)

$$\varphi(t - n^{-1/2}\gamma y/\sigma) = \varphi(t)[1 + o(n^0)]$$

we obtain that the term in (5) is equal to

$$n^{-1/2}(\gamma/\sigma)\varphi(t)\int_{0}^{v(n)} \exp(-y)\,dy(1 + o(n^0)). \tag{7}$$

Notice that the relations above hold uniformly in ρ and t. Now (1) is immediate.   □

Notice that for $\gamma = 0$ we again get the expansion of length two of the normalized d.f. of $X_{r(n):n}$ as given in P.4.5. Moreover, for $\gamma = 0$ and for $r(n)$ replaced by $r(n) + 1$ we get the same expansion as for $\rho = 1$ and $r(n)$.

If $q = \frac{1}{2}$, $f'(F^{-1}(1/2)) = 0$, $n = 2m$, and $r = m$ then

$$P\{[(2m)^{1/2}f(F^{-1}(1/2))/2][(X_{m:2m} + X_{m+1:2m})/2 - F^{-1}(1/2)] \le t\}$$
$$= \Phi(t) + o(n^{-1/2}). \tag{6.2.12}$$

Thus, the sample median for even sample sizes is asymptotically normal with a remainder term of order $o(n^{-1/2})$. For odd sample sizes the corresponding result was proved in Section 4.2.

Remark 6.2.5. Let $q_0 \in (0, 1)$. Assume that F has three bounded derivatives on a neighborhood of $F^{-1}(q_0)$ and that $f(F^{-1}(q_0)) > 0$. Then a short examination of the proof to Theorem 6.3.4 reveals that the assertion holds uniformly over all q in a sufficiently small neighborhood of q_0 and $r(n) \equiv r(q, n)$ such that $\sup_q |r(q, n)/n - q| = o(n^{-1/2})$. This yields the version of Theorem 6.3.4 as cited in Pfanzagl (1985).

The Meanwhile Classical Theory of Linear Combinations of Order Statistics

The central idea of the classical approach is to use weight functions to represent a linear function of order statistic in an elegant way.

Linear combinations of order statistics of the form

$$T_n = n^{-1} \sum_{i=1}^{n} J\left(\frac{i}{n+1}\right) X_{i:n} \tag{6.2.13}$$

are estimators of the functional

$$\mu(F) = \int_0^1 J(s) F^{-1}(s) \, ds. \tag{6.2.14}$$

Notice that according to (1.2.13) and (1.2.14)

$$\mu(F) = \int x J(F(x)) \, dF(x) \tag{6.2.15}$$

for continuous d.f.'s F.

The following theorem is due to Helmers (1981). The proof of Theorem 6.2.6 (see also Helmers (1982, Theorem 3.1.2)) is based on the calculus of characteristic functions.

Theorem 6.2.6. *Suppose that $E|\xi_1|^3 < \infty$ and*

$$\sigma^2(F) := \int \int J(F(x)) J(F(y)) (\min(F(x), F(y)) - F(x)F(y)) \, dx \, dy > 0.$$

Moreover, let the weight function J satisfy a Lipschitz condition of order 1 on $(0, 1)$. Then,

$$\sup_x \left| P\left\{ \frac{n^{1/2}}{\sigma(F)} (T_n - \mu(F)) \le x \right\} - \Phi(x) \right| = O(n^{-1/2}).$$

The smoothness condition imposed on J can be weakened by imposing appropriate smoothness conditions on F.

6.3. Bahadur Approximation

In Section 1.1 we have seen that the d.f. of an order statistic—and thus that of the sample q.f.—can be represented by means of the sample d.f. It was observed by R.R. Bahadur (1966) that an amazingly accurate stochastic approximation of the sample d.f. to the sample q.f. holds.

Motivation

To get some insight into the nature of this approximation let us consider the special case of i.i.d. $(0, 1)$-uniformly distributed r.v.'s $\eta_1, \eta_2, \ldots, \eta_n$. Denote by G_n and $U_{i:n}$ the pertaining sample d.f. and the ith order statistics. We already know that the distributions of

$$G_n(r/n) = n^{-1} \sum_{i=1}^{n} 1_{(-\infty, r/n]}(\eta_i)$$

and of $U_{r:n}$ are concentrated about r/n. Moreover, relation (1.1.6) shows that pointwise

$$U_{r:n} - \frac{r}{n} \le 0 \quad \text{iff} \quad G_n\left(\frac{r}{n}\right) - \frac{r}{n} \ge 0. \tag{6.3.1}$$

Thus, it is plausible that the distribution of

$$(U_{r:n} - r/n) + (G_n(r/n) - r/n)$$

is more closely concentrated about zero than each of the distributions of $U_{r:n} - r/n$ and $G_n(r/n) - r/n$. Instead of $(U_{r:n} - r/n) + (G_n(r/n) - r/n)$, the so-called Bahadur statistic

$$(G_n^{-1}(q) - q) + (G_n(q) - q), \qquad q \in (0, 1), \tag{6.3.2}$$

may apparently be studied as well.

Recall that $G_n^{-1}(q) = U_{r(q):n}$ where $r(q) = nq$ if nq is an integer and $r(q) = [nq] + 1$ otherwise.

In the general case of order statistics $X_{i:n}$ from n i.i.d. random variables ξ_i with common d.f. F and derivative $f(F^{-1}(q))$ the Bahadur statistic is given by

$$f(F^{-1}(q))(F_n^{-1}(q) - F^{-1}(q)) + (F_n(F^{-1}(q)) - q), \qquad q \in (0, 1), \tag{6.3.3}$$

where F_n and F_n^{-1} are the sample d.f. and sample q.f. based on the r.v.'s ξ_i.

The connection between (6.3.2) and (6.3.3) becomes obvious by noting that the transformation technique yields

$$\begin{aligned} f(F^{-1}(q))(F_n^{-1}(q) &- F^{-1}(q)) + (F_n(F^{-1}(q)) - q) \\ &\overset{d}{=} f(F^{-1}(q))(F^{-1}(G_n^{-1}(q)) - F^{-1}(q)) + (G_n(F(F^{-1}(q))) - q). \end{aligned} \tag{6.3.4}$$

If $F^{-1}(q)$ is a continuity point of F then $F(F^{-1}(q))$ can be replaced by q and, moreover, if F^{-1} has a bounded second derivative then

$$f(F^{-1}(q))(F^{-1}(G_n^{-1}(q)) - F^{-1}(q)) = G_n^{-1}(q) - q + O(G_n^{-1}(q) - q)^2$$

and hence results for the Bahadur statistic in the uniform case can easily be extended to continuous d.f.'s F.

Probabilities of Moderate Deviation

Since we are interested in the Bahadur statistic as a technical tool we shall confine our attention to a result concerning moderate deviations. The upper bound for the accuracy of the stochastic approximation will be non-uniform in q.

Theorem 6.3.1. *For every s > 0 there exists a constant C(s) such that*

$$P\{|(G_n^{-1}(q) - q) + (G_n(q) - q)| > (\log n)/n)^{3/4}\delta(q, s, n) \quad \text{for some}$$

$$q \in (0, 1)\} \le C(s)n^{-s}$$

where $\delta(q, s, n) = 7(s + 3)\max\{(q(1 - q))^{1/4}, (7(s + 3)(\log n)/n)^{1/2}\}.$

Before proving Theorem 6.3.1 we make some comments and preparations. Theorem 6.3.1 is sufficient as a technical tool in statistical applications, however, one should know that sharp results concerning the stochastic behavior of the Bahadur statistic exist in literature. The following limit theorem is due to Kiefer (1969a): For every $t > 0$,

$$P\left\{\sup_{q \in (0, 1)} |(G_n^{-1}(q) - q) + (G_n(q) - q)| > \frac{(\log n)^{1/2}t}{n^{3/4}}\right\}$$

$$\to 2\sum_m (-1)^{m+1} e^{-2m^2t^4}$$

as $n \to \infty$ where the summation runs over all positive integers m.

Kiefer's result indicates that Theorem 6.3.1 is sharp in so far that the $((\log n)/n)^{3/4}$ cannot be replaced by some term of order $o[((\log n)/n)^{3/4}]$.

To prove Theorem 6.3.1 we shall use a simple result concerning the oscillation of the sample d.f. For this purpose define the sample probability measure Q_n by

$$Q_n(A) = n^{-1} \sum_{i=1}^{n} 1_A(\eta_i)$$

where the η_i are i.i.d. random variables with common uniform distribution Q_0 on $(0, 1)$. Recall that the Glivenko–Cantelli theorem yields

$$\sup_{I \in \mathscr{I}} |Q_n(I) - Q_0(I)| \to 0, \qquad n \to \infty, \text{ w.p. 1}, \tag{6.3.5}$$

where $\mathscr{I}$ is the system of all intervals in $(0, 1)$.

Lemma 6.3.2 will indicate the rate of convergence in (6.3.5); moreover, this result will show that the rate is better for those intervals I for which $\sigma^2(I) = Q_0(I)(1 - Q_0(I))$ is small.

Lemma 6.3.2. *For every s > 0 there exists a constant A(s) such that for every n:*

$$P\left\{\sup_{I \in \mathscr{I}} \frac{n^{1/2}|Q_n(I) - Q_0(I)|}{\max\{\sigma(I), ((\log n)/n)^{1/2}\}} \ge (s + 3)(\log n)^{1/2}\right\} < A(s)n^{-s}.$$

PROOF. Given $\varepsilon, \rho > 0$ we shall prove that

$$K(s, n) := P\left\{\sup_{I \in \mathscr{I}} \frac{n^{1/2}|Q_n(I) - Q_0(I)|}{\max\{\sigma(I), \rho/n^{1/2}\}} \ge \varepsilon\right\}$$

$$\le (n + 2)^2 \exp\left[-\varepsilon\rho + \frac{3}{4}\rho^2 + \frac{13}{2}\right]. \tag{6.3.6}$$

Then, an application of (6.3.6) to $\rho = (\log n)^{1/2}$ and $\varepsilon = (s + 3)(\log n)^{1/2}$ yields the assertion.

Put $\mathscr{I}_0 = \{(i/n, j/n]: 0 \le i < j \le n\}$. Straightforward calculations yield

$$K(s, n) \le P\left\{\sup_{I \in \mathscr{I}_0} \frac{n^{1/2}|Q_n(I) - Q_0(I)| + 2n^{-1/2}}{\max\{\sigma^2(I) - 2/n - 4/n^2, \rho^2/n\}^{1/2}} \ge \varepsilon\right\} \tag{1}$$

$$\le \sum_{I \in \mathscr{I}_0} P\{n^{1/2}|Q_n(I) - Q_0(I)| \ge \varepsilon(I)\}$$

where $\varepsilon(I) = \varepsilon \max\{\sigma^2(I) - 2/n - 4/n^2, \rho^2/n\}^{1/2} - 2n^{-1/2}$. Let $I \in \mathscr{I}_0$ be fixed. Assume w.l.g. that $\sigma(I) > 0$ and $\varepsilon\rho \ge 7/2$ so that $\varepsilon(I) > 0$. Using the exponential bound (3.1.1) with $t = \sigma(I)/\max\{(\sigma^2(I) - 2/n - 4/n^2)/\rho^2, 1/n\}^{1/2}$ we obtain

$$P\{n^{1/2}|Q_n(I) - Q_0(I)| \ge \varepsilon(I)\} \le 2 \exp\left[-\frac{\varepsilon(I)}{\sigma(I)}t + \frac{3}{4}t^2\right] \tag{2}$$

$$\le 2 \exp\left[-\varepsilon\rho + \frac{3}{4}\rho^2 + \frac{7}{2} + \frac{3}{n}\right].$$

Now, (1) and (2) yield (6.3.6). The proof is complete. $\qquad\square$

Remark 6.3.3. Lemma 6.3.2 holds for any i.i.d. random variables (with arbitrary common distribution Q in place of Q_0). The general case be reduced to the special case of Lemma 6.3.2 by means of the quantile transformation.

Lemma 6.3.2 together with the Borel–Cantelli lemma yields

$$\limsup_n \sup_{I \in \mathscr{I}_n} \frac{n^{1/2}|Q_n(I) - Q_0(I)|}{\sigma(I)(\log n)^{1/2}} \le 5 \quad \text{w.p. 1} \tag{6.3.7}$$

where $\mathscr{I}_n = \{I \in \mathscr{I}: \sigma^2(I) = Q_0(I)(1 - Q_0(I)) \ge (\log n)/n\}$. In this context, we mention a result of Stute (1982) who proved a sharp result concerning the almost sure behavior of the oscillation of the sample d.f.:

$$\limsup_n \sup_{I \in \mathscr{I}_n^*} \frac{n^{1/2}|Q_n(I) - Q_0(I)|}{(2Q_0(I)\log a_n^{-1})^{1/2}} = 1 \quad \text{w.p. 1} \tag{6.3.8}$$

where $\mathscr{I}_n^* = \{I \in \mathscr{I}: I = (a, b], \alpha a_n \le Q_0(I) \le \beta a_n\}$ with $0 < \alpha < \beta < \infty$, and a_n has the properties $a_n \downarrow 0$, $na_n \uparrow \infty$, $\log a_n^{-1} = o(na_n)$ and $(\log a_n^{-1})/(\log\log n) \to \infty$ as $n \to \infty$. Note that (6.3.8) shows that the rate in (6.3.7) is sharp.

Theorem 6.3.1 will be an immediate consequence of Lemma 6.3.2 and Lemma 3.1.5 which concerns the maximum deviation of the sample q.f. G_n^{-1} from the $(0, 1)$-uniform q.f.

PROOF OF THEOREM 6.3.1. Since $|G_n(G_n^{-1}(q)) - q| \le 1/n$ we obtain

$$|G_n^{-1}(q) - q + (G_n(q) - q)| \le |G_n^{-1}(q) - G_n(G_n^{-1}(q)) + (G_n(q) - q)| + 1/n$$

$$\le \sup_{|x-q| \le \kappa} |x - G_n(x) + G_n(q) - q|$$

$$= \sup_{I(q)} |Q_n(I(q)) - Q_0(I(q))|$$

whenever $|G_n^{-1}(q) - q| \leq \kappa$ and $I(q)$ runs over all intervals $(x, q]$ and $(q, x]$ with $|x - q| \leq \kappa$. Thus, by Lemma 6.3.2 and Lemma 3.1.5 applied to $\kappa = \kappa(q, s, n)$, we get

$$P\{|G_n^{-1}(q) - q + (G_n(q) - q)| \leq \delta(q, s, n), \quad q \in (0, 1)\}$$

$$\geq P\left\{\sup_{I(q)} |Q_n(I(q)) - Q_0(I(q))| \leq \right.$$

$$\left. (s + 3)((\log n)/n)^{1/2} \kappa(q, s, n)^{1/2}, \quad q \in (0, 1)\right\} - B(s)n^{-s}$$

$$\geq 1 - [A(s) + B(s)]n^{-s}$$

where $A(s)$ and $B(s)$ are the constants of Lemma 6.3.2 and Lemma 3.1.5. The proof is complete. $\square$

6.4. Bootstrap Distribution Function of a Quantile

In this section we give a short introduction to Efron's bootstrap technique and indicate its applicability to problems concerning order statistics.

Introduction

Since the sample d.f. F_n is a natural nonparametric estimator of the unknown underlying d.f. F it is plausible that the statistical functional $T(F_n)$ is an appropriate estimator of $T(F)$ for a large class of functionals T.

In connection with covering probabilities and confidence intervals one is interested in the d.f.

$$T_n(F, t) = P_F\{T(F_n) - T(F) \leq t\}$$

of the centered statistic $T(F_n) - T(F)$.

The basic idea of the bootstrap approach is to estimate the d.f. $T_n(F, \cdot)$ by means of the bootstrap d.f. $T_n(F_n, \cdot)$. Thus, the underlying d.f. F is simply replaced by the sample d.f. F_n.

Let us touch on the following aspects:

(a) the calculation of the bootstrap d.f. by enumeration or alternatively, by Monte Carlo resampling,
(b) the validity of the bootstrap approach,
(c) the construction of confidence intervals for $T(F)$ via the bootstrap approach.

Evaluation of Bootstrap D.F.: Enumeration and Monte Carlo

Hereafter, let the observations $x_1, \ldots, x_n$ be generated according to n i.i.d. random variables with common d.f. F. Denote by F_n^x the corresponding realization of the sample d.f. F_n; thus, we have

$$F_n^{\mathbf{x}}(t) = 1/n \sum_{i=1}^{n} 1_{(-\infty, t]}(x_i), \qquad \mathbf{x} = (x_1, \ldots, x_n).$$

Since $F_n^{\mathbf{x}}$ is a discrete d.f. it is clear that the realization $T_n(F_n^{\mathbf{x}}, \cdot)$ of the bootstrap d.f. $T_n(F_n, \cdot)$ can be calculated by enumeration:

If $x_i \neq x_j$ for $i \neq j$ then $T_n(F_n^{\mathbf{x}}, t)$ is the relative frequency of vectors $\mathbf{z} \in \{x_1, \ldots, x_n\}^n$ which satisfy the condition

$$T(F_n^{\mathbf{z}}) - T(F_n^{\mathbf{x}}) \leq t. \tag{6.4.1}$$

Notice that inequality (6.4.1) has to be checked for n^n vectors $\mathbf{z}$.

A Monte Carlo approximation to $T_n(F_n^{\mathbf{x}}, t)$ is given by the relative frequency of pseudo-random vectors $\mathbf{z}_1, \ldots, \mathbf{z}_m$ satisfying (6.4.1) where $\mathbf{z}_i = (z_{i,1}, \ldots, z_{i,n})$. The values $z_{1,1}, \ldots, z_{1,n}, z_{2,1}, \ldots, z_{m,n}$ are pseudo-random numbers generated according to the d.f. $F_n^{\mathbf{x}}$. The sample size m should be large enough so that the deviation of the Monte Carlo approximation from $T_n(F_n^{\mathbf{x}}, t)$ is negligible.

The 3σ-rule leads to a crude estimate of the necessary sample size. It says that the absolute deviation of the Monte Carlo approximation from $T_n(F_n^{\mathbf{x}}, t)$ is smaller than $3/(2m^{1/2})$ with a probability $\geq .99$. Thus, if e.g. a deviation of 0.005 is negligible then one should take $m = 90\,000$.

These considerations show that the Monte Carlo procedure is preferable to the exact calculation of the bootstrap estimate by enumeration if m is small compared to n^n (which will be the case if $n \geq 10$). In special cases it is possible to represent the bootstrap estimate by some analytical expression (see (6.4.2)).

A Counterexample: Sample Minima

Next, we examine the statistical performance of bootstrap estimates in the particular cases of sample minima. This problem will serve as an example where the bootstrap approach is not valid.

Let again $\alpha(F) = \inf\{x: F(x) > 0\}$ denote the left endpoint of the d.f. F. The corresponding statistical functional $\alpha(F_n)$ is the sample minimum $X_{1:n}$. If $\alpha(F) > -\infty$ then according to (1.3.3),

$$T_n(F, t) = P\{X_{1:n} - \alpha(F) \leq t\} = 1 - [1 - F(\alpha(F) + t)]^n$$

and

$$T_n(F_n, t) = 1 - [1 - F_n(X_{1:n} + t)]^n.$$

If F is continuous then w.p. 1,

$$T_n(F_n, 0) - T_n(F, 0) = 1 - \left(1 - \frac{1}{n}\right)^n \to 1 - \exp(-1), \qquad n \to \infty.$$

Hence the bootstrap method leads to an inconsistent sequence of estimators.

Sample Quantiles: Exact Evaluation of Bootstrap D.F.

Monte Carlo simulations provide some knowledge about the accuracy of the bootstrap procedure for a fixed sample size. Further insight into the validity of the bootstrap method is obtained by asymptotic considerations.

The consistency of $T_n(F_n, \cdot)$ holds if e.g. the normalized d.f.'s $T_n(F_n, \cdot)$ and $T_n(F, \cdot)$ have the same limit, as n goes to infinity. Then the accuracy of the bootstrap approximation will be determined by the rates of convergence of the two sequences of d.f.'s to the limiting d.f. As an example we study the bootstrap approximation to the d.f. of the sample q-quantile.

If $T(F) = F^{-1}(q)$ then $T(F_n) = F_n^{-1}(q) = X_{m(n):n}$ where $m(n) = nq$ if nq is an integer, and $m(n) = [nq] + 1$, otherwise. By Lemma 1.3.1,

$$T_n(F, t) = P\{X_{m(n):n} - F^{-1}(q) \le t\}$$

(6.4.2)

$$= \sum_{i=m(n)}^{n} \binom{n}{i} (F(F^{-1}(q) + t))^i (1 - F(F^{-1}(q) + t))^{n-i}$$

and the same representation holds for $T_n(F_n, t)$ with F^{-1} replaced by F_n^{-1}.

From Theorem 4.1.4 we know that $T_n(F, t)$, suitably normalized, approaches the standard normal d.f. Φ as $n \to \infty$. The normalized version of $T_n(F, t)$ is given by

$$T_n^*(F, t) = T_n(F, (q(1 - q))^{1/2} t/n^{1/2} \varphi(\Phi^{-1}(q)))$$

(6.4.3)

if $F = \Phi$.

To prove that the bootstrap d.f. $T_n(F_n, \cdot)$ is a consistent estimator of $T_n(\Phi, \cdot)$ one has to show that, $T_n^*(F_n, t) \to \Phi(t)$, $n \to \infty$, for every t, w.p. 1.

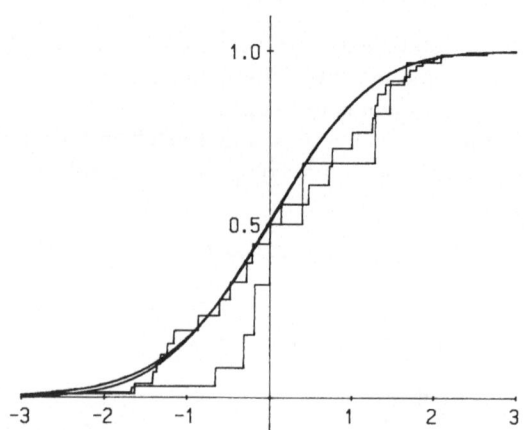

Figure 6.4.1. Normalized d.f. $T_n^*(\Phi, \cdot)$ of sample q-quantile and bootstrap d.f. $T_n^*(F_n, \cdot)$ for $q = 0.4$ and $n = 20, 200$.

The numerical calculations above were carried out by using the normal approximation to the d.f. of the sample quantile of i.i.d. $(0, 1)$-uniformly distributed r.v.'s. Otherwise, the computation of the binomial coefficients would cause numerical difficulties. Computations for the sample size $n = 20$ showed that the error of this approximation is negligible.

From Figure 6.4.1 we see that $T_{20}^*(\Phi, \cdot)$ and $T_{200}^*(\Phi, \cdot)$ are close together (and, by the way, close to Φ) indicating a quick convergence. The bootstrap d.f. $T_n^*(F_n, \cdot)$ is a step function which slowly approaches Φ. Next, we indicate this rate of convergence.

Asymptotic Investigations

The further analysis will be simplified by using the normal approximation to the d.f. of the sample q-quantile of n i.i.d. $(0, 1)$-uniformly distributed r.v.'s. From Corollary 1.2.7 and (4.2.1), applied to $m = 1$, we deduce

$$T_n(F_n, t/n^{1/2}) - T_n(F, t/n^{1/2})$$

$$= \Phi\left[\frac{n^{1/2}(F_n(F_n^{-1}(q) + t/n^{1/2}) - q)}{(q(1 - q))^{1/2}}\right]$$

$$- \Phi\left[\frac{n^{1/2}(F(F^{-1}(q) + t/n^{1/2}) - q)}{(q(1 - q))^{1/2}}\right] + O(n^{-1/2}) \qquad (6.4.4)$$

$$= \Phi[tg_{n,t}/(q(1 - q))^{1/2}] - \Phi[tf(F^{-1}(q))/(q(1 - q))^{1/2}] + o(1)$$

uniformly over t [where the second relation holds if F has a derivative, say, $f(F^{-1}(q))$ at $F^{-1}(q)$]. Moreover, the function $g_{n,t}$ is defined by

$$g_{n,t} = \frac{F_n(F_n^{-1}(q) + t/n^{1/2}) - F_n(F_n^{-1}(q))}{t/n^{1/2}}, \qquad t \neq 0, \qquad (6.4.5)$$

and $= 0$ if $t = 0$.

The auxiliary function $g_{n,t}$ is a "naive" estimator of the density at the random point $F_n^{-1}(q)$. Thus, the stochastic behavior of the bootstrap error $T_n(F_n, t/n^{1/2}) - T_n(F, t/n^{1/2})$ is closely related to that of a density estimator. We have

$$\sup_t |T_n(F_n, t) - T_n(F, t)| \to 0, \qquad n \to \infty, \text{ w.p. } 1, \qquad (6.4.6)$$

that is, the bootstrap estimator is strongly consistent, if w.p. 1 for every $t \neq 0$,

$$g_{n,t} \to f(F^{-1}(q)), \qquad n \to \infty. \qquad (6.4.7)$$

Let us assume that F has a derivative, say, f near $F^{-1}(q)$ and f is continuous at $F^{-1}(q)$. From Lemma 3.1.7(ii) and the Borel–Cantelli lemma it follows that, w.p. 1, for every $t \neq 0$,

$$g_{n,t} = \frac{F(F_n^{-1}(q) + t/n^{1/2}) - F(F_n^{-1}(q))}{t/n^{1/2}}$$

$$+ O\left[\left(\frac{F(F_n^{-1}(q) + t/n^{1/2}) - F(F_n^{-1}(q))}{t/n^{1/2}}\right)^{1/2} \frac{(\log n)^{1/2}}{n^{1/4}} + \frac{\log n}{n^{1/2}}\right]$$

$$= f(F^{-1}(q) + \theta_{n,t} t/n^{1/2}) + O\left[(f(F_n^{-1}(q) + \theta_{n,t} t/n^{1/2}))^{1/2} \frac{(\log n)^{1/2}}{n^{1/4}} + \frac{\log n}{n^{1/2}}\right]$$

eventually, for some $\theta_{n,t} \in (0, 1)$.

Thus, (6.4.7) holds because $F_n^{-1}(q)$ is a strongly consistent estimator of $F^{-1}(q)$ under the present conditions (compare with Lemma 1.2.9).

It is easy to see that the proof, developed above, also leads to a bound of the rate of convergence which is, roughly speaking, of order $O(n^{-1/4})$ under slightly stronger conditions imposed on F.

An exact answer to the question concerning the accuracy of the bootstrap approximation can e.g. be obtained by a law of the iterated logarithm as proved by Singh (1981):

If F has a bounded second derivative near $F^{-1}(q)$ and $f(F^{-1}(q)) > 0$ then

$$\limsup_n \frac{n^{1/4}}{(\log\log n)^{1/2}} \sup_t |T_n(F_n, t) - T_n(F, t)| = K_{q,F} > 0 \qquad \text{w.p. } 1$$

where K is a constant depending on q and F only.

The accuracy of the bootstrap approach is also described in a theorem due to Falk and Reiss (1989) which concerns the weak convergence of the process Z_n defined by

$$Z_n(t) = c(t)n^{1/4}[T_n(F_n, t/n^{1/2}) - T_n(F, t/n^{1/2})]$$

where   (6.4.8)

$$c(t) = [(q(1 - q))/f(F^{-1}(q))]^{1/2}/\varphi[f(F^{-1}(q))t/(q(1 - q))^{1/2}]$$

and $\varphi = \Phi'$.

Theorem 6.4.1. *Assume that F is a continuous d.f. having a derivative f near $F^{-1}(q)$ which satisfies a local Lipschitz-condition of order $\delta > 1/2$ and that $f(F^{-1}(q)) > 0$. Then, Z_n weakly converges to a process Z defined by*

$$Z(t) = \begin{cases} B_1(-t) & \text{if} \quad t \le 0 \\ B_2(t) & \quad t > 0 \end{cases}$$

where B_1 and B_2 are independent standard Brownian motions on $[0, \infty)$.

We refer to Falk and Reiss (1989) for a detailed proof of Theorem 6.4.1 and for a definition of the weak convergence on the set of all right continuous functions on the real line having left-hand limits.

The basic idea of the proof is to examine the expressions in (6.4.3) and (6.4.5) conditioned on the sample q-quantile $F_n^{-1}(q)$. Notice that the r.v.'s $g_{n,t}$ only depend on order statistics smaller (larger) than $F_n^{-1}(q)$ if $t \le 0$ (if $t > 0$). Thus, it follows from Theorem 1.8.1 that, conditioned on $F_n^{-1}(q)$, the processes

$$(g_{n,t})_{t \le 0} \quad \text{and} \quad (g_{n,t})_{t > 0}$$

are conditionally independent. Theorem 6.4.1 reveals that we get the unconditioned independence in the limit.

The Maximum Deviation

Let $T_n^*(F_n, \cdot)$ be the normalized bootstrap d.f. as defined in (6.4.3). Denote by H_n the normalized d.f. of the maximum deviation of the bootstrap d.f. $T_n^*(F_n, t)$ from $T_n^*(\Phi, t)$ over $|t| \le 3$. More precisely, we have

$$H_n(s) = P\{n^{1/4} \max_{|t| \le 3} |T_n^*(F_n, t) - T_n^*(\Phi, t)| \le s\}. \tag{6.4.9}$$

We present a Monte Carlo result based on a sample of size $N = 5000$.

Figure 6.4.2 shows that the asymptotic result in Theorem 6.4.1 is of relevance for small and moderate sample sizes.

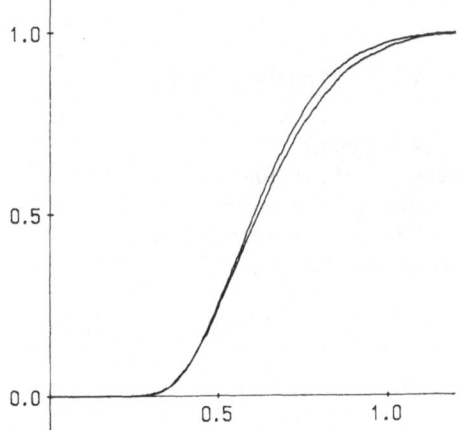

Figure 6.4.2. Normalized d.f. H_n of maximum bootstrap error for $q = 0.5$ and $n = 200$, 2000 with $H_{200} \ge H_{2000}$.

Confidence Bounds

Next, we consider the problem of setting two-sided confidence bounds for the unknown parameter $T(F)$. First, let us look at the problem from the point of view of a practitioner. One has to find a random variable $c_n(\alpha)$ such that

$$P_F\{|T(F_n) - T(F)| \le c_n(\alpha)\} \simeq 1 - \alpha. \tag{6.4.10}$$

The bootstrap solution is to take $c_n(\alpha)$ such that the bootstrap d.f. satisfies

$$T_n(F_n, c_n(\alpha)) - T_n(F_n, -c_n(\alpha)) \simeq 1 - \alpha.$$

The validity of (6.4.10) can be made plausible by the argument that uniformly over all t

$$P_F\{|T(F_n) - T(F)| \le t\} \simeq (T_n(F_n, t) - T_n(F_n, -t)).$$

This idea will be made rigorous in the particular case of the q-quantile via asymptotic considerations.

If F has a derivative f near $F^{-1}(q)$ and f is continuous at $F^{-1}(q)$ then we know that

$$P_F\{|F_n^{-1}(q) - F^{-1}(q)| \le u_n(q)\} \to 1 - \alpha, \qquad n \to \infty,$$

where $u_n(\alpha) = \Phi^{-1}(1 - \alpha/2)(q(1 - q)/n)^{1/2}/f(F^{-1}(q))$. Moreover, by using the fact that

$$\sup_t |T_n(F_n, t) - T_n(F, t)| \to 0, \qquad n \to \infty, \text{ w.p. } 1,$$

we obtain $c_n(\alpha)/u_n(\alpha) \to 1$, $n \to \infty$, w.p. 1. Hence, Slutzky's lemma yields (6.4.10).

For a continuation of this topic we refer to Section 8.4 where the smooth bootstrap is examined.

P.6. Problems and Supplements

1. (Gram–Charlier series of type A)

   Let φ denote the density of the standard normal d.f. Φ. The Chebyshev–Hermite polynomials $H_i = (-1)^i \varphi^{(i)}/\varphi$ are orthonormal w.r.t. the inner product $(h, g) = \int h(x)g(x)\varphi(x)\,dx$ (see Kendall and Stuart (1958), page 155). Write $H_i = \sum_{j=0}^i e_{j,i} x^j$. Denote by P_n the distribution and by $\mu_{n,j}$ the jth moment of

   $$\frac{n^{1/2} f(F^{-1}(q))}{(q(1 - q))^{1/2}} (X_{r(n):n} - F^{-1}(q)).$$

   Prove, under the conditions of Theorem 6.1.1, that

   $$\sup_B \left| P_n(B) - \int_B \varphi(x)\left(1 + \sum_{i=1}^{3(m-1)} (-1)^i \frac{c_{n,i}}{i!} H_i(x)\right) dx \right| = O(n^{-m/2})$$

   where $c_{n,i} = (-1)^i \sum_{j=0}^i e_{j,i} \mu_{n,j}$.

   (Reiss, 1974b)

2. Under the conditions of Theorem 6.1.2, with $q = 1/2, m = 3$ and odd sample sizes n,

   $$P\left\{X_{[n/2]+1:n} > F^{-1}(1/2) + \frac{1}{2f_0 n^{1/2}}\left(\lambda_\alpha - \frac{1}{4n^{1/2}}\frac{f_1}{f_0^2}\lambda_\alpha^2\right.\right.$$

   $$\left.\left. - \frac{1}{4n}\left(\lambda_\alpha + \lambda_\alpha^3\left(1 - \frac{f_1^2}{2f_0^4} + \frac{f_2}{6f_0^3}\right)\right)\right)\right\} = \alpha + O(n^{-3/2})$$

   where $\lambda_\alpha = \Phi^{-1}(1 - \alpha)$ and $f_i = f^{(i)}(F^{-1}(1/2))$.

   (F.N. David and N. L. Johnson, 1954)

3. Let $X_{i:n}$ be the ith order statistic of n i.i.d. exponential r.v.'s $\xi_1, \ldots, \xi_n$. Show that

$$\sum_{i=1}^{n} a_{i,n}(X_{i:n} - EX_{i:n}) \stackrel{\text{d}}{=} \sum_{i=1}^{n} b_{i,n}(\xi_i - 1)$$

where $b_{j,n} = (n - j + 1)^{-1} \sum_{i=j}^{n} a_{i,n}$. Moreover, with $\tau_n^2 = \sum_{i=1}^{n} b_{i,n}^2$, we have

$$P\left\{\tau_n^{-1} \sum_{i=1}^{n} a_{i,n}(X_{i:n} - EX_{i:n}) \le t\right\} \to \Phi(t), \qquad n \to \infty,$$

if, and only if,

$$\max_{j=1}^{n+1} \tau_n^{-1}|b_{j,n}| \to 0, \qquad n \to \infty.$$

4. Prove an expansion of length 2 in Lemma 6.2.3(ii).
5. Show that the accuracy of the bootstrap approximation can be improved by treating the standardized version

$$T_n(F_n, t/n^{1/2} g_{n,t}) - T_n(F, t/n^{1/2} f(F^{-1}(q))).$$

[Hint: Use (6.4.4).]

Bibliographical Notes

An approach related to that in Theorem 6.1.1 was adopted by Hodges and Lehmann (1967) for expanding the variance of the sample median (without rigorous proof). These investigations led to the famous paper by Hodges and Lehmann (1970) concerning the second order efficiency (deficiency).

Concerning limit theorems for moments of extremes we refer to Pickands (1968), Polfeldt (1970), Ramachandran (1984), and Resnick (1987).

Concerning linear combinations of order statistics we already mentioned the book of Helmers (1982). A survey of other approaches for deriving limit theorems for linear combinations of order statistics is given in the book of Serfling (1980). A more recent result concerning linear combinations of order statistics is due to van Zwet (1984): A representation as a symmetric statistics leads to a Berry–Esséen type theorem that is essentially equivalent to Theorem 6.2.6.

Limit laws for sums of extremes and intermediate order statistics have attained considerable attention in the last years. This problem is related to that of weak convergence of sums of i.i.d. random variables to a stable law (see Feller (1972)). Concerning weak laws we refer to the articles of M. Csörgő et al. (1986), S. Csörgő and D.M. Mason (1986), and S. Csörgő et al. (1986). A. Janssen (1988) proved a corresponding limit law w.r.t. the variational distance. An earlier notable article pertaining to this is that of Teugels (1981), among others.

Spacings and functions of spacings (understood in the greater generality of m-step spacings) are dealt with in several parts of the book as e.g. in the context of estimating the quantile density function. We did not make any attempt to cover this field to its full extent. For a comprehensive treatment of spacings see Pyke (1965, 1972). Several test statistics in nonparametric statistics are based on spacings. In the present context, the most interesting ones are perhaps those based on m-step spacings. For a survey of recent results we refer to the article of Jammalamadaka S. Rao and M. Kuo (1984). Interesting results concerning "systematic" statistic (including χ^2-test) are given by Miyamoto (1976).

A first improvement of Bahadur's original result in 1966 was achieved by Kiefer (1967), namely a law of the iterated logarithm analogue for the Bahadur approximation evaluated at a single point. Limit theorems like that stated in Section 6.3 are contained in the article of Kiefer (1969a). Further extensions concern (a) the weakening of conditions imposed on the underlying r.v.'s (see e.g. Sen, 1972) and (b) non-uniform bounds for the remainder term of the Bahadur approximation (e.g. Singh, 1979).

It was observed by Bickel and Freedman (1981) that bootstrapping leads to inconsistent estimators in case of extremes. An interesting recent survey of various techniques related to bootstrap was given by Beran (1985). We refer to Klenk and Stute (1987) for an application of the bootstrap method to linear combinations of order statistics.

CHAPTER 7

Approximations in the Multivariate Case

The title of this chapter should be regarded more as a program than as a description of the content (in view of the declared aims of this book).

In Section 7.1 we shall give an outline of the present state-of-the-art of the asymptotic treatment of multivariate central order statistics.

Contrary to the field of central order statistics a huge amount of literature exists concerning the asymptotic behavior of multivariate extremes. For an excellent treatment of this subject we refer to Galambos (1987) and Resnick (1987). In Section 7.2 we shall present some elementary results concerning the rate of convergence in the weak sense. Our interest will be focused on maxima where the marginals are asymptotically independent. As an example we shall compute the rate at which the marginal maxima of normal random vectors become independent.

7.1. Asymptotic Normality of Central Order Statistics

Throughout this section, we assume that $\xi_1, \xi_2, \xi_3, \ldots$ is a sequence of i.i.d. random vectors of dimension d with common d.f. F. Let $X_{r:n}^{(j)}$ be the rth order statistic in the jth component as defined in (2.1.4).

For $j = 1, \ldots, d$, let $I(j) \subset \{1, \ldots, n\}$. If F statisfies some mild regularity conditions then it is plausible that a collection of order statistics

$$X_{r(j):n}^{(j)}, \qquad j = 1, \ldots, d, r(j) \in I(j) \tag{7.1.1}$$

is jointly asymptotically normal if for each $j = 1, \ldots, d$ the order statistics

$$X_{r(j):n}^{(j)}, \qquad r(j) \in I(j), \tag{7.1.2}$$

have this property. We do not know whether this idea can be made rigorous, though.

The asymptotic normality of order statistics can be proved via the device of Section 2.1, namely, to represent the d.f. of order statistics as the d.f. of a sum of i.i.d. random vectors. To simplify the writing let us study the 2-dimensional case. According to Section 2.1 we have

$$P\{X^{(1)}_{r(1,n):n} \le t_{1,n}, X^{(2)}_{r(2,n):n} \le t_{2,n}\}$$

$$= P\left\{\sum_{i=1}^{n} (1_{(-\infty,t_{1,n}]}(\xi_{i,1}), 1_{(-\infty,t_{2,n}]}(\xi_{i,2})) \ge \mathbf{r}(n)\right\} \tag{7.1.3}$$

where $\xi_i = (\xi_{i,1}, \xi_{i,2})$ and $\mathbf{r}_n = (r(1,n), r(2,n))$. On the right-hand side we are given the distribution of a sum of i.i.d. random vectors whence the multidimensional central limit theorem is applicable.

Let $0 < q_1, q_2 < 1$ be fixed and assume that

$$n^{1/2}(r(i,n)/n - q_i) \to 0, \qquad n \to \infty, i = 1, 2. \tag{7.1.4}$$

According to the univariate case the appropriate choice of constants $\mathbf{t}_n = (t_{1,n}, t_{2,n})$ is

$$t_{i,n} = F_i^{-1}(q_i) + x_i/n^{1/2} f_i, \qquad i = 1, 2, \tag{7.1.5}$$

where F_i is the ith marginal d.f. of F and $f_i = F_i'(F_i^{-1}(q_i))$. Let us rewrite the right-hand side of (7.1.3) by

$$P\left\{n^{-1/2}\sum_{i=1}^{n}\boldsymbol{\eta}_{i,n} \le \mathbf{x}_n\right\} \tag{7.1.6}$$

where the random vectors $\boldsymbol{\eta}_{i,n}$ are given by

$$\boldsymbol{\eta}_{i,n} = -[(1_{(-\infty,t_{1,n}]}(\xi_{i,1}), 1_{(-\infty,t_{2,n}]}(\xi_{i,2})) - (F_1(t_{1,n}), F_2(t_{2,n}))], \tag{7.1.7}$$

and

$$\mathbf{x}_n = (x_{1,n}, x_{2,n}) = n^{-1/2}[n(F_1(t_{1,n}), F_2(t_{2,n})) - \mathbf{r}(n)]. \tag{7.1.8}$$

Obviously, $\boldsymbol{\eta}_{i,n}$, $i = 1, 2, \ldots, n$, are bounded i.i.d. random vectors with mean vector zero and covariance matrix $\Sigma_n = (\sigma_{i,j,n})$ given by

$$\sigma_{i,i,n} = F_i(t_{i,n})(1 - F_i(t_{i,n})), \qquad i = 1, 2$$

and   (7.1.9)

$$\sigma_{1,2,n} = F(\mathbf{t}_n) - F_1(t_{1,n})F_2(t_{2,n}).$$

Theorem 7.1.1. *Assume that F is continuous at the point $(F_1^{-1}(q_1), F_2^{-1}(q_2))$. Moreover, for $i = 1, 2$, let F_i be differentiable at $F_i^{-1}(q_i)$ with $f_i = F_i'(F_i^{-1}(q_i)) > 0$. Define $\Sigma = (\sigma_{i,j})$ by*

$$\sigma_{i,i} = q_i(1 - q_i), \qquad i = 1, 2,$$

and   (7.1.10)

$$\sigma_{1,2} = F(F_1^{-1}(q_1), F_2^{-1}(q_2)) - q_1 q_2.$$

If $\det(\Sigma) \neq 0$ *and condition (7.1.4) holds then for every* (x_1, x_2):

$$P\{n^{1/2} f_i(X^{(i)}_{r(i,n):n} - F_i^{-1}(q_i)) \leq x_i, i = 1, 2\} \rightarrow \Phi_\Sigma(x_1, x_2), \qquad n \rightarrow \infty, \quad (7.1.11)$$

where Φ_Σ *is the bivariate normal d.f. with mean vector zero and covariance matrix* Σ.

PROOF. Let Σ_n and $\boldsymbol{\eta}_{i,n}$ be as in (7.1.9) and (7.1.7). Since $\Sigma_n \rightarrow \Sigma$, $n \rightarrow \infty$, we may assume w.l.g. that $\det(\Sigma_n) \neq 0$. Let T_n be a matrix such that $T_n^2 = \Sigma_n^{-1}$ [compare with Bhattacharya and Rao (1976), (16.3), and (16.4)]. Then, according to a Berry–Esséen type theorem (see Bhattacharya and Rao (1976), Corollary 18.3) we get

$$\sup_z \left| P\left\{ n^{-1/2} \sum_{i=1}^n \boldsymbol{\eta}_{i,n} \leq \mathbf{z} \right\} - \Phi_{\Sigma_n}(\mathbf{z}) \right| \leq cn^{-1/2} E\|T_n \boldsymbol{\eta}_{1,n}\|_2^3 = O(n^{-1/2})$$

$$(7.1.12)$$

for some constant $c > 0$. Here $\|\ \|_2$ denotes the Euclidean norm.

The differentiability of F_i at $F_i^{-1}(q_i)$ and condition (7.1.4) yield that $x_{i,n} \rightarrow x_i$, $n \rightarrow \infty$, and hence

$$\Phi_{\Sigma_n}(x_{1,n}, x_{2,n}) \rightarrow \Phi_\Sigma(x_1, x_2), \qquad n \rightarrow \infty. \quad (7.1.13)$$

Combining (7.1.3), (7.1.6), (7.1.12), and (7.1.13) we obtain (7.1.11). $\qquad\square$

The error rates in (7.1.11) can easily be computed under slightly stronger regularity conditions imposed on F.

The condition $\det(\Sigma) \neq 0$ is rather a mild one. If $\xi_i = (\xi_i, \xi_i)$ are random vectors having the same r.v. in both components then $\det(\Sigma) = 0$ if $q_1 = q_2$ and $\det(\Sigma) \neq 0$ if $q_1 \neq q_2$. It is clear that the two procedures of taking two order statistics $X_{r:n}, X_{s:n}$ according to $\xi_1, \ldots, \xi_n$ or order statistics $X^{(1)}_{r:n}, X^{(2)}_{s:n}$ according to $\xi_1, \ldots, \xi_n$ are identical. Thus, the situation of Section 4.5 can be regarded as a special case of the multivariate one.

Next we give a straightforward generalization of Theorem 7.1.1 to the case $d \geq 3$. We take one order statistic $X^{(i)}_{r(i,n):n}$ out of each of the d components.

Theorem 7.1.2. *Let* $\xi_1, \xi_2, \ldots$ *be a sequence of d-variate i.i.d. random vectors with common d.f. F. Denote by* F_i *and* $F_{i,j}$ *the univariate and bivariate marginal d.f.'s of F. Let* $0 < q_i < 1$ *for* $i = 1, \ldots, d$. *Assume that* $F_{i,j}$ *is continuous at the point* $(F_i^{-1}(q_i), F_j^{-1}(q_j))$ *for* $i, j = 1, \ldots, d$. *Moreover, for* $i = 1, \ldots, d$, *let* F_i *be differentiable at* $F_i^{-1}(q_i)$ *with* $f_i = F_i'(F_i^{-1}(q_i)) > 0$. *Assume that*

$$n^{1/2}(r(i,n)/n - q_i) \rightarrow 0, \qquad n \rightarrow \infty, i = 1, \ldots, d. \quad (7.1.14)$$

Define $\Sigma = (\sigma_{i,j})$ *by*

$$\sigma_{i,i} = q_i(1 - q_i), \qquad i = 1, \ldots, d,$$

and $\qquad\qquad\qquad\qquad\qquad\qquad\qquad\qquad\qquad\qquad\qquad\qquad\qquad\qquad\qquad$ (7.1.15)

$$\sigma_{i,j} = F_{i,j}(F_i^{-1}(q_i), F_j^{-1}(q_j)) - q_i q_j, \qquad i \neq j.$$

If $\det(\Sigma) \neq 0$, *then for every* $\mathbf{x} = (x_1, \ldots, x_d)$,

$$P\{n^{1/2} f_i[X^{(i)}_{r(i,n):n} - F_i^{-1}(q_i)] \leq x_i, i = 1, \ldots, d\} \to \Phi_{\Sigma}(\mathbf{x}), \qquad n \to \infty, \quad (7.1.16)$$

where Φ_{Σ} *is the d-variate normal d.f. with mean vector zero and covariance matrix* Σ.

7.2. Multivariate Extremes

In this section, we shall deal exclusively with maxima of d-variate i.i.d. random vectors $\xi_{1,n}, \ldots, \xi_{n,n}$ with common d.f. F_n. It is assumed that F_n has identical univariate marginals $F_{n,i}$. Thus,

$$F_{n,1} = \cdots = F_{n,d}.$$

It will be convenient to denote the d-variate maximum by

$$M_n = (M_{n,1}, \ldots, M_{n,d})$$

where $M_{n,1}, \ldots, M_{n,d}$ are the identically distributed univariate marginal maxima (compare with (2.1.8)) with common d.f. $F_{n,1}^n$. Recall that F_n^n is the d.f. of M_n.

Weak Convergence

The weak convergence is again the pointwise convergence of d.f.'s if the limiting d.f. is continuous which will always be assumed in this section. The weak convergence of d-variate d.f.'s implies the weak convergence of the univariate marginal d.f.'s (since the projections are continuous). In particular, if F_n^n weakly converges to G_0 then the univariate marginal d.f.'s $F_{n,1}^n$ also converge weakly to the univariate marginal $G_{0,1}$ of G_0. Notice that G_0 also has identical univariate marginals. If $G_{0,1}$ is nondegenerate then the results of Chapter 5 already give some insight into the present problem.

Recall from Section 2.2 that the d-variate d.f.'s $\mathbf{x} \to \prod_{i=1}^d G_{0,1}(x_i)$ and $\mathbf{x} \to G_{0,1}(\min(x_1, \ldots, x_d))$ represent the case of independence and complete dependence.

Lemma 7.2.1. *Assume that the univariate marginals* $F_{n,1}^n$ *converge pointwise to the d.f.* $G_{0,1}$.
(i) *Then, for every* $\mathbf{x}$,

$$\prod_{i=1}^d G_{0,1}(x_i) \leq \liminf_n F_n^n(\mathbf{x}) \leq \limsup_n F_n^n(\mathbf{x}) \leq G_{0,1}(\min(x_1, \ldots, x_d)).$$

(ii) *If* F_n^n *converges pointwise to some right continuous function* G *then* G *is a d.f.*

PROOF. Ad (i): Check that $F_n^n(\mathbf{x}) \leq F_{n,1}^n(\min(x_1, \ldots, x_d))$. Now, the upper bound is obvious.

Secondly, Bonferroni's inequality (see P.2.5(iv)) yields

$$F_n^n(\mathbf{x}) \geq \exp\left[-\sum_{j=1}^d n(1 - F_{n,1}(x_j)) \right] + o(1)$$

$$= \prod_{j=1}^d \exp[-n(1 - F_{n,1}(x_j))] + o(1) = \prod_{j=1}^d G_{0,1}(x_j) + o(1).$$

Therefore, the lower bound also holds.

Ad (ii): Use (i) to prove that G is a normed function. Moreover, the pointwise convergence of F_n^n to G implies that G is Δ-monotone (see (2.2.19)). ☐

It is immediate from Lemma 7.2.1 that max-stable d.f.'s G_0 have the property

$$\prod_{i=1}^d G_{0,1}(x_i) \leq G_0(\mathbf{x}) \leq G_{0,1}(\min(x_1,\ldots,x_d)). \qquad (7.2.1)$$

Let $\boldsymbol{\xi} = (\xi_1,\ldots,\xi_d)$ be a random vector with d.f. F. Recall from P.2.5 that for some universal constant $C > 0$,

$$\sup_{\mathbf{t}} \left| F^n(\mathbf{t}) - \exp\left(\sum_{j=1}^d (-1)^j n h_j(\mathbf{t}) \right) \right| \leq C n^{-1} \qquad (7.2.2)$$

where

$$h_j(\mathbf{t}) = \sum_{1 \leq i_1 < \cdots < i_j \leq d} P\{\xi_{i_1} > t_{i_1},\ldots,\xi_{i_j} > t_{i_j}\}, \qquad j = 1,\ldots,d. \quad (7.2.3)$$

Combining Lemma 7.2.1 and (7.2.2) we obtain

Corollary 7.2.2. *Let $\boldsymbol{\xi}_n$ be a d-variate random vector with d.f. F_n. Define $h_{n,j}$ in analogy to h_j in (7.2.3) with $\boldsymbol{\xi}$ replaced by $\boldsymbol{\xi}_n$. Suppose that the univariate marginals $F_{n,1}^n$ converge pointwise to a d.f. Moreover, for every $j = 1,\ldots,d$,*

$$n h_{n,j} \to h_{0,j}, \qquad n \to \infty, \text{ pointwise,}$$

where $h_{0,j}$, $j = 1,\ldots,d$, are right continuous functions. Then,

(i)
$$G_0 = \exp\left(\sum_{j=1}^d (-1)^j h_{0,j} \right) \qquad \text{is a d.f.,}$$

and

(ii)
$$F_n^n(\mathbf{x}) \to G_0(\mathbf{x}), \qquad n \to \infty, \text{ for every } \mathbf{x}.$$

The formulation of Lemma 7.2.1 and Corollary 7.2.2 is influenced by a recent result due to Hüsler and Reiss (1989) where maxima under multivariate normal vectors, with correlation coefficients $\rho(n)$ tending to 1 as $n \to \infty$, are studied. In the bivariate case the following result holds: If

$$(1 - \rho(n))\log n \to \lambda^2, \qquad n \to \infty,$$

then the normalized distributions of maxima weakly converge to a d.f. H_λ defined by

$$H_\lambda(x, y) = \exp\left[-\Phi\left(\lambda + \frac{x - y}{2\lambda}\right)e^{-y} - \Phi\left(\lambda + \frac{y - x}{2\lambda}\right)e^{-x}\right] \quad (7.2.4)$$

with

$$H_0 = \lim_{\lambda \downarrow 0} H_\lambda \quad \text{and} \quad H_\infty = \lim_{\lambda \uparrow \infty} H_\lambda.$$

If $\lambda = 0$, the marginal maxima are asymptotically completely dependent; if $\lambda = \infty$, we have asymptotic independence. Notice that H_λ is max-stable and thus belongs to the usual class of multivariate extreme value d.f.'s.

Next (7.2.2) will be specialized to the bivariate case. Let (ξ_n, η_n) be a random vector with d.f. F_n. The identical marginal d.f.'s are again denoted by $F_{n,1}$ and $F_{n,2}$. According to (7.2.2),

$$\sup_{(x,y)} |F_n^n(x, y) - \exp(-n(1 - F_{n,1}(x)) - n(1 - F_{n,1}(y)) + nL_n(x, y))| \leq Cn^{-1}$$

$$(7.2.5)$$

where

$$L_n(x, y) = P\{\xi_n > x, \eta_n > y\}$$

is the bivariate survivor function. Assume that

$$F_{n,1}^n(x) \to G_{0,1}(x), \qquad n \to \infty, \tag{7.2.6}$$

for every x, where $G_{0,1}$ is a d.f. Then,

$$F_n^n(x, y) = \exp[-n(1 - F_{n,1}(x)) - n(1 - F_{n,1}(y)) + nL_n(x, y)] + O(n^{-1})$$

$$= G_{0,1}(x)G_{0,1}(y)\exp[nL_n(x, y)] + o(1). \tag{7.2.7}$$

Therefore, the asymptotic behavior of the bivariate survivor function is decisive for the asymptotic behavior of the bivariate maximum. The convergence rate in the univariate case and the convergence rate of the survivor functions determine the convergence rate for the bivariate maxima.

Asymptotic (Quadrant-) Independence

We discuss the particular situation where the term $nL_n(x, y)$ in (7.2.7) goes to zero as $n \to \infty$. The following result is a trivial consequence of (7.2.7).

Lemma 7.2.3. *Assume that (7.2.6) holds. For every (x, y) with $G_{0,1}(x)G_{0,1}(y) > 0$ the following equivalence holds*:

$$F_n^n(x, y) \to G_{0,1}(x)G_{0,1}(y), \qquad n \to \infty,$$

if, and only if,

$$nL_n(x, y) \to 0, \qquad n \to \infty. \tag{7.2.8}$$

Thus under condition (7.2.8) the marginal maxima $M_{n,1}$ and $M_{n,2}$ are asymptotically independent in the sense that $(M_{n,1}, M_{n,2})$ converge in distribution to a random vector with independent marginals.

Corollary 7.2.4. *Let ξ and η be r.v.'s with common d.f. F such that $F^n(b_n + a_n \cdot) \to$ G weakly.*

Then, the pertaining normalized maxima $a_n^{-1}(M_{n,i} - b_n)$, $i = 1, 2$, are asymptotically independent if

$$\lim_{x \uparrow \omega(F)} P(\xi > x | \eta > x) = 0. \tag{7.2.9}$$

PROOF. Notice that $(b_n + a_n x) \uparrow \omega(F)$ and $n(1 - F(b_n + a_n x)) \to -\log G(x)$, $n \to \infty$, for $\alpha(G_0) < x < \omega(G_0)$ and hence the assertion is immediate from Lemma 7.2.3 applied to $\xi_n = a_n^{-1}(\xi - b_n)$ and $\eta_n = a_n^{-1}(\eta - b_n)$. □

It is well known that (7.2.9) is also necessary for the asymptotic independence. Moreover, Corollary 7.2.4 can easily be extended to the d-variate case (see Galambos (1987), page 301, and Resnick (1987), Proposition 5.27).

Next, Lemma 7.2.3 will be applied to prove that, for multivariate extremes, the asymptotic pairwise independence of the marginal maxima implies asymptotic independence.

Theorem 7.2.5. *Assume that $(M_{n,1}, \ldots, M_{n,d})$ converge in distribution to a d-variate random vector with d.f. G_0. Then, the asymptotic pairwise independence of the marginal maxima implies the asymptotic independence.*

PROOF. The Bonferroni inequality (see P.2.4 and P.2.5) implies that

$$P\{M_n \le \mathbf{x}\}$$

$$\begin{cases} \le \exp\left(-\sum_{i=1}^{d} n(1 - F_{n,1}(x_i)) + \sum_{1 \le i < j \le d} nL_{n,i,j}(x_i, x_j)\right) \\ \ge \exp\left(-\sum_{i=1}^{d} n(1 - F_{n,1}(x_i))\right) + o(1) \end{cases}$$

$$\begin{cases} \le \left(\prod_{i=1}^{d} G_{0,1}(x_i)\right) \exp\left(\sum_{1 \le i < j \le d} nL_{n,i,j}(x_i, x_j)\right) + o(1) \\ \ge \left(\prod_{i=1}^{d} G_{0,1}(x_i)\right) + o(1) \end{cases}$$

where

$$L_{n,i,j}(x_i, x_j) = P\{\xi_i > x_i, \xi_j > x_j\}.$$

It remains to prove that

$$\exp\left(\sum_{1 \le i < j \le d} nL_{n,i,j}(x_i, x_j)\right) \to 1, \qquad n \to \infty,$$

for every **x** with $\prod_{i=1}^{d} G_{0,1}(x_i) > 0$. This, however, is obvious from the fact that according to Lemma 7.2.3 the pairwise independence implies

$$nL_{n,i,j}(x_i, x_j) \to 0, \qquad n \to \infty,$$

for every $1 \leq i < j \leq d$.   □

As an immediate consequence of Theorem 7.2.5 one gets

Theorem 7.2.6. *Let* $\xi = (\xi_1, \ldots, \xi_d)$ *have a max-stable d.f. Then, the pairwise indepencence of* $\xi_1, \ldots, \xi_d$ *implies the independence.*

In fact a much stronger result holds as pointed out to me by J. Hüsler. If the r.v.'s $\xi_1, \ldots, \xi_d$ are uncorrelated and jointly have a max-stable d.f. then they are mutually independent (see P.7.2).

Rates for the Distance from Independence

For notational convenience we shall only study the bivariate case. From (7.2.5) and by noting that

$$\sup_x |F_{n,1}^n(x) - \exp(-n(1 - F_{n,1}(x)))| \leq Cn^{-1} \qquad (7.2.10)$$

(compare with P.2.5(ii)) we get

$$F_n^n(x, y) = F_{n,1}^n(x)F_{n,1}^n(y)\exp[nL(x, y)] + O(n^{-1}). \qquad (7.2.11)$$

From (7.2.11) we see that the term $nL_n(x, y)$ determines the rate at which the independence of the marginal maxima is attained.

It is apparent from the proof of Theorem 7.2.5 that (7.2.11) can easily be extended to the case $d \geq 2$.

Next (7.2.11) will be specialized to bivariate normal vectors. It was observed by Sibuya (1960) that the marginal maxima of i.i.d. normal random vectors are asymptotically independent. In the following example we shall calculate the rate at which the marginal maxima become quadrant-independent.

EXAMPLE 7.2.7. Let F be the d.f. of a normal vector (ξ, η) where ξ and η are standard normal r.v.'s. Let ρ denote the covariance of ξ and η where $-1 < \rho < 1$. Put $u_n(x) = b_n + b_n^{-1}x$ where again $b_n = n\varphi(b_n)$. Then, for every x, y,

$$F^n(u_n(x), u_n(y))$$
$$= \Phi^n(u_n(x))\Phi^n(u_n(y))[1 + O(n^{-(1-\rho)/(1+\rho)}(\log n)^{-\rho/(1+\rho)})] + O(n^{-1}). \qquad (7.2.12)$$

According to (7.2.11) we have to prove that

$$nL_n(x, y) = nL(u_n(x), u_n(y)) = O(n^{-(1-\rho)/(1+\rho)}(\log n)^{-\rho/(1+\rho)}).$$

It is well known that the normal distribution $N_{(\rho z, 1-\rho^2)}$ is the conditional distribution of ξ given $\eta = z$. Thus,

$$nL(u_n(x), u_n(y)) = n \int_{u_n(y)}^{\infty} (1 - N_{(\rho z, 1-\rho^2)}(-\infty, u_n(x)]) \varphi(z) \, dz$$

$$= \int_y^{\infty} (1 - \Phi[(u_n(x) - \rho u_n(z))/(1-\rho^2)^{1/2}]) \exp[-(z + z^2/b_n^2)] \, dz$$

$$= O(b_n^{-1} \varphi(b_n)^{(1-\rho)/(1+\rho)})$$

where the final step is carried out by using the inequality $1 - \Phi(x) \leq \varphi(x)/x$, $x > 0$. We remark that for $\rho > 0$ the integration over z with $y \leq (u_n(x) - \rho u_n(z))/(1-\rho^2)^{1/2} \leq b_n$ has to be dealt with separately. Since $b_n = O((\log n)^{1/2})$ the proof can easily be completed.

Final Remarks

If one confines the attention to asymptotically independent r.v.'s then it is natural to replace, in a first step, the original marginal r.v.'s by some independent versions. The calculation of an upper bound of the Hellinger distance between the distribution of a multivariate maximum and the joint distribution of the independent versions of the marginals is an open problem. In a second step one could apply Lemma 3.3.10 and the results of Section 5.2 to obtain an upper bound of the Hellinger distance between the original distribution and a limit distribution.

If we analyze the density of the normalized bivariate maximum, with normalizing constants $a_n > 0$ and b_n, in the form as given in (2.2.8), we find that the decisive condition for the asymptotic independence, in the strong sense, is that the conditional d.f.'s $F_1(b_n + a_n x | b_n + a_n y)$ and $F_2(b_n + a_n y | b_n + a_n x)$ converge to 1 as $n \to \infty$. Recall that the related condition (7.2.9) yields the asymptotic independence in the weak sense.

In case of asymptotic independence the statistical results in the univariate case carry over to the multivariate case. If the marginals are asymptotically dependent then new statistical problems have to be solved (see e.g. P.2.11 and Bibliographical Notes).

P.7. Problems and Supplements

1. Denote by $M_{r,s,n}$ the number of random vectors ξ_i in the random quadrant $(-\infty, X_{r:n}^{(1)}) \times (-\infty, X_{s:n}^{(2)})$. Under the conditions of Theorem 7.1.1 the random vectors $(M_{r(n),s(n),n}, X_{r(n):n}^{(1)}, X_{s(n):n}^{(2)})$ are asymptotically normal.

(Siddiqui, 1960)

2. (i) Let $\xi = (\xi_1, \ldots, \xi_d)$ have a max-stable d.f. Then,

$$\xi_1, \ldots, \xi_d \text{ are associated,}$$

that is, $\text{cov}(g(\xi), f(\xi)) \geq 0$ for all componentwise nondecreasing, real-valued functions f, g, whenever the relevant expectations exist.

(Marshall and Olkin, 1983; for an extension see Resnick, 1987)

(ii) If $\xi_1, \ldots, \xi_d$ are associated and uncorrelated then they are mutually independent.

(Joag-Dev, 1983)

Bibliographical Notes

Under slightly stronger conditions than those stated in Theorems 7.1.1 and 7.1.2, Weiss (1964) proved the asymptotic normality of the d.f.'s of multivariate central order statistics. The proof is based on the normal approximation of the multinomial distribution.

The asymptotic normality of multivariate central order statistics was already proved by Mood (1941), in the special case of sample medians, and by Siddiqui (1960). In both articles the exact densities of the order statistics are computed. By using the normal approximation of the multinomial distribution it is then shown that the densities converge pointwise to the normal density. Thus, according to the Schéffe lemma, one also gets the convergence in the variational distance. Kuan and Ali (1960) verified the joint asymptotic normality of multivariate order statistics, including the case where several order statistics are taken from each component. It is evident that such ordered values define a grid in the Euclidean d-space. The frequencies of sample points in the cells of the grid define further r.v.'s. Weiss (1982) proved the joint asymptotic normality of multivariate order statistics and such associated cell frequencies.

The research work on multivariate maxima of i.i.d. random vectors started with the articles of J. Tiago de Oliveira (1958), J. Geffroy (1958/59), and M. Sibuya (1960). In literature, further reference is given to Finkelstein (1953). From the beginning much attention was focused on the case where the marginal maxima are asymptotically independent. It was observed by S.M. Berman (1961) that for the components of an extreme value vector the independence is equivalent to the pairwise independence. In this context one also has to note that the marginal maxima are asymptotically, mutually (quadrant-) independent when, and only when, this is true for each pair of marginal maxima [see e.g. Galambos (1987, Corollary 5.3.1) or Resnick (1987, Proposition 5.27)].

If measurements of a certain phenomenon are made at places close together then there will be a certain dependence between the observations which, in the present context, are supposed to be maxima. From the results of Section 2.2 it is apparent that the family of max-stable distributions is large enough to serve as a model for this situation. One may argue that this model is even so large that the problem has to be tackled of finding a smaller nonparametric or a parametric model. If one has some knowledge of the mechanism underlying the maxima, then, speaking in mathematical terms, a limit theorem for maxima will single out certain max-stable distributions.

However, one has to face the difficulty of finding "attractive" multivariate distributions under which the asymptotic distribution of maxima reflects the dependence of the observed marginal maxima. In this context, the result obtained by Hüsler and Reiss (1989) (see (7.2.4)) looks promising: Distributions of bivariate maxima are studied under normal distributions where the correlation coefficient $\rho(n)$ varies as the sample size increases. In the limit one obtains a family of max-stable distributions describing situations between independence and complete dependence.

We refer to Tiago de Oliveira (1984) for a review of parametric submodels of bivariate max-stable d.f.'s. The nonparametric approach of Pickands for estimating the dependence function (see P.2.11) has been pursued further by Smith (1985b) by introducing the smoothing technique to multivariate extreme value theory. This work has been continued in Smith et al. (1987) where the kernel method is applied to the estimation of max-stable d.f.'s.

STATISTICAL MODELS
AND PROCEDURES

CHAPTER 8

Evaluating the Quantile and Density Quantile Function

In this chapter

(a) we start with the "pure" nonparametric, statistical model,
(b) introduce smoothness conditions.

In Chapter 9 this discussion will be continued by studying

(c) semi-parametric models,
(d) parametric extreme value models.

As pointed out in the Introduction the sample q.f. F_n^{-1} is the natural estimator of the underlying q.f. F^{-1}. In Section 8.1 some results will be collected which concern the statistical performance of sample quantiles. It will be shown, in particular, that statistical procedures built on sample quantiles are optimal if the model is large enough.

Given the information that the unknown q.f. F^{-1} is a smooth function one should not use step functions generated by the sample q.f. as estimates of the q.f. Consequently, two different classes of kernel type estimators will be introduced in Section 8.2.

The first class of estimators is obtained by smoothing the sample q.f. by means of a kernel. The second class of estimators is established in analogy of the construction of the sample q.f. as the "inverse" of the sample d.f.: Take the "inverse" of the kernel type estimator of the d.f. Derivatives of the kernel type estimates of the q.f. will be appropriate estimates of the density quantile function $(F^{-1})' = 1/f(F^{-1})$ where f is the density of F.

8.1. Sample Quantiles

In this section we shall primarily study results for a fixed sample size. The statistical procedures for evaluating the unknown q-quantile will be optimal if the underlying model is large enough. The test, estimation, and confidence procedures have to be randomized to satisfy the usual requirements in an exact way (e.g. attainment of a level or median unbiasedness).

One-Sided Test of Quantiles

Let $X_{i:n}$ be the ith order statistic of n i.i.d. random variables $\xi_1, \xi_2, \ldots, \xi_n$ with common continuous d.f. F. A basic problem is to test the null-hypothesis

$$F^{-1}(q) \leq u \text{ against } F^{-1}(q) > u.$$

We shall briefly summarize some well-known facts concerning tests based on sample quantiles.

Given $\alpha, q \in (0, 1)$ and a positive integer n, let $r(\alpha) \equiv r(\alpha, q, n)$ be the largest integer $r \in \{0, \ldots, n\}$ such that

$$\sum_{j=0}^{r-1} \binom{n}{j} q^j (1 - q)^{n-j} \leq \alpha. \tag{8.1.1}$$

Notice that the left-hand side of (8.1.1) is equal to $P\{X_{r:n} > F^{-1}(q)\}$.

Keep in mind that $r(\alpha)$ also depends on q and n. Put $X_{0:n} = -\infty$ and $X_{n+1:n} = \infty$.

It is clear that $\{X_{r(\alpha):n} > u\}$ is a critical region of level α for testing

$$F^{-1}(q) \leq u \text{ against } F^{-1}(q) > u,$$

however, the level α will not be attained on the null-hypothesis except in those cases where equality holds in (8.1.1).

To define a test which is similar on $\{F: F^{-1}(q) = u\}$ we introduce a randomized test procedure based on two order statistics. Define the critical function φ by

$$\varphi = \begin{cases} 1 & X_{r(\alpha):n} > u \\ \gamma(\alpha) & \text{if } X_{r(\alpha):n} \leq u, X_{r(\alpha)+1:n} > u \\ 0 & X_{r(\alpha)+1:n} \leq u \end{cases} \tag{8.1.2}$$

where $\gamma(\alpha) \equiv \gamma(\alpha, q, n)$ is the unique solution of the equation

$$\sum_{j=0}^{r(\alpha)-1} \binom{n}{j} q^j (1 - q)^{n-j} + \gamma \binom{n}{r(\alpha)} q^{r(\alpha)} (1 - q)^{n-r(\alpha)} = \alpha \tag{8.1.3}$$

with $0 \leq \gamma < 1$.

Simple calculations show that the left-hand side of (8.1.3) is equal to

$$E_F \varphi = P\{X_{r(\alpha):n} > u\} + \gamma P\{X_{r(\alpha):n} \leq u, X_{r(\alpha)+1:n} > u\}.$$

We have

$$E_F \varphi \begin{array}{l} \leq \alpha \\ = \alpha \end{array} \text{ if } \begin{array}{l} F^{-1}(q) \leq u \\ F^{-1}(q) = u. \end{array} \tag{8.1.4}$$

Moreover,

$$E_F\varphi = \alpha \quad \text{if} \quad F(u) = q.$$

The critical function φ as defined in (8.1.2) is uniformly most powerful for the testing $F^{-1}(q) \le u$ against $F^{-1}(q) > u$. To prove this consider the simple testing problem F_0 against F_1 where F_1 is a d.f. with $0 < q_1 := F_1(u) < q$; notice that $F_1(u) < q$ is equivalent to $F_1^{-1}(q) > u$. Define the d.f. F_0 via f_0 by $F_0(t) = \int_{-\infty}^{t} f_0 \, dF_1$ where

$$f_0 = \frac{q}{q_1} 1_{(-\infty, u]} + \frac{1-q}{1-q_1} 1_{(u, \infty)}. \tag{8.1.5}$$

Denote by Q_i the probability measures belonging to F_i. Then, f_0 is the Q_1-density of Q_0. Easy calculations show that $F_0(u) = q$ and hence $F_0^{-1}(q) \le u$. It will turn out that F_0 is a "least-favorable" null-hypothesis.

Lemma 8.1.1. φ *as defined in (8.1.2) is a most powerful, critical function of level* α *for testing* F_0 *against* F_1.

PROOF. In view of the Fundamental Lemma of Neyman and Pearson it suffices to prove that

$$\varphi = \begin{matrix} 1 \\ 0 \end{matrix} \quad \text{iff} \quad 1 \begin{matrix} > \\ < \end{matrix} c \prod_{i=1}^{n} f_0(\xi_i)$$

for some $c > 0$. Put $S_n = \sum_{i=1}^{n} 1_{(-\infty, u]}(\xi_i)$.
  We have

$$\prod_{i=1}^{n} f_0(\xi_i) = \left(\frac{q(1-q_1)}{q_1(1-q)} \right)^{S_n} \left(\frac{1-q}{1-q_1} \right)^n$$

and hence

$$1 \begin{matrix} > \\ < \end{matrix} c \prod_{i=1}^{n} f_0(\xi_i) \quad \text{iff} \quad r(\alpha) \begin{matrix} > \\ < \end{matrix} S_n$$

where $c > 0$ is defined by the equation

$$r(\alpha) = \left[-\log c - n \log \frac{1-q}{1-q_1} \right] \Big/ \log \frac{q(1-q_1)}{q_1(1-q)}.$$

From (1.1.8) we know that

$$r(\alpha) \begin{matrix} > \\ < \end{matrix} S_n \quad \text{iff} \quad \begin{matrix} X_{r(\alpha):n} > u \\ X_{r(\alpha)+1:n} \le u \end{matrix}$$

and hence φ is of the desired form. $\square$

Corollary 8.1.2. *The critical function* φ_n *defined in (8.1.2) is uniformly most powerful of level* α *for testing* $F^{-1}(q) \le u$ *against* $F^{-1}(q) > u$.

PROOF. Obvious from (8.1.4) and Lemma 8.1.1 since the d.f. F_0 defined in Lemma 8.1.1 is continuous.   □

For $k = 1, 2, 3, \ldots$ or $k = \infty$ we define $\mathscr{F}_k$ as the family of all d.f.'s F which possess the following properties:

(i) F has a (Lebesgue) density f,
(ii) $f > 0$ on $(\alpha(F), \omega(F))$,   (8.1.6)
(iii) f has k bounded derivatives on $(\alpha(F), \omega(F))$.

The crucial point of the conditions above is that the derivatives above need not be uniformly bounded over the given model.

Lemma 8.1.3. *Let $k = 1, 2, \ldots$ or $k = \infty$ be fixed.*
Then, φ as defined in (8.1.2) is a uniformly most powerful critical function of level α for testing $F^{-1}(q) \leq u$ against $F^{-1}(q) > u$ with $F \in \mathscr{F}_k$.

PROOF. Notice that F_0 (see the line before (8.1.5)) does not belong to $\mathscr{F}_k$. If f_1 is the density of $F_1 \in \mathscr{F}_k$ then F_0 has the density

$$f_0 = f_1 \left(\frac{q}{q_1} 1_{(-\infty, u]} + \frac{1-q}{1-q_1} 1_{(u, \infty)} \right).$$ (8.1.7)

Since $q_1 < q$ it is clear that f_0 has a jump at u, thus $F_0 \notin \mathscr{F}_k$. To make Lemma 8.1.1 applicable to the case $k \geq 1$ one can choose d.f.'s $G_m \in \mathscr{F}_k$ with $G_m^{-1}(q) = u$ having densities g_m such that $g_m(x) \to f_0(x)$ as $m \to \infty$ for every $x \neq u$. Then, applying Fatou's lemma, one can prove that every critical function ψ of level α on $\{F \in \mathscr{F}_k : F^{-1}(q) \leq u\}$ has the property $E_{F_0} \psi \leq \alpha$. Thus, Lemma 8.1.1 yields $E_{F_1} \psi \leq E_{F_1} \varphi$ and hence, φ is uniformly most powerful.   □

Randomized Estimators of Quantiles

Whereas randomized test procedures expressed in the form of critical functions are widely accepted in statistics this cannot be said of randomized estimators. Therefore, we keep our explanations here as short as possible. Nevertheless, we hope that the following lines and some further details in the Supplements will create some interest.

Recall that the randomized sample median was defined in (1.7.19) as the Markov kernel

$$M_n(\cdot \mid \cdot) = (\varepsilon_{X_{[(n+1)/2]:n}} + \varepsilon_{X_{[(n+1)/2]+1:n}})/2$$ (8.1.8)

where ε_x again denotes the Dirac measure with mass 1 at x.

In Lemma 1.7.10 it was proved that M_n is median unbiased; that is, the median of the underlying distribution is a median of the distribution of the Markov kernel M_n. In analogy to (8.1.8) one can also construct a randomized sample q-quantile which is a median unbiased estimator of the unknown q-quantile.

Given $q \in (0, 1)$ and the sample size n let

$$r \equiv r(1/2, q, n) \quad \text{and} \quad \gamma \equiv \gamma(1/2, q, n)$$

be defined as in (8.1.1) and (8.1.3). Define the randomized estimator Q_n by

$$Q_n(\cdot \mid \cdot) = (1 - \gamma)\varepsilon_{X_{r:n}} + \gamma \varepsilon_{X_{r+1:n}} \tag{8.1.9}$$

where $X_{r:n}$ is the rth order statistic of n i.i.d. random variables with common continuous d.f. F.

From the results concerning test procedures one can deduce by routine calculations that the randomized sample q-quantile is an optimal estimator of the q-quantile in the class of all randomized, median unbiased estimators which are equivariant under translations. Non-randomized estimators will be studied at the end of this section.

Randomized One-Sided Confidence Procedures

Another relevant source is Chapter 12 in Pfanzagl (1985). There the quantiles serve as an example of an irregular functional in the sense that the standard theory of 2nd order efficiency is not applicable. This is due to the fact that for this particular functional a certain 2nd derivative does not exist. Hence, a direct approach is necessary to establish upper bounds for the 2nd order efficiency of the relevant statistical procedures.

Randomized statistics of the form (8.1.9) with $r \equiv r(1 - \beta, q, n)$ and $\gamma \equiv \gamma(1 - \beta, q, n)$ also define randomized, one-sided confidence procedures where the lower confidence bound is $X_{r:n}$ with probability $1 - \gamma$ and $X_{r+1:n}$ with probability γ. These confidence procedures are optimal under all procedures that exactly attain the confidence level β. Pfanzagl proves that the asymptotic efficiency still holds within an error bound of order $o(n^{-1/2})$ in the class of all confidence procedures attaining the confidence level $\beta + o(n^{-1/2})$ uniformly in a local sense (compare with Pfanzagl (1985), Proposition 12.3.3). A corresponding result can be proved for test and estimation procedures.

Estimator Based on a Convex Combination of Two Consecutive Order Statistics

For some fixed $q \in (0, 1)$ define

$$\hat{q}_n = (1 - \gamma(n))X_{r(n):n} + \gamma(n)X_{r(n)+1:n} \tag{8.1.10}$$

where $r(n) \equiv r(q, n) \in \{1, \dots, n\}$ and $\gamma(n) \in [0, 1)$ satisfy the equation

$$nq - r - \gamma + (1 + q)/3 = 0. \tag{8.1.11}$$

Put $\sigma^2 = q(1 - q)$. Under the conditions of Theorem 6.2.4 we get

$$\sup_t \left| P \left\{ \frac{n^{1/2} f(F^{-1}(q))}{\sigma} (\hat{q}_n - F^{-1}(q)) \le t \right\} \right.$$
$$\left. - \left(\Phi(t) - n^{-1/2} \varphi(t) \left[\frac{1 - 2q}{3\sigma} - \frac{\sigma f'(F^{-1}(q))}{2f(F^{-1}(q))^2} \right] t^2 \right) \right| = o(n^{-1/2}). \tag{8.1.12}$$

It is immediate that $\hat{q}_n$ is median unbiased of order $o(n^{-1/2})$.

Moreover, notice that $\hat{q}_n$ is equivariant under translations; that is, shifting the observations amounts to the same as shifting the distribution of $\hat{q}_n$. One can prove that $\hat{q}_n$ is optimal in the class of all estimators that are equivariant under translations and median unbiased of order $o(n^{-1/2})$. The related result for confidence intervals is proved in Pfanzagl (1985), Proposition 12.3.9.

In the present section the statistical procedures are, roughly speaking, based on the sample q-quantile. These procedures possess an optimality property because the class of competitors was restricted by strong conditions like exact median unbiasedness or median unbiasedness of order $o(n^{-1/2})$. If these conditions are weakened then one can find better procedures. We refer to Section 8.3 for a continuation of this discussion.

8.2. Kernel Type Estimators of Quantiles

Recall that the sample q-quantile $F_n^{-1}(q)$ is given by

$$F_n^{-1}(q) = X_{i:n} \text{ if } (i - 1)/n < q \le i/n \text{ and } q \in (0, 1) \text{ for } i = 1, \dots, n.$$

Thus, F_n^{-1} generates increasing step functions which have jumps at the points i/n for $i = 1, \dots, n - 1$. Throughout we define $F_n^{-1}(0) = F_n^{-1}(0^+) = X_{1:n}$ and $F_n^{-1}(1) = F_n^{-1}(1^-) = X_{n:n}$.

If the underlying q.f. F^{-1} is continuous or differentiable then it is desirable to construct functions as estimates which share this property. Moreover, the information that F^{-1} is a smooth curve should be utilized to obtain estimators of a better statistical performance than that of the sample q.f. F_n^{-1}. The key idea will be to average over the order statistics close to the sample q-quantile for every $q \in (0, 1)$.

The Polygon

In a first step we construct a piecewise linear version of the sample q.f. F_n^{-1} by means of linear interpolation. Thus, given a predetermined partition $0 = q_0 < q_1 < \cdots < q_k < q_{k+1} = 1$ we get an estimator of the form

$$F_n^{-1}(q_{j-1}) + \frac{q - q_{j-1}}{q_j - q_{j-1}} [F_n^{-1}(q_j) - F_n^{-1}(q_{j-1})], \qquad q_{j-1} < q \le q_j. \tag{8.2.1}$$

For $j = 2, \ldots, k$ we may take values q_j such that $q_j - q_{j-1} = \beta$ for some appropriate "bandwidth" $\beta > 0$. This estimator evaluated at q is equal to the sample q-quantile if $q = q_j$ and equal to $[F_n^{-1}(q - \beta/2) + F_n^{-1}(q + \beta/2)]/2$ if $q = (q_{j-1} + q_j)/2$ for $j = 2, \ldots, k$. Notice that the derivative of the polygon is equal to

$$[F_n^{-1}(q_j) - F_n^{-1}(q_{j-1})]/(q_j - q_{j-1}), \qquad q_{j-1} < q < q_j.$$

Moving Scheme

This gives reason to construct another estimator of F^{-1} by using a "moving scheme." For every $q \in (0, 1)$ define the estimator of $F^{-1}(q)$ by

$$[F_n^{-1}(q - \beta(q)) + F_n^{-1}(q + \beta(q))]/2 \qquad (8.2.2)$$

where the "bandwidth function" $\beta(q)$ has to be defined in such a way that $0 \leq q - \beta(q) < q + \beta(q) \leq 1$. Given a predetermined value $\beta \in (0, 1/2)$ the bandwidth function $\beta(q)$ can e.g. be defined by

$$\beta(q) = \begin{matrix} q & 0 < q < \beta \\ \beta & \text{if} \quad \beta \leq q \leq 1 - \beta \\ 1 - q & 1 - \beta < q < 1. \end{matrix} \qquad (8.2.3)$$

Another reasonable choice of a bandwidth function is

$$\beta(q) = \begin{matrix} q - q^2/4\beta & 0 < q < 2\beta \\ \beta & \text{if} \quad 2\beta \leq q \leq 1 - 2\beta \\ (1 - q) - (1 - q)^2/4\beta & 1 - 2\beta < q < 1 \end{matrix} \qquad (8.2.4)$$

where it is assumed that $\beta \leq 1/4$. Notice that the bandwidth function in (8.2.4) is differentiable.

The use of bandwidths depending on q can be justified by the following arguments:

Since $F_n^{-1}(q)$ is the natural, nonparametric estimator of $F^{-1}(q)$ it is clear that (8.2.2) defines an estimator of $[F^{-1}(q - \beta(q)) + F^{-1}(q + \beta(q))]/2$ which in turn is approximately equal to $F^{-1}(q)$ if F^{-1} is a smooth function near q and if $\beta(q)$ is not too large. However, if q is close to one of the endpoints of the domain of F^{-1}, then one has to be cautious. If q or $1 - q$ is small than the usual q.f.'s (e.g. normal or exponential) do not fulfill the required smoothness condition. Thus, without further information about the form of the q.f. at the endpoints of $(0, 1)$ a statistician should again adopt the sample q.f. or any estimator close to the sample q.f. This aim is achieved by using bandwidths as defined above.

The use of variable bandwidths also enters the scene when a pointwise optimal bandwidth (depending on the underlying d.f.) is estimated from the data. In this case the bandwidth is random and depends on the given argument q.

The polygon (Figure 8.2.1) and the moving scheme (Figure 8.2.2) are based on $n = 50$ pseudo standard exponential random numbers. F^{-1} is the standard exponential q.f.

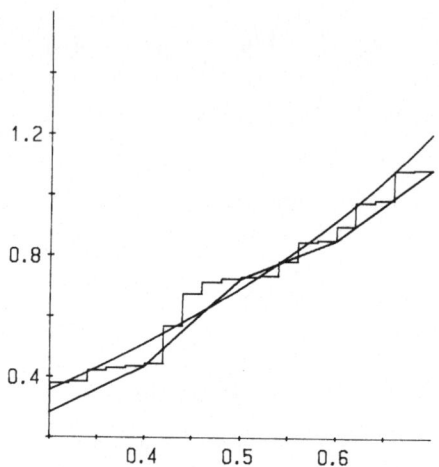

Figure 8.2.1. F^{-1}, F_n^{-1}, polygon with $n = 50$, $\beta = 0.1$.

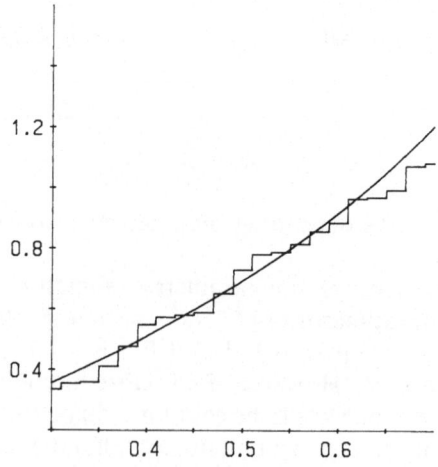

Figure 8.2.2. F^{-1}, moving scheme with $n = 50$, $\beta = 0.1$.

Quasi-Quantiles and Trimmed Means

The estimator in (8.2.2) can be written

$$(X_{r(q):n} + X_{s(q):n})/2. \tag{8.2.5}$$

If $q - \beta(q)$ and $q + \beta(q)$ are not integers then we have $r(q) = \max(1, [n(q - \beta(q))] + 1)$ and $s(q) = \min(n, [n(q + \beta(q))] + 1)$.

Another ad hoc estimator of the q-quantile is a certain "trimmed mean" defined by

$$(s(q) - r(q) + 1)^{-1} \sum_{i=r(q)}^{s(q)} X_{i:n}. \qquad (8.2.6)$$

To extend the class of estimators of the q-quantile we introduce estimators of the form

$$F_{n,0}^{-1}(q) = \sum_{i=1}^{n} a_{i,n}(q) X_{i:n} \qquad (8.2.7)$$

where the scores $a_{i,n}(q)$ satisfy the condition

$$\sum_{i=1}^{n} a_{i,n}(q) = 1. \qquad (8.2.8)$$

Within this class of estimators we shall study those where the scores are defined by a kernel. The "trimmed mean" will be closely related to a kernel estimator which is based on a uniform kernel.

The Kernel Method

Since we shall also need a kernel estimator $F_{n,0}$ of the d.f. F we discuss the method of smoothing a function via a kernel within a general framework.

Notice that the q.f. of $F_{n,0}$ will be another competitor of the sample q.f. as an estimator of the underlying q.f. F^{-1}.

Hereafter, let H be a real-valued function with domain (a, b). Particular cases are q.f.'s and d.f.'s with domain $(0, 1)$ and, respectively, the real line.

We say that a real-valued function k with domain $(a, b) \times (a, b)$ is a kernel if for every $x \in (a, b)$,

$$\int_{a}^{b} k(x, y) \, dy = 1. \qquad (8.2.9)$$

Given an initial estimator H_n define

$$H_{n,0}(x) = \int_{a}^{b} k(x, y) H_n(y) \, dy. \qquad (8.2.10)$$

By partial integration we get the representation

$$H_{n,0}(x) = \int_{a}^{b} K(x, y) \, dH_n(y) + H_n(a^+) \qquad (8.2.11)$$

if $H_n(a^+)$ and $H_n(b^-)$ exist and are finite where the function K is defined by

$$K(x, z) = \int_{z}^{b} k(x, y) \, dy. \qquad (8.2.12)$$

We shall study special kernels of the form

$$k(x, y) = \frac{1}{\beta(x)} u\left(\frac{x - y}{\beta(x)}\right).$$ (8.2.13)

The function u is again called a kernel.

Kernel Estimators of Q.F.

If $H_n = F_n^{-1}$ then

$$F_{n,0}^{-1}(q) = \int_0^1 k(q, y) F_n^{-1}(y) \, dy = \sum_{i=1}^n a_{i,n}(q) X_{i:n}$$ (8.2.14)

where the score functions $a_{i,n}$ are given by

$$a_{i,n}(q) = \int_{(i-1)/n}^{i/n} k(q, y) \, dy.$$

Obviously, condition (8.2.9) implies that the scores $a_{i,n}(q)$ satisfy condition (8.2.8).

Let u have the properties $\int u(y) \, dy = 1$ and $u(x) = 0$ for $|x| > 1$. Moreover, assume that the bandwidth function β satisfies the condition $\beta(q) \le \min(q, 1 - q)$; e.g. the bandwidth functions in (8.2.3) and (8.2.4) satisfy this condition. Then the kernel k defined in (8.2.13) satisfies (8.2.9). Now, $F_{n,0}^{-1}$ can be written in the form

$$F_{n,0}^{-1}(q) = \int_{-1}^1 F_n^{-1}(q - \beta(q)y) u(y) \, dy.$$ (8.2.15)

For $\beta(q)$, defined in (8.2.3) and (8.2.4), the function $q \to q - \beta(q)y$ is non-decreasing for every $|y| \le 1$ showing that $F_{n,0}^{-1}$ is nondecreasing if $u \ge 0$. Thus, $F_{n,0}^{-1}$ is in fact a q.f. Moreover, this construction has the favorable property that the range of $F_{n,0}^{-1}$ is a subset of the support of the underlying d.f. F.

Writing $U(z) = \int_{-1}^z u(y) \, dy$ we have

$$F_{n,0}^{-1}(q) = \sum_{i=1}^n \left[U\left(\frac{q - (i - 1)/n}{\beta(q)}\right) - U\left(\frac{q - i/n}{\beta(q)}\right) \right] X_{i:n}.$$ (8.2.16)

It is easy to verify that the coefficients are equal to zero if $i \le n(q - \beta(q))$ or $i \ge n(q + \beta(q)) + 1$.

Kernel Estimators of D.F.

The kernel estimators of the d.f. are of the form

$$F_{n,0}(x) = \int K(x, y) \, dF_n(y) = n^{-1} \sum_{i=1}^n K(x, \xi_i),$$ (8.2.17)

or, alternatively,

$$F_{n,0}(x) = n^{-1} \sum_{i=1}^{n} U((x - \xi_i)/\beta) \qquad (8.2.18)$$

where $U(z) = \int_{-\infty}^{z} u(y)\,dy$ and u is a function such that $\int u(y)\,dy = 1$. If $u \geq 0$ then $F_{n,0}$ generates d.f.'s hence by constructing the corresponding q.f.'s we obtain a further estimator $(F_{n,0})^{-1}$ of the q.f. F^{-1}.

Density Estimation

The kernel method enables us to construct differentiable functions as estimates of the d.f. F and the q.f. F^{-1}, although the initial estimates are step functions. Thus, we get estimators of the density $f = F'$ and the density quantile function $(F^{-1})' = 1/f(F^{-1})$ as well.

From (8.2.18) we obtain

$$F_{n,1}(x) = F'_{n,0}(x) = (n\beta)^{-1} \sum_{i=1}^{n} u((x - \xi_i)/\beta). \qquad (8.2.19)$$

A corresponding formula holds for $F_{n,1}^{-1} = (F_{n,0}^{-1})'$. If the bandwidth function is defined as in (8.2.3) then $F_{n,0}^{-1}$ is differentiable on the interval $(\beta, 1 - \beta)$. We get

$$F_{n,1}^{-1}(q) = \beta^{-1} \sum_{i=1}^{n} \left[u\left(\frac{q - (i - 1)/n}{\beta} \right) - u\left(\frac{q - i/n}{\beta} \right) \right] X_{i:n} \qquad (8.2.20)$$

for $\beta \leq q \leq 1 - \beta$. With $\beta(q)$ as in (8.2.4) the same representation of $F_{n,1}^{-1}$ holds for $2\beta \leq q \leq 1 - 2\beta$. However, now $F_{n,1}^{-1}$ also exists on $(0, 1)$ and can easily be computed.

Some Illustrations

In this sequel, we shall apply the Epanechnikov kernel defined by

$$u(x) = (3/4)(1 - x^2)1_{[-1,1]}(x). \qquad (8.2.21)$$

Notice that

$$U(x) = \begin{cases} 0 & x < -1 \\ 1/2 + 3x/4 - x^3/4 & \text{if} \quad -1 \leq x \leq 1 \\ 1 & x > 1. \end{cases}$$

In Figures 8.2.3 and 8.2.4 the kernel q.f. $F_{n,0}^{-1}$ and the q.f. $(F_{n,0})^{-1}$ of the kernel d.f. $F_{n,0}$ are based on $n = 100$ pseudo standard exponential random numbers.

For q bounded away from 0 and 1 one realizes that $F_{n,0}^{-1}$ and $(F_{n,0})^{-1}$ have about the same performance. Near to 0 and 1 the estimate taken from $(F_{n,0})^{-1}$

has the unpleasant property that (a) it is inaccurate and (b) it attains values which do not belong to the support of the exponential d.f. The second property is of course not very surprising. To avoid this unpleasant behavior of $(F_{n,0})^{-1}$ one should modify $F_{n,0}(x)$ in such a way that the bandwidth depends on x.

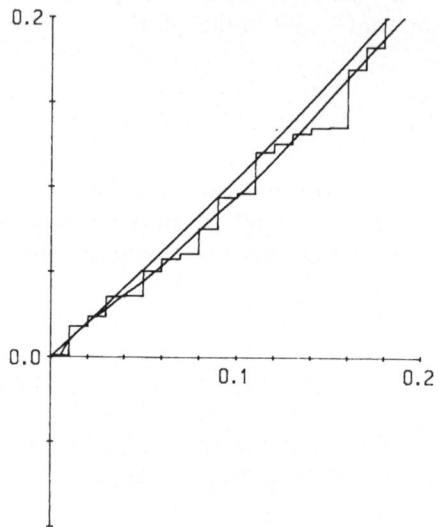

Figure 8.2.3. F^{-1}, F_n^{-1}, and $F_{n,0}^{-1}$ with $n = 100$, $\beta = 0.08$.

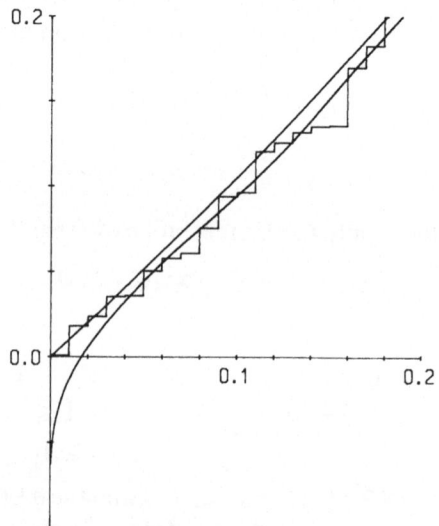

Figure 8.2.4. F^{-1}, F_n^{-1}, and $(F_{n,0})^{-1}$ with $n = 100$, $\beta = 0.08$.

Figures 8.2.3 and 8.2.4 show clearly that the kernel estimates reduce the random fluctuation of the "natural" estimates thus, also reducing the maximum deviation from the underlying d.f.

Next $F_{n,0}^{-1}$ and $(F_{n,0})^{-1}$ will be evaluated at the right end of the domain. Again $F_{n,0}^{-1}$ is defined with the bandwidth function in (8.2.4).

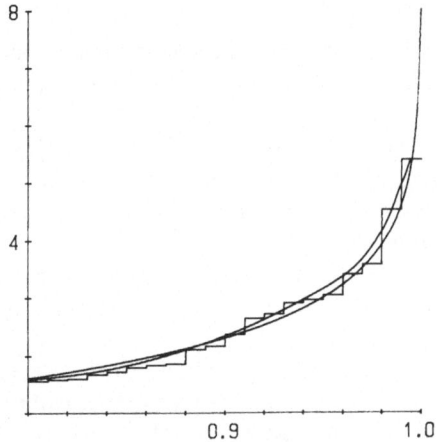

Figure 8.2.5. F^{-1}, F_n^{-1}, and $F_{n,0}^{-1}$ with $n = 100$, $\beta = 0.08$.

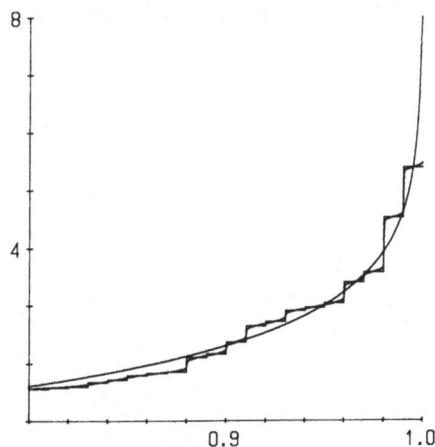

Figure 8.2.6. F^{-1}, F_n^{-1} and $(F_{n,0})^{-1}$ with $n = 100$, $\beta = 0.08$.

At the first moment I thought there was an error in the computer program when the graph in Figure 8.2.6 appeared on the screen. The graph of $(F_{n,0})^{-1}$ can hardly be distinguished from the sample q.f. The explanation for $(F_{n,0})^{-1}$

being close to the sample q.f. is that the largest order statistics are not close to each other, and so the kernel d.f. with the bandwidth $\beta = 0.08$ does not smooth the sample d.f.

Parametric versus Nonparametric Estimation

Finally, we examine the estimation of the standard normal q.f. This situation is related to estimating the exponential q.f. near the right endpoint of the domain.

In addition to the smoothed sample q.f. $F_{n,0}^{-1}$, we shall take the estimator $\mu_n + \sigma_n \Phi^{-1}$ where (μ_n, σ_n) is the maximum likelihood (m.l.) estimator of the location and scale parameter of the normal d.f. The kernel q.f. is again defined by means of the Epanechnikov kernel.

In Figures 8.2.7 and 8.2.8 the observations are sampled according to the standard normal d.f. We remark that the m.l. estimate (μ_n, σ_n) of (μ, σ) has the value $(0.028, 1.032)$.

The performance of the estimators is bad near the endpoints 0 and 1 of the domain. This is not surprising since the parametric estimate $\mu_n + \sigma_n \Phi^{-1}$ does not converge to Φ^{-1} uniformly over $(0, 1)$. Notice that

$$\mu_n + \sigma_n \Phi^{-1} - \Phi^{-1} = \mu_n + (\sigma_n - 1)\Phi^{-1}$$

is an unbounded function whenever $\sigma_n \neq 1$. Thus, Figure 8.2.8 is misleading to some extent.

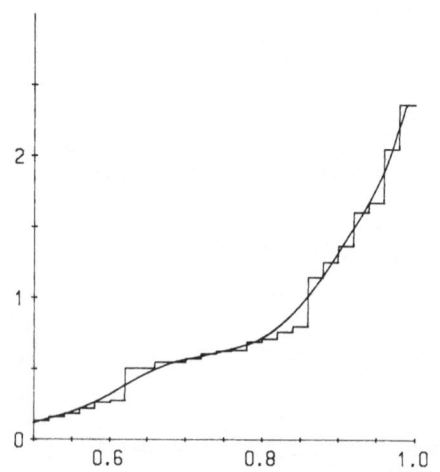

Figure 8.2.7. F_n^{-1}, $F_{n,0}^{-1}$ with $n = 100$, $\beta = 0.08$.

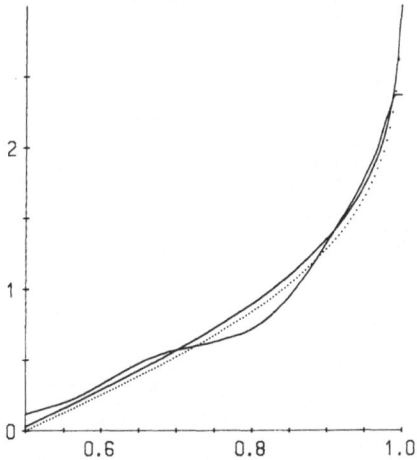

Figure 8.2.8. Φ^{-1} (dotted curve), $\mu_n + \sigma_n \Phi^{-1}$, $F_{n,0}^{-1}$ with $n = 100$, $\beta = 0.08$.

Fitting a Density to Data

For the visual comparison of two different d.f.'s the probability paper plays a dominant role (see e.g. Gumbel's book or Barnett (1975)). For this purpose the "theoretical" d.f. is transformed to a straight line. When applying the same transformation to the sample d.f., a deviation of the transformed sample d.f. from the straight line can easily be detected.

It can be advisable to compare distributions by their densities. One advantage is that one can see the original form of the distribution. The data will visually be represented by means of the kernel density $f_n = F_{n,1}$ as introduced in (8.2.19). In a second step an extreme value density is fitted to the kernel density. We suppose that the graphs given in Sections 1.3 and 5.1 have already sensitized the reader for extreme value densities.

We shall examine the monthly and annual maxima of the temperature at De Bilt (Netherlands). Data of 133 years (1849–1981) are available and have been first studied by M.A.J. van Montfort (1982). The plot of the annual maxima on a normal probability paper shows an excellent fit of a normal distribution. Van Montfort points out the resemblance of normal distributions and certain "symmetric" Weibull distributions (compare also Figure 1.3.4). The author is grateful to van Montfort for a translation of his paragraph 8.3 (written in Dutch) and for providing the data. Despite of van Montfort's remark, Sneyers (1984) considers this as an "... example of an extreme value distribution following not a Fisher–Tippett asymptote...".

Below Weibull densities with location, scale, and shape parameters μ, σ, and α are fitted to kernel densities based on monthly maxima of the temperature. The kernel density is defined with the Epanechnikov kernel and the

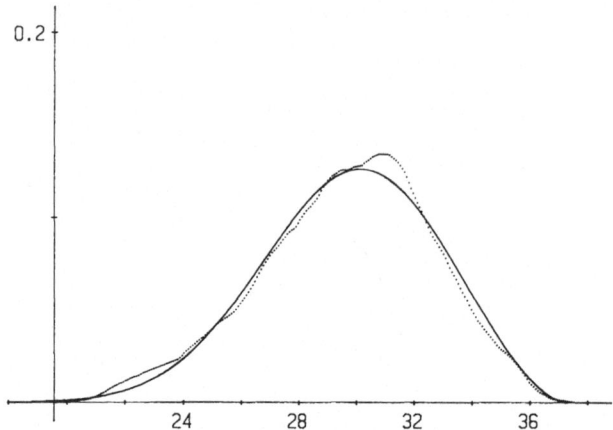

Figure 8.2.9. July: Kernel density and Weibull density with parameters $\mu = 37.3$, $\sigma = 8.5$, $\alpha = 2.7$.

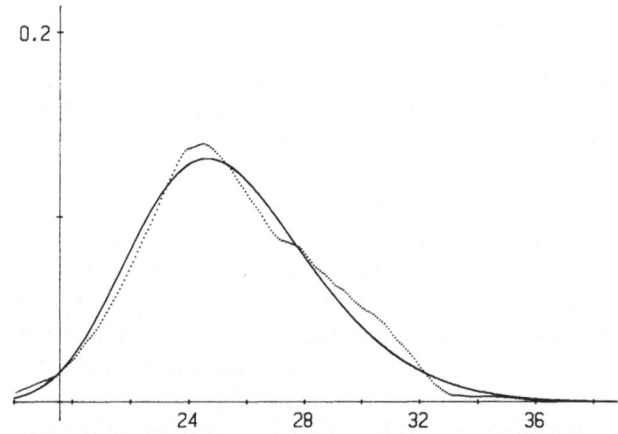

Figure 8.2.10. September: Kernel density and Weibull density with parameters $\mu = 44.0$, $\sigma = 19.8$, $\alpha = 7.0$.

bandwidth $\beta = 2.0$. A better fit can be achieved by more smoothing, that is, for a larger bandwidth.

We see that the densities of the maxima of temperature in July (Fig. 8.2.9) are skewed to the left; those for September (Fig. 8.2.10) are skewed to the right.

Below we also include the corresponding Weibull densities for June and August which are close together. That for June is nearly symmetric and that for August is slightly skewed to the right. The kernel density for annual maxima is nearly symmetric.

The largest observed values of monthly maxima within 133 years are

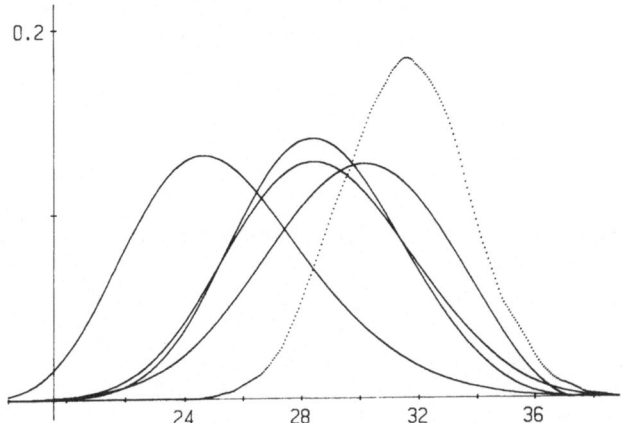

Figure 8.2.11. Kernel density for annual maxima; Weibull densities: June: $\mu = 38.2$, $\sigma = 10.6$, $\alpha = 3.9$; July: $\mu = 37.3$, $\sigma = 8.5$, $\alpha = 2.7$; August: $\mu = 39.8$, $\sigma = 12.2$, $\alpha = 4.1$; September: $\mu = 44.0$, $\sigma = 19.8$, $\alpha = 7.0$.

(a) 36.8 in June 1947, (b) 35.6 in July 1911, (c) 35.8 in August 1857, and (d) 34.2 in September 1949.

We suggest to classify the annual maximum as a maximum of independent, not identically distributed Weibull r.v.'s according to the maxima in June, July, August, and September. According to (1.3.4), the calculation of the d.f. and the density of the maximum of not identically distributed r.v.'s creates no difficulties. The resulting density shows an excellent fit to the kernel density of the annual maxima as given in Figure 8.2.11.

The choice of the Weibull density was accomplished by some visual, subjective judgment. To obtain an automatic procedure one should fix a distance between densities like the maximum deviation, χ^2-distance, Hellinger distance, or some other distance. Then, take that parameter (μ, σ, α) which minimizes the distance between the kernel density and the Weibull density. From the foregoing remarks it becomes obvious that our estimates are produced by some kind of minimum distance method. By using this method we are getting larger estimates of the unknown right endpoint than by taking the sample maximum. Recall that the "minimum distance" estimates are 38.2, 37.3, 39.8, and 44.0 compared to the sample maxima 36.8, 35.6, 35.8, and 34.2. The difference is particularly significant in those cases where the density is skewed to the right.

Hosking (1985) developed a modified Newton–Raphson iteration algorithm for solving the maximum likelihood equation in the 3-parameter extreme value model (given by the von Mises parametrization). This algorithm seems to work if $|\beta| < 0.5$. When using the "minimum distance" estimates, given in Figure 8.2.11, as initial estimates, then one obtains the following estimates:

June: $\mu = 39.1$, $\sigma = 11.4$, $\alpha = 4.3$;   July: $\mu = 36.2$, $\sigma = 7.4$, $\alpha = 2.4$;
August: $\mu = 38.7$, $\sigma = 11.1$, $\alpha = 4.0$;   September: $\mu = 38.8$, $\sigma = 14.4$, $\alpha = 5.2$.

The densities pertaining to the maximum likelihood estimates show again an excellent fit to the kernel (sample) densities.

8.3. Asymptotic Performance of Quantile Estimators

The kernel estimator of the q.f. is given by

$$F_{n,0}^{-1}(q) = \beta^{-1} \int_0^1 u\left(\frac{q-y}{\beta}\right) F_n^{-1}(y)\, dy \tag{8.3.1}$$

if $2\beta < q < 1 - 2\beta$. Notice that under appropriate regularity conditions the ith derivative $F_{n,i}^{-1}$ of $F_{n,0}^{-1}$ is given by

$$F_{n,i}^{-1}(q) = \beta^{-(i+1)} \int_0^1 u^{(i)}\left(\frac{q-y}{\beta}\right) F_n^{-1}(y)\, dy. \tag{8.3.2}$$

Moderate Deviations

Our first aim will be to deduce rough bounds for the rate of convergence of kernel estimators of the q.f. and its derivatives. For this purpose we shall study again the oscillation property of the sample q.f.

The basic tool for the following considerations will be Lemma 3.1.7(ii) which describes the stochastic behavior of

$$n^{1/2}|F_n^{-1}(q_2) - F_n^{-1}(q_1) - [F^{-1}(q_2) - F^{-1}(q_1)]|/(q_2 - q_1)^{1/2} \tag{8.3.3}$$

uniformly over q_1, q_2 with $0 < p_1 \leq q_1 < q_2 \leq p_2 < 1$.

In the sequel we shall assume that the kernel u satisfies the following regularity conditions:

Condition 8.3.1. Let m be a positive integer. Assume that

(i) u has the support $[-1, 1]$.
(ii) u has $m + 1$ derivatives.
(iii) $\int u(y)\, dy = 1$.

Integration by parts yields

$$\int u^{(i)}(y) y^i\, dy = i!, \qquad i = 0, \ldots, m + 1$$

and

$$\int u^{(i)}(y) y^j\, dy = 0, \qquad 0 \leq j < i \leq m + 1.$$

(8.3.4)

Condition 8.3.2. Let k be a positive integer. Assume that

$$\int u(y)y^j\,dy = 0, \qquad j = 1, \ldots, k.$$

Under Conditions 8.3.1 and 8.3.2 we get, by means of integration by parts, that

$$\int u^{(i)}(y)y^{i+j}\,dy = 0, \qquad i = 0, \ldots, m+1 \text{ and } j = 1, \ldots, k. \qquad (8.3.5)$$

The following representation of $F_{n,i}^{-1}$ will be useful:

$$F_{n,i}^{-1}(q) = \begin{cases} F_n^{-1}(q) & i = 0 \\ (F^{-1})^{(i)}(q) \\ 0 & \end{cases} + R_{i,n}(q) \quad \text{for} \quad \begin{matrix} \\ i = 1, \ldots, m \\ i = m+1 \end{matrix} \qquad (8.3.6)$$

where the remainder term is given by

$$R_{i,n}(q) = \beta^{-i}\int u^{(i)}(y)[F_n^{-1}(q-\beta y) - F_n^{-1}(q) - (F^{-1}(q-\beta y) - F^{-1}(q))]\,dy$$

$$+ \beta^{-i}\int u^{(i)}(y)\left[F^{-1}(q-\beta y) - F^{-1}(q) - \sum_{j=1}^{k+i}\frac{(-\beta y)^j}{j!}(F^{-1})^{(j)}(q)\right]dy$$
$$(8.3.7)$$

if again $2\beta < q < 1 - 2\beta$ and if the derivatives of F^{-1} at q exist.

We remark that (8.3.7) always holds for $k = 0$.

The representation above shows that $R_{i,n}(q)$ splits up (a) into a random part which is governed by the oscillation behavior of the sample q.f. and (b) into a non-random part which depends on the remainder term of a Taylor expansion of F^{-1} about q.

It is evident that a similar representation holds for the sample d.f. F_n in place of the sample q.f. F_n^{-1}. Recall that the oscillation behavior of F_n was studied in Remark 6.3.3.

The histograms with random or non-random cells are based on terms of the form

$$(p_2 - p_1)/(F_n^{-1}(p_2) - F_n^{-1}(p_1))$$

or

$$(F_n(t_2) - F_n(t_1))/(t_2 - t_1).$$

Thus, the oscillation behavior of the sample q.f. and the sample d.f. can be regarded as a property which summarizes the properties of histograms.

The representation (8.3.6) shows that the stochastic behavior of kernel estimators of the q.f. is exhaustively determined by the oscillation behavior of the sample q.f.

Next, we give a technical result which concerns the moderate deviation of the kernel q.f. from the underlying q.f.

Lemma 8.3.3. *Suppose that Conditions 8.3.1 and 8.3.2 hold for some $m \geq 1$ and $k = m - 1$. Moreover, assume that the q.f. F^{-1} has $m + 1$ bounded derivatives on a neighborhood of the interval (p_1, p_2).*

Then, for every $s > 0$ and every sufficiently small $\beta \geq (\log n)/n$ there exist constants $B, C > 0$ (being independent of β and n) such that

(i) $P\left\{ \sup_{p_1 \leq q \leq p_2} |F_{n,0}^{-1}(q) - F_n^{-1}(q)| > C\left[\left(\frac{\beta \log n}{n} \right)^{1/2} + \beta^m \right] \right\} < Bn^{-s}$,

(ii) $P\left\{ \sup_{p_1 \leq q \leq p_2} |F_{n,i}^{-1}(q) - (F^{-1})^{(i)}(q)| > C\left[\left(\frac{\log n}{\beta^{2i-1} n} \right)^{1/2} + \beta^{m-i+1} \right] \right\} < Bn^{-s}$

for $i = 1, \ldots, m$, and

(iii) $P\left\{ \sup_{p_1 \leq q \leq p_2} |F_{n,m+1}^{-1}(q)| > C\left[\left(\frac{\log n}{\beta^{2m+1} n} \right)^{1/2} + 1 \right] \right\} < Bn^{-s}$.

PROOF. Immediate from Lemma 3.1.7(ii) and (8.3.6). $\qquad \square$

It is easy to see that Lemma 8.3.3(i) holds with $(\log n)/n$ in place of $(\beta(\log n)/n)^{1/2}$ if $0 < \beta \leq (\log n)/n$. This yields that for every $\varepsilon > 0$:

$$P\left\{ \sup_{p_1 \leq q \leq p_2} |n^{1/2}(F_{n,0}^{-1}(q) - F^{-1}(q)) - n^{1/2}(F_n^{-1}(q) - F^{-1}(q))| \geq \varepsilon \right\} \to 0 \quad (8.3.8)$$

as $n \to \infty$ for every sequence of bandwidths $\beta \equiv \beta_n$ with $n\beta_n^{2m} \to 0$, $n \to \infty$.

This means that the quantile process $n^{1/2}(F_n^{-1} - F^{-1})$ and the smooth quantile process $n^{1/2}(F_{n,0}^{-1} - F^{-1})$ have the same asymptotic behavior on the interval (p_1, p_2). Lemma 8.3.3 shows that, with high probability, the kernel estimates of the q.f. are remarkably smooth. This fact is basic for the considerations of Section 8.4.

Kernel Estimators Evaluated at a Fixed Point

The results above do not enable us to distinguish between the asymptotic performance of the sample q.f. and the kernel estimator of the q.f. This is possible if a limit theorem together with a bound for the remainder term is established. The first theorem, taken from Reiss (1981c), concerns the estimation of the d.f.

Theorem 8.3.4. *Let $F_{n,0}$ be the kernel estimator of the d.f. as given in (8.2.18). Suppose that the kernel u satisfies Conditions 8.3.1(i), (ii), and 8.3.2 for some $k \geq 1$. Moreover, let F have $k + 1$ derivatives on a neighborhood of the fixed point t such that $|F^{(k+1)}| \leq A$.*

Then, uniformly over the bandwidths $\beta \in (0, 1)$,

$$\left| E(F_{n,0}(t) - F(t))^2 - E(F_n(t) - F(t))^2 + 2(\beta/n)F'(t) \int xu(x)U(x)\,dx \right|$$

$$\leq (\beta^{k+1} A \int |u(x)x^{k+1}|dx/(k+1)!)^2 + O(\beta^2/n). \tag{8.3.9}$$

This result enables us to compare the mean square error $E(F_{n,0}(t) - F(t))^2$ of $F_{n,0}(t)$ and the variance $E(F_n(t) - F(t))^2 = F(t)(1 - F(t))/n$ of the sample d.f. $F_n(t)$ evaluated at t. If $F'(t) > 0$ and the bandwidth β is chosen so that the right-hand side of (8.3.9) is sufficiently small then the term $\int xu(x)U(x)\,dx$ can be taken as a measure of performance of $F_{n,0}(t)$. If

$$\int xu(x)U(x)\,dx > 0 \tag{8.3.10}$$

then, obviously, $F_{n,0}(t)$ is of a better performance than $F_n(t)$.

If u is a non-negative, symmetric kernel then

$$\int xu(x)U(x)\,dx = \int xu(x)[2U(x) - 1]\,dx > 0$$

since the integrand on the right-hand is non-negative. Notice that a non-negative kernel u satisfies Condition 8.3.2 only if $k = 1$.

From (8.3.9) we see that $F_{n,0}(t)$ and $F_n(t)$ have the same asymptotic efficiency, however, $F_n(t)$ is asymptotically deficient w.r.t. $F_{n,0}(t)$. The concept of deficiency was introduced by Hodges and Lehmann (1970). Define

$$i(n) = \min\{m: E(F_m(t) - F(t))^2 \leq E(F_{n,0}(t) - F(t))^2\}. \tag{8.3.11}$$

Thus, $i(n)$ is the smallest integer m such that $F_m(t)$ has the same or a better performance than $F_{n,0}(t)$. Since $i(n)/n \to 1$, $n \to \infty$, we know that $F_{n,0}(t)$ and $F_n(t)$ have the same asymptotic efficiency. However, the relative deficiency $i(n) - n$ of $F_n(t)$ w.r.t. $F_{n,0}(t)$ quickly tends to infinity as $n \to \infty$. In short, we may say that the relative deficiency $i(n) - n$ is the number of observations that are wasted if we use the sample d.f. instead of the kernel estimator.

The comparison of $F_n(t)$ and $F_{n,0}(t)$ may as well be based on covering probabilities. The Berry–Esséen theorem yields

$$P\{(n^{1/2}/\sigma)|F_n(t) - F(t)| \leq y\} = 2\Phi(y) - 1 + O(n^{-1/2}) \tag{8.3.12}$$

where $\sigma^2 = F(t)(1 - F(t))$. The Berry–Esséen theorem, Theorem 8.3.4, and P.8.6 lead to the following theorem.

Theorem 8.3.5. *Under the conditions of Theorem 8.3.4 we get, uniformly over $\beta > 0$,*

$$P\{(n^{1/2}/\sigma)|F_{n,0}(t) - F(t)| \leq y\}$$
$$= 2\Phi\left[y\left(\frac{3}{2} - \frac{E(F_{n,0}(t) - F(t))^2}{2E(F_n(t) - F(t))^2}\right) \right] + O(n^{-1/2} + (\beta + n\beta^{2(m+1)})^{3/2}). \tag{8.3.13}$$

We see that the performance of $F_{n,0}(t)$ again depends on the mean square error. A modified definition of the relative deficiency, given w.r.t. covering probabilities, leads to the same conclusion as in the case of the mean square error.

In analogy to the results above, one may compare the performance of the sample q-quantile $F_n^{-1}(q)$ and a kernel estimator $F_{n,0}^{-1}(q)$. If the comparison is based on the mean square error, one has to impose appropriate moment conditions. To avoid this, we restrict our attention to covering probabilities.

Recall from Section 4.2 that under weak regularity conditions,

$$P\{(n^{1/2}/\sigma_0)|F_n^{-1}(q) - F^{-1}(q)| \le y\} = 2\Phi(y) - 1 + O(n^{-1/2}) \quad (8.3.14)$$

with $\sigma_0^2 = (q(1 - q))/[f(F^{-1}(q))^2]$ and f denoting the derivative of F.

The following lemma is taken from Falk (1985a, Proposition 1.5).

Lemma 8.3.6. Let $F_{n,0}^{-1}$ be the kernel estimator of the q.f. as given in (8.3.1). Suppose that the kernel u satisfies Conditions 8.3.1(i), (ii). Suppose that the q.f. F^{-1} has a bounded second derivative on a neighborhood of the fixed point $q \in (0, 1)$, and that $f(F^{-1}(q)) > 0$.

Then, if $\beta \equiv \beta(n) \to 0$, $n \to \infty$, we have,

$$P\{(n^{1/2}/\sigma_n)(F_{n,0}^{-1}(q) - \mu_n) \le y\} = \Phi(y) + O(\log(n)n^{-1/4}) \quad (8.3.15)$$

where

$$\mu_n = \int_0^1 F^{-1}(x)\beta^{-1}u((q - x)/\beta)\,dx \quad (8.3.16)$$

and

$$\sigma_n^2 = \int_0^1 \left(\int u(x)[q - \beta x - 1_{(0,q-\beta x]}(y)](F^{-1})'(q - \beta x)\,dx \right)^2 dy. \quad (8.3.17)$$

Moreover,

$$\sigma_n \to \sigma_0, \qquad n \to \infty. \quad (8.3.18)$$

Thus, from Lemma 8.3.6 we know that $F_{n,0}^{-1}(q)$ is asymptotically normal with mean value μ_n and variance σ_n^2/n. The proof of Lemma 8.3.6 is based on a Bahadur approximation argument. (8.3.18) indicates that $F_{n,0}^{-1}(q)$ and $F_n^{-1}(q)$ have the same asymptotic efficiency. It would be of interest to know whether the remainder term in (8.3.14) is of order $O(n^{-1/2})$. Applying P.8.6 we obtain as a counterpart of Theorem 8.3.5 the following result.

Under the conditions of Lemma 8.3.6,

$$P\{(n^{1/2}/\sigma_0)|F_{n,0}^{-1}(q) - F^{-1}(q)| \le y\}$$
$$= 2\Phi\left[y\left(\frac{3}{2} - \frac{\sigma_n^2/n + (\mu_n - F^{-1}(q))^2}{2\sigma_0^2/n}\right) \right] \quad (8.3.19)$$
$$+ O(\log(n)n^{-1/4} + [\max(|\sigma_n - \sigma_0|, n(\mu_n - F^{-1}(q))^2)]^{3/2}).$$

This shows that the performance of $F_{n,0}^{-1}(q)$ depends on the "mean square error" $\sigma_n^2/n + (\mu_n - F^{-1}(q))^2$. As in Falk (1985a, proof of Theorem 2.3) we may prove that

$$\sigma_n^2 = \sigma_0^2 - 2\beta(n) \int xu(x)U(x)\,dx + O(\beta(n)^2) \qquad (8.3.20)$$

and

$$|\mu_n - F^{-1}(q)| \leq O(\beta(n)^{k+1}) \qquad (8.3.21)$$

if F^{-1} has $k + 1$ derivatives on a neighborhood of q and the kernel u satisfies Condition 8.3.2 for k. Thus, the results for the q-quantile are analogous to that for the sample d.f.

8.4. Bootstrap via Smooth Sample Quantile Function

In Section 6.4 we introduced the bootstrap d.f. $T_n(F_n, \cdot)$ as an estimator of the d.f.

$$T_n(F, \cdot) = P_F\{T(F_n) - T(F) \leq \cdot\}.$$

Thus, $T_n(F, \cdot)$ is the centered d.f. of the statistical functional $T(F_n)$. Then, in the next step, the bootstrap d.f. $T_n(F_n, \cdot)$ is the statistical functional of $T_n(F, \cdot)$.

For the q-quantile (which is the functional $T(F) = F^{-1}(q)$) it was indicated that the bootstrap error $T_n(F_n, t) - T_n(F, t)$ is of order $O(n^{-1/4})$.

Thus, the rate of convergence of the bootstrap estimator is very slow. We also refer to the illustrations in Section 6.4 which reveal the poor performance for small sample sizes. Another unpleasant feature of the bootstrap estimate was that it is a step function.

In the present section we shall indicate that under appropriate regularity conditions the bootstrap estimator based on a smooth version of the sample d.f. has a better performance.

The Smooth Bootstrap D.F.

Let again $F_{n,0}^{-1}$ denote the kernel q.f. as defined in Section 8.2. We have

$$F_{n,0}^{-1}(q) = \int_0^1 \frac{1}{n\beta(q)} u\left(\frac{q-y}{\beta(q)}\right) F_n^{-1}(y)\,dy \qquad (8.4.1)$$

where the kernel u satisfies the conditions $u \geq 0$, $u(x) = 0$ for $|x| > 1$, and $\int u(x)\,dx = 1$. Moreover, the bandwidth function $\beta(q)$ is defined as in (8.2.3) or (8.2.4). Denote by $F_{n,0}$ the smooth sample d.f. which is defined as the inverse of the kernel q.f. $F_{n,0}^{-1}$.

By plugging $F_{n,0}$ into $T_n(\cdot, t)$ (instead of F_n) we get the smooth bootstrap d.f. $T_n(F_{n,0}, \cdot)$.

We remark that one may also use the kernel estimator of the d.f. as introduced in Section 8.2.

Since $F_{n,0}$ is absolutely continuous one can expect that the smooth bootstrap d.f. $T_n(F_{n,0}, \cdot)$ is also absolutely continuous. This will be illustrated in the particular case of the q-quantile.

Illustration

Given n i.i.d. random variables with standard normal d.f. Φ define again, as in Section 6.4, the normalized d.f. of the sample q-quantile by

$$T_n^*(F, t) = T_n(F, (q(1 - q))^{1/2} t / n^{1/2} \varphi(\Phi^{-1}(q))).$$

For a sample of size $n = 20$ (Figure 8.4.1) and $n = 200$ (Figure 8.4.2) we

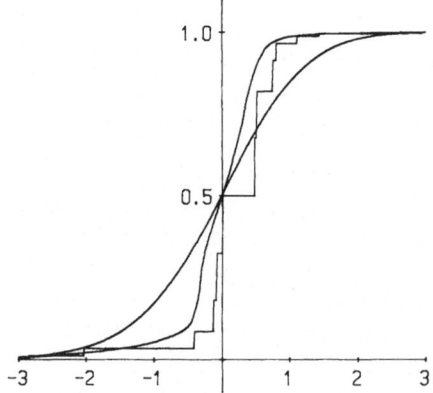

Figure 8.4.1. $T_n^*(\Phi, \cdot)$, $T_n^*(F_n, \cdot)$, $T_n^*(F_{n,0}, \cdot)$ for $q = .4$, $n = 20$.

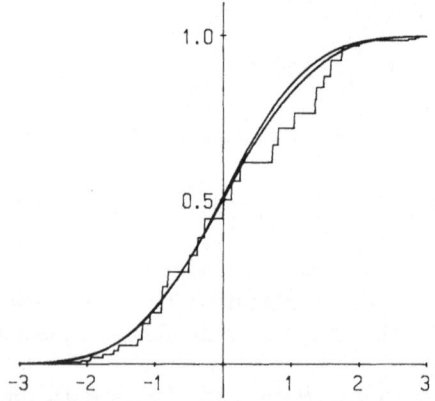

Figure 8.4.2. $T_n^*(\Phi, \cdot)$, $T_n^*(F_n, \cdot)$, $T_n^*(F_{n,0}, \cdot)$ for $q = .4$, $n = 200$.

compare the normalized d.f. $T_n^*(F, \cdot)$ of the sample q-quantile, the normalized bootstrap d.f. $T_n^*(F_n, \cdot)$, and the normalized smooth bootstrap d.f. $T_n^*(F_{n,0}, \cdot)$. The kernel q.f. $F_{n,0}^{-1}$ is defined with the bandwidth function in (8.2.4) with $\beta = 0.07$. Moreover, u is the Epanechnikov kernel.

Smooth Bootstrap Error Process

In this sequel, let us again use the same symbol for the d.f. and the corresponding probability measure. Write

$$T_n(F, B) = P_F\{(T(F_n) - T(F)) \in B\} \tag{8.4.2}$$

for Borel sets B.

Define the bootstrap error process $\mu_n(F, \cdot)$ by

$$\mu_n(F, B) = T_n(F_{n,0}, B) - T_n(F, B). \tag{8.4.3}$$

Notice that $\mu_n(F, \cdot)$ is the difference of two random probability measures and thus a random signed measure. Below we shall study the stochastic behavior of $\mu_n(F, \cdot)$ as $n \to \infty$ in the particular case of the q-quantile $T(F) = F^{-1}(q)$.

Let $\mathscr{S}$ be a system of Borel sets. We shall study the asymptotic behavior of

$$\sup_{B \in \mathscr{S}} |\mu_n(F, B)|$$

in the particular case of the functional $T(F) = F^{-1}(q)$ for some fixed $q \in (0, 1)$. Put

$$\sigma_n = (q(1 - q))^{1/2}(F^{-1})'(q)n^{-1/2}$$

and

$$v_n(F, B) = \int_B [1 - (x/\sigma_n)]^2 \, dN_{(0,\sigma_n^2)}(x). \tag{8.4.4}$$

Straightforward calculations show that

$$\sup_t |v_n(F, (-\infty, t])| = (2\pi e)^{-1/2}$$

and $\hfill (8.4.5)$

$$\sup_{t>0} |v_n(F, [-t, t]| = \sup_B |v_n(F, B)| = (2/\pi e)^{1/2}.$$

Notice that these expressions do not depend on the underlying d.f. F.

Theorem 8.4.1. *Assume that*

(a) *F^{-1} has a bounded second derivative near q and that $(F^{-1})'(q) > 0$ for some fixed $q \in (0, 1)$,*
(b) *the bandwidth β_n satisfies the conditions $n\beta_n^3 \to 0$ and $n\beta_n^2 \to \infty$ as $n \to \infty$,*
(c) *the kernel u has a bounded second derivative.*

Then,

$$P_F\left\{\left(n\beta_n \middle/ \int u^2(y)\,dy\right)^{1/2} \sup_{B \in \mathscr{S}} |\mu_n(F,B)| \middle/ \sup_{B \in \mathscr{S}} |v_n(F,B)| \leq t\right\} \to 2\Phi(t) - 1$$

(8.4.6)

as $n \to \infty$ *for every* $t \geq 0$ *whenever* $\sup_{B \in \mathscr{S}} |v_n(F,B)| > 0$.

The key idea of the proof is to compute the asymptotic normality of the sample q-quantile of i.i.d. random variables with common q.f. $F_{n,0}^{-1}$. According to Lemma 8.3.3 such q.f.'s satisfy the required smoothness conditions with high probability.

A version of Theorem 8.4.1, with $\mathscr{S} = \{(-\infty, t]\}$ and $F_{n,0}$ being the smooth sample d.f., is proved in Falk and Reiss (1989). A detailed proof of the present result will be given somewhere else.

If $\beta_n = n^{-1/3}$ then the accuracy of the bootstrap approximation is, roughly speaking, of order $O(n^{-1/3})$. The choice of $\beta_n = n^{-1/2}$ leads to a bootstrap estimator related to that of Section 6.4 as far as the rate of convergence is concerned.

Under stronger regularity conditions it is possible to construct bootstrap estimates of a higher accuracy. Assume that F^{-1} has three bounded derivatives near q and that the kernel u has three bounded derivatives. Moreover, assume that $\int u(x)x\,dx = 0$. Notice that nonnegative, symmetrical kernels u satisfy this condition. Then, the condition $n\beta_n^3 \to 0$ in Theorem 8.4.1 can be weakened to $n\beta_n^5 \to 0$ as $n \to \infty$. This yields that the rate of convergence of the smooth bootstrap d.f. is, roughly speaking, of order $O(n^{-2/5})$ for an appropriate choice of β_n.

P.8. Problems and Supplements

1. (Randomized sample quantiles)
   (i) Define a class of median unbiased estimators of the q-quantile by choosing $X_{r:n}$ with probability $p(r)$ where $\sum_{r=0}^n p(r) = 1$ and

$$\sum_{r=0}^n \sum_{k=r}^n \binom{n}{k} q^k (1-q)^{n-k} p(r) = 1/2.$$

(Pfanzagl, 1985, page 435)

   (ii) Establish a representation corresponding to that in P.1.28 for the randomized sample median.

2. (Testing the q-quantile)
   (i) Let f_0 and f_1 be the densities in (8.1.7). Construct d.f.'s $G_m \in \mathscr{F}_k$ with densities g_m for $m = 1, 2, 3, \ldots$ such that $G_m^{-1}(q) = u$ and $g_m(x) \to f_0(x)$ as $m \to \infty$ for every $x \neq u$.
   (ii) Let φ and $\mathscr{F}_k$ be as in Lemma 8.1.3. Prove that for every critical function ψ of level α such that $E_F\psi = \alpha$ if $F \in \mathscr{F}_k$ and $F^{-1}(q) = u$ the following relations hold:

$$E_F\psi \underset{\geq}{\overset{\leq}{}} E_F\varphi \qquad \text{if } F \in \mathscr{F}_k \text{ and } F^{-1}(q) \underset{>}{\overset{<}{}} u.$$

3. Let φ and $\mathscr{F}_k$ be as in Lemma 8.1.3 and let $\mathscr{G}$ be a sub-family of $\mathscr{F}_k$. For $\varepsilon > 0$ define a "ε-neighborhood" $\mathscr{G}_\varepsilon$ of $\mathscr{G}$ by

$$\mathscr{G}_\varepsilon = \{F \in \mathscr{F}_k : |f - g| \le \varepsilon g \text{ for some } G \in \mathscr{G}\}$$

where f and g denote the differentiable densities of F and G.

  Then for every critical function ψ which has the property

$$E_F \psi \le \alpha \qquad \text{if } F \in \mathscr{G}_\varepsilon \text{ and } F^{-1}(q) \le u$$

we have

$$E_F \psi \le E_F \varphi \qquad \text{if } F \in \mathscr{G}_{\varepsilon/2} \text{ and } q - \frac{\varepsilon q(1 - q)}{4(1 + \varepsilon)} \le F(u) < q.$$

4. (Stochastic properties of kernel density estimator)
   Find conditions under which the density estimator $f_n \equiv F_{n,1}$ [see (8.2.19)] has the following properties:
   (i) $\int f_n(y)\,dy = 1$.
   (ii) $Ef_n(x) = \int u(y)f(x + \beta y)\,dy$.
   (iii) $Ef_n(x) \to f(x)$ as $\beta \to 0$.
   (iv) $|Ef_n(x) - f(x) - \beta^2 f^{(2)}(x) \int u(y)y^2\,dy/2| = O(\beta^2)$.
   (v) $|E[f_n(x) - Ef_n(x)]^2 - (n\beta)^{-1}f(x)\int u^2(y)\,dy| = O(n^{-1})$.
   (vi) Let $\int u(y)y^2\,dy > 0$. Show that

$$\beta = n^{-1/5}\left[f(x)\int u^2(y)\,dy\right]^{1/5}\Bigg/\left[f^{(2)}(x)\int u(y)y^2\,dy\right]^{2/5}$$

minimizes the term

$$\left[\beta^2 f^{(2)}(x)\int u(y)y^2\,dy/2\right]^2 + (n\beta)^{-1}f(x)\int u^2(y)\,dy.$$

For this choice of β, the mean square error of $f_n(x)$ satisfies the relation

$$E[f_n(x) - f(x)]^2$$

$$= n^{-4/5}\frac{5}{4}\left[f(x)\int u^2(y)\,dy\right]^{4/5}\left[f^{(2)}(x)\int u(y)y^2\,dy\right]^{2/5} + o(n^{-4/5}).$$

5. (Orthogonal series estimator)
   For $x \in [0, 1]$ define $e_0(x) = 1$ and

$$e_{2j-1}(x) = 2^{1/2}\cos(2\pi jx)$$

$$e_{2j}(x) = 2^{1/2}\sin(2\pi jx), \qquad j = 1, 2, 3, \ldots.$$

   (i) (a) $e_0, e_1, e_2, \ldots$ are orthonormal [w.r.t. the inner product

$$(f, g) = \int_0^1 f(x)g(x)\,dx].$$

     (b) Let

$$f = 1 + \sum_{j=1}^s a_j e_j\left(= 1 + \sum_{j=1}^s (f, e_j)e_j\right)$$

be a probability density and $\xi_1, \ldots, \xi_n$ i.i.d. random variables with common density f. Then, for every $x \in [0, 1]$,

$$\hat{f}_n(x) = 1 + \sum_{j=1}^{s} \left(n^{-1} \sum_{i=1}^{n} e_j(\xi_i) \right) e_j(x)$$

is an expectation unbiased estimator of $f(x)$ having the integrated variance

$$\int_0^1 \text{Var}(\hat{f}_n(x)) \, dx = n^{-1} \sum_{i=0}^{s} \int_0^1 e_i^2(x) f(x) \, dx - n^{-1} \left(1 + \sum_{i=1}^{s} a_i^2 \right) = O(s/n)$$

(see Prakasa Rao, 1983, Example 2.2.1)

(ii) (Problem) Investigate the asymptotic performance of

$$\int_0^1 (\hat{f}_n(x) - 1)^2 \, dx$$

as a test statistic for testing the uniform distribution on $(0, 1)$ against alternatives as given in (i) (b) with $s \equiv s(n) \to \infty$ as $n \to \infty$.

(Compare with Example 10.4.1.)

6. There exists a constant $C(\rho) > 0$, only depending on $\rho > 0$, such that

$$|N_{(\mu_n, v_n^2)}(\mu - \sigma y n^{-1/2}, \mu + \sigma y n^{-1/2}) - 2\Phi(y[1 + \{1 - (n/\sigma^2)(v_n^2 + (\mu_n - \mu)^2)\}/2]) - 1|$$

$$\leq C(\rho)(\max(|n^{1/2} v_n - \sigma|, n(\mu_n - \mu)^2))^{3/2}$$

for every $y \geq 0$, $v_n > 0$, $\sigma \geq \rho$, $-\infty < \mu_n, \mu < \infty$ and positive integers n.

(Reiss, 1981c)

7. Denote by G_n^{-1} the sample q.f. if $F_{n,0}$ is the underlying d.f. Prove that

$$P_{F_{n,0}}\{(G_n^{-1}(q) - F_{n,0}^{-1}(q))/F_{n,1}^{-1}(q) \in B\}$$

is a more accurate approximation to

$$P_F\{(F_n^{-1}(q) - F^{-1}(q))/(F^{-1})'(q) \in B\}$$

than the bootstrap distribution $T_n(F_{n,0}, \cdot)$ to $T_n(F, \cdot)$.

8. (Generating pseudo-random variables)
   Generate pseudo-random numbers according to the kernel q.f. $F_{n,0}^{-1}$ and the kernel d.f. $F_{n,0}$.

Bibliographical Notes

It was proved by Pfanzagl (1975) that the sample q-quantile (including the sample median) is an asymptotically efficient estimator of the q-quantile (the median) in the class of an asymptotically median unbiased estimators. It is well known that for symmetric densities one can find nonparametric estimators of the symmetry point which are as efficient as parametric estimators; according to Pfanzagl's result a corresponding procedure is not possible if there is even the slightest violation of the symmetry condition.

In Section 8.2 we studied special topics belonging to nonparametric density estimation or, in other words, nonparametric curve estimation. We refer to the book of Prakasa Rao (1983) for a comprehensive account of this field. In data analysis extensive use of histograms, that are closely related to kernel estimators, has been made for a long time. As early as 1944, Smirnov established an interesting mathematical result concerning the maximum deviation of the histogram from the underlying density. Since the celebrated articles of Rosenblatt (1956) and Parzen (1962) much research work has been done in this field.

The kernel estimator of the d.f. was studied by Nadaraya (1964), Yamato (1973), Winter (1973), and Reiss (1981c). It was proved by Falk (1983) that kernels u exist which satisfy condition (8.3.10) as well as Condition 8.3.2 for $k > 1$. Falk (1983) and Mammitzsch (1984) solved the question of optimal choice of kernels in the context of estimating d.f.'s and q.f.'s.

The basic idea behind the kernel estimator of the q-quantile is to average over order statistics close to the sample q-quantile. The most simple case is given by quasi-quantiles which are built by two, or more general by a fixed number k, of order statistics. In the nonparametric context, quasi-quantiles were used by Hodges and Lehmann (1967) in order to estimate the center of a symmetric distribution and by Reiss (1980, 1982) to estimate and test q-quantiles. The kernel estimator of the q.f. was introduced by Parzen (1979) and, independently, by Reiss (1982). The asymptotic performance of the kernel estimator of the q.f. was investigated by Falk (1984a, 1985a). Other notable articles pertaining to this are Brown (1981), Harrell and Davis (1982), and Yang (1985), among others.

The derivative of the q.f. ($\equiv$ quantile density function) can easily be estimated by means of the difference of two order statistics. An estimator of the quantile density function may e.g. be applied to construct confidence bounds for the q-quantile. The estimation of the quantile density function is closely related to the estimation of the density by means of histograms with random cell boundaries. Such histograms were dealt with by Siddiqui (1960), Bloch and Gastwirth (1968), van Ryzin (1973), Tusnády (1974), and Reiss (1975a, 1978). A confidence band, based on the moving scheme (see (8.2.2)), was established in Reiss (1977b) by applying a result for kernel density estimators due to Bickel and Rosenblatt (1973) and a Bahadur approximation result like Theorem 6.3.1.

Another example of the kernel method is provided by smoothing the log survivor function and taking the derivative which leads to a kernel estimator of the hazard function (see Rice and Rosenblatt, 1976). A related estimator of the hazard function was earlier investigated by Watson and Leadbetter (1964a, 1964b).

Sharp results for the almost sure behavior of kernel density estimators were proved by Stute (1982) by applying the result concerning the oscillation of the sample d.f. A notable article pertaining to this is Reiss (1975b).

CHAPTER 9

Extreme Value Models

This chapter is devoted to parametric and nonparametric extreme value models. The parametric models result from the limiting distributions of sample extremes, whereas the nonparametric models contain actual distributions of sample extremes. The statistical inference within the nonparametric framework will be carried out by applying the parametric results.

The importance of parametric statistical procedures for the nonparametric set-up (see also Section 10.4) may possibly revive the interest in parametric problems. However, it is not our intention to give a detailed, exhaustive survey of the various statistical procedures concerning extreme values.

The central idea of our approach will be pointed out by studying the simple—nevertheless important—problem of estimating a parameter α which describes the shape of the distribution in the parametric model and the domain of attraction in the nonparametric model.

In Section 9.1 we give an outline of some important statistical ideas which are basic for our considerations. In particular, we explain in detail the straight-forward and widely adopted device of transforming a given model in order to simplify the statistical inference. A continuation of this discussion can be found in Section 10.1 where the concept of "sufficiency" is included into our considerations.

Sections 9.2 and 9.3 deal with the sampling of independent maxima. Section 9.4 introduces the parametric model which describes the sampling of the k largest order statistics. It is shown that in important cases the given model can be transformed into a model defined by independent observations. The nonparametric counterpart is treated in Section 9.5.

A comparison of the results of Sections 9.3 and 9.5 is given in Section 9.6. The 3-parameter extreme value family contains regular and non-regular sub-families and hence the statistical inference can be intricate. However, the classical model is of a rather limited range; it can be enlarged by adding further parameters as it will be indicated in Section 9.6.

In Section 9.7 we continue our research concerning the evaluation of the unknown q.f. The information that the underlying d.f. belongs to the domain of attraction of an extreme value d.f. is used to construct a competitor of the sample q.f. near the endpoints.

9.1. Some Basic Concepts of Statistical Theory

In the present section we shall recall some simple facts from statistical theory. The first part mainly concerns the estimation of an unknown parameter as e.g. the shape parameter of an extreme value distribution. The second part deals with the comparison of statistical models.

Remarks about Estimation Theory

Consider the fairly general estimation problem where a sequence $\xi_1, \xi_2, \ldots$ of r.v.'s (with common distribution P_θ, $\theta \in \Theta$) is given which enables us to construct a consistent estimator of a real-valued parameter θ as the sample size k tends to infinity.

In applications the sample size k will be predetermined or chosen by the statistician so that the estimation procedure attains a certain accuracy. Then, one faces two problems, namely that of measuring the accuracy of estimators and in a second step that of finding an optimal estimator in order not to waste observations (although in some cases it may be preferable to use quick estimators in order not to waste time).

For an estimator $\theta_k^* \equiv \theta_k^*(\xi_1, \ldots, \xi_k)$ of the parameter θ a widely accepted measure of accuracy is the mean square error

$$E_\theta(\theta_k^* - \theta)^2. \tag{9.1.1}$$

If necessary the expectation is denoted by E_θ etc. instead of E in order to indicate that the r.v.'s $\xi_1, \ldots, \xi_k$ and, thus, the expectation as well depends on the parameter θ. Since

$$E_\theta(\theta_k^* - \theta)^2 = E_\theta(\theta_k^* - E_\theta\theta_k^*)^2 + (E_\theta\theta_k^* - \theta)^2 \tag{9.1.2}$$

we know that the mean square error is the variance if θ_k^* is expectation unbiased.

In general, the accuracy of the estimator can be measured by the expected loss

$$E_\theta L(\theta_k^* | \theta) \tag{9.1.3}$$

where L is an appropriate loss function. Note that A. Wald in his supreme wisdom decided to call $E_\theta L(\theta_k^* | \theta)$ risk instead of expected loss. For a detailed discussion of the problem of comparing estimators and of the definitions of optimal estimators we refer to Pfanzagl (1982), pages 151–154. We indicate some basic facts.

There does not exist a canonical criterion for the selection of an optimal estimator. However, one basic idea for any definition of optimality is to exclude degenerated estimators as e.g. an estimator which is a constant.

An estimator θ_k^* is optimal w.r.t. the global minimax criterion if

$$\sup_\theta E_\theta L(\theta_k^*|\theta) = \inf \sup_\theta E_\theta L(\theta_k^{**}|\theta) \qquad (9.1.4)$$

where the inf is taken over the given class of estimators θ_k^{**}. Notice that (9.1.4) can be modified to a local minimax criterion by taking the sup over a neighborhood of θ_0 for each $\theta_0 \in \Theta$.

The Bayes risk of an estimator θ_k^{**} w.r.t. a "prior distribution λ" is given by the weighted risk

$$\int E_\theta L(\theta_k^{**}|\theta)d\lambda(\theta)$$

where λ is a probability measure on the parameter space Θ equipped with a σ-field. The optimum estimator is now the Bayes estimator θ_k^* which minimizes the Bayes risk; that is,

$$\int E_\theta L(\theta_k^*|\theta)d\lambda(\theta) = \inf \int E_\theta L(\theta_k^{**}|\theta)d\lambda(\theta) \qquad (9.1.5)$$

where the inf is taken over the given class of estimators θ_k^{**}. In certain applications one also considers generalized Bayes estimators where λ is a measure; this generalization e.g. leads to Pitman estimators (compare with (10.1.23)). For a detailed treatment of Bayes and minimax procedures we refer to Ibragimov and Has'minskii (1981) and Witting (1985).

Alternatively, one can try to find an optimal estimator within a class of estimators which satisfy an additional regularity condition. Recall that if the estimators are assumed to be expectation unbiased then the use of (9.1.1) leads to the famous Cramér–Rao bound as a lower bound for the variance. In the nonparametric context (e.g. when estimating a density) one has to admit a certain amount of bias of the estimator to gain a smaller mean square error.

The extension of the concept above to randomized estimators (Markov kernels having their distributions on the parameter space Θ) is straightforward. Notice that $E_\theta L(\theta_k^*|\theta) = \int L(\cdot|\theta)dQ_\theta$ where Q_θ is the distribution of θ_k^*. The extension is easily obtained by putting the distribution of the randomized estimator in place of Q_θ.

A different restriction is obtained by the requirement that the estimator θ_k^* is median unbiased or asymptotically median unbiased (compare with Section 8.1).

Moreover, we shall base our calculations on covering probabilities of the form

$$P\{-t' \le \theta_k^* - \theta \le t''\} \tag{9.1.6}$$

which measure the concentration of the estimator θ_k^* about θ.

Let $L(\theta_1|\theta_2)$ be of the form $L(\theta_1 - \theta_2)$. An estimator θ_k^* which is maximally concentrated about the true parameter θ will also minimize the risk $E_\theta L(\theta_k^* - \theta)$ for bounded, negative unimodal loss functions L having the mode at zero [that is, L is nonincreasing on $(-\infty, 0]$ and nondecreasing on $[0, \infty)$]. This can easily be deduced from P.3.5.

Comparison of Statistical Models

Next we describe the simplest version of the fundamental operation of replacing a given statistical model by another one which might be more accessible to the statistician. The model

$$\mathscr{P} = \{P_\theta: \theta \in \Theta\} \tag{9.1.7}$$

will be replaced by

$$\mathscr{Q} = \{Q_\theta: \theta \in \Theta\}. \tag{9.1.8}$$

The two models can be compared by means of a map T or, in general, by a Markov kernel (the latter case will be dealt with in Chapter 10). The crucial point is that the map T is independent of the parameter θ.

Given $\theta \in \Theta$ and a r.v. ξ with distribution P_θ let $\eta = T(\xi)$ be distributed according to Q_θ.

Then, obviously, for any estimator $\hat{\theta}(\eta)$ operating on $\mathscr{Q}$ [or in greater generality, a statistical procedure] we find an estimator operating on $\mathscr{P}$, namely,

$$\theta^*(\xi) = \hat{\theta}(T(\xi))$$

having the same distribution as $\hat{\theta}(\eta)$.

In terms of risks this yields that for every loss function L

$$E_\theta L(\theta^*(\xi)|\theta) = E_\theta L(\hat{\theta}(\eta)|\theta), \qquad \theta \in \Theta. \tag{9.1.9}$$

An extension of the framework above is needed in Section 9.3 where $\mathscr{Q}$ and $\mathscr{P}$ have different parameter sets. Let $\mathscr{Q}$ be as in (9.1.8) and

$$\mathscr{P} = \{P_{\theta,h}: \theta \in \Theta, h \in H(\theta)\}. \tag{9.1.10}$$

Let T be a map such that for every r.v. ξ with distribution $P_{\theta,h}$ and r.v. η with distribution Q_θ,

$$\sup_B |P\{T(\xi) \in B\} - P\{\eta \in B\}| \le \varepsilon(\theta, h). \tag{9.1.11}$$

This implies (compare with P.3.5) that with $\theta^*(\xi) = \hat{\theta}(T(\xi))$,

$$|E_{\theta,h} L(\theta^*(\xi)|\theta) - E_\theta L(\hat{\theta}(\eta)|\theta)| \le \varepsilon(\theta, h) \sup_t L(t|\theta) \tag{9.1.12}$$

for every loss function $L(\cdot|\theta)$.

For every procedure acting on $\mathcal{2}$, we found a procedure on $\mathcal{P}$ with the same performance (within a certain error bound). Until now we have not excluded the possibility that there exists a procedure on $\mathcal{P}$ which is superior to those carried over from $\mathcal{2}$ to $\mathcal{P}$. However, if T is a one-to-one map (as e.g. in Example 9.1.1), one may interchange the role of $\mathcal{2}$ and $\mathcal{P}$ by taking the inverse T^{-1} instead of T. Thus, the optimal procedure on $\mathcal{P}$ can be regained from the corresponding one on $\mathcal{2}$.

In connection with loss functions the parameter θ is not necessarily real-valued. The extension of the concept to functional parameters is obvious.

EXAMPLE 9.1.1. Section 9.2 will provide a simple example for the comparison of two models. Here, with $\theta = (\sigma, \alpha)$, P_θ is the Fréchet distribution with scale parameter σ and shape parameter $1/\alpha$, and Q_θ is the Gumbel distribution with location parameter $\log \sigma$ and scale parameter α. The transformation T is given by $T = \log$.

Moreover, given a sample of size k one has to take the transformation

$$T(x_1, \ldots, x_k) = (\log x_1, \ldots, \log x_k).$$

A continuation of this discussion can be found in Section 10.1.

9.2. Efficient Estimation in Extreme Value Models

Given a d.f. G denote by $G^{(\mu, \sigma)}$ the corresponding d.f. with location parameter μ and scale parameter σ; thus, we have

$$G^{(\mu, \sigma)}(x) = G((x - \mu)/\sigma).$$

Fréchet and Gumbel Model

The starting point is the scale and shape parameter family of Fréchet d.f.'s $G_{1, 1/\alpha}^{(0, \sigma)}$. We have

$$G_{1, 1/\alpha}^{(0, \sigma)}(x) = \exp(-(x/\sigma)^{-1/\alpha}), \qquad x \geq 0. \tag{9.2.1}$$

The usual procedure of treating the estimation problem is to transform the given model to the location and scale parameter family of Gumbel d.f.'s $G_3^{(\theta, \alpha)}$ where $\theta = \log \sigma$. Notice that if ξ is a r.v. with d.f. $G_{1, 1/\alpha}^{(0, \sigma)}$ then $\eta = \log \xi$ is a r.v. with d.f. $G_3^{(\theta, \alpha)}$.

The density of $G_3^{(\theta, \alpha)}$ will be denoted by $g_3^{(\theta, \alpha)}$.

Gumbel Model: Fisher Information Matrix

For the calculation of the Fisher information matrix within the location and scale parameter family of Gumbel d.f.'s we need the first two moments of the distributions. The following two formulas are well known (see e.g. Johnson and Kotz (1970)):

$$\int x \, dG_3(x) = \int_0^\infty (-\log x) e^{-x} \, dx = \gamma \qquad (9.2.2)$$

where $\gamma = 0.5772 \ldots$ is Euler's constant. Moreover,

$$\int x^2 \, dG_3(x) = \gamma^2 + \pi^2/6. \qquad (9.2.3)$$

From (9.2.2) and (9.2.3) it is obvious that a r.v. η with d.f. $G_3^{(\theta,\alpha)}$ has the expectation

$$E\eta = \theta + \alpha\gamma \qquad (9.2.4)$$

and variance

$$\operatorname{Var} \eta = (\alpha\pi)^2/6. \qquad (9.2.5)$$

The Fisher information matrix can be written as

$$I(\theta_1, \theta_2) = \left[\int \left[\frac{\partial}{\partial \theta_i} \log g_3^{(\theta_1, \theta_2)}(x) \right] \left[\frac{\partial}{\partial \theta_j} \log g_3^{(\theta_1, \theta_2)}(x) \right] dG_3^{(\theta_1, \theta_2)}(x) \right]_{i,j}.$$

By partial integration one can easily deduce from (9.2.4) and (9.2.5) that

$$I(\theta, \alpha) = \alpha^{-2} \begin{bmatrix} 1 & (\gamma - 1) \\ (\gamma - 1) & \pi^2/6 + (1 - \gamma)^2 \end{bmatrix}. \qquad (9.2.6)$$

Check that the inverse matrix $I(\theta, \alpha)^{-1}$ of $I(\theta, \alpha)$ is given by

$$I(\theta, \alpha)^{-1} = (6\alpha^2/\pi^2) \begin{bmatrix} \pi^2/6 + (1 - \gamma)^2 & (1 - \gamma) \\ (1 - \gamma) & 1 \end{bmatrix}. \qquad (9.2.7)$$

Gumbel Model: The Maximum Likelihood Estimator

The maximum likelihood (m.l.) estimator $(\hat{\theta}_k, \hat{\alpha}_k)$ of the location and scale parameters in the Gumbel model is asymptotically normal with mean vector (θ, α) and covariance matrix $k^{-1} I(\theta, \alpha)^{-1}$. The rate of convergence to the limiting normal distribution is of order $O(k^{-1/2})$ (proof!).

In the sequel, the estimators will be written in a factorized form: If the m.l. estimator is based on k i.i.d. random variables $\eta_1, \ldots, \eta_k$ we shall write $\hat{\alpha}_k(\eta_1, \ldots, \eta_k)$ instead of $\hat{\alpha}_k$.

If the r.v.'s $\eta_1, \ldots, \eta_k$ have the common d.f. $G_3^{(\theta,\alpha)}$ then we obtain according to (9.2.7) that

$$P\{(k/V(\alpha))^{1/2} (\hat{\alpha}_k(\eta_1, \ldots, \eta_k) - \alpha) \le t\} \to \Phi(t), \qquad n \to \infty, \qquad (9.2.8)$$

where $V(\alpha) = 6\alpha^2/\pi^2$.

Given the observations $x_1, \ldots, x_k$ the m.l. estimate $\hat{\alpha}_k(x_1, \ldots, x_k)$ is the solution of the two log-likelihood-equations

$$\sum_{i=1}^{k} e^{-(x_i - \theta)/\alpha} = k \qquad (9.2.9)$$

and

$$\sum_{i=1}^{k} (x_i - \theta)[1 - e^{-(x_i - \theta)/\alpha}] = k\alpha. \qquad (9.2.10)$$

Notice that (9.2.9) is equivalent to the equation

$$\theta = -\alpha \log\left[k^{-1} \sum_{i=1}^{k} e^{-x_i/\alpha}\right] \qquad (9.2.11)$$

so that by inserting the expression for θ in (9.2.10) we get the equation

$$g(\alpha) = 0 \qquad (9.2.12)$$

with g defined by

$$g(\alpha) = \alpha - k^{-1} \sum_{i=1}^{k} x_i + \left[\sum_{i=1}^{k} x_i e^{-x_i/\alpha}\right]\bigg/\left[\sum_{i=1}^{k} e^{-x_i/\alpha}\right]. \qquad (9.2.13)$$

Observe that the solution $\hat{\alpha}_k(x_1, \ldots, x_k)$ of the equation (9.2.12) has the following property: For reals θ and $\alpha > 0$ we have

$$\hat{\alpha}_k(\theta + \alpha x_1, \ldots, \theta + \alpha x_k) = \alpha \hat{\alpha}_k(x_1, \ldots, x_k). \qquad (9.2.14)$$

This property yields that there exist correction terms which make the m.l. estimator of α median unbiased. The corresponding result also holds w.r.t. the expectation unbiasedness.

Equation (9.2.12) has to be solved numerically; however, this can hardly be regarded as a serious drawback in the computer era. Approximate solutions can be obtained by the Newton–Raphson iteration procedure. Notice that

$$(6^{1/2}/\pi)s_k(\eta_1, \ldots, \eta_k)$$

may serve as an initial estimator of α where

$$s_k^2(x_1, \ldots, x_k) = (k - 1)^{-1} \sum_{i=1}^{k} \left[x_i - k^{-1} \sum_{j=1}^{k} x_j\right]^2$$

is the sample variance. The asymptotic performance of $(6^{1/2}/\pi)s_k$ is indicated in P.9.2. We remark that the first iteration leads to

$$\alpha_k^* = (6^{1/2}/\pi)s_k - g[(6^{1/2}/\pi)s_k]/g'[(6^{1/2}/\pi)s_k]. \qquad (9.2.15)$$

The estimator $\alpha_k^*(\eta_1, \ldots, \eta_k)$ has the same asymptotic performance as the m.l. estimator. Further iterations may improve the finite sample properties of the estimator.

From (9.2.11) we know that the m.l. estimator of the location parameter is given by

$$\hat{\theta}_k(\eta_1, \ldots, \eta_k) = -\hat{\alpha}_k(\eta_1, \ldots, \eta_k)\log\left[k^{-1} \sum_{i=1}^{k} e^{-x_i/\hat{\alpha}_k(\eta_1, \ldots, \eta_k)}\right]. \qquad (9.2.16)$$

Efficient Estimation of α

Let us concentrate on estimating the parameter α.

(9.2.14) yields that (9.2.8) holds uniformly over the location and scale parameters θ and α. A further consequence is that the m.l. estimator is asymptotically efficient in the class of all estimators $\alpha_k^*(\eta_1, \ldots, \eta_k)$ which are asymptotically median unbiased in a locally uniform way. For such estimators we get for every $t', t'' > 0$,

$$P\{-t'k^{-1/2} \le \alpha_k^*(\eta_1, \ldots, \eta_k) - \alpha \le t''k^{-1/2}\}$$
$$\le P\{-t'k^{-1/2} \le \hat{\alpha}_k(\eta_1, \ldots, \eta_k) - \alpha \le t''k^{-1/2}\} + o(k^0). \tag{9.2.17}$$

We return to the Fréchet model of d.f.'s $G_{1,1/\alpha}^{(0,\sigma)}$ with scale parameter σ and shape parameter $1/\alpha$. The results above can easily be made applicable to the Fréchet model.

If $\xi_1, \ldots, \xi_k$ are i.i.d. random variables with common d.f. $G_{1,1/\alpha}^{(0,\sigma)}$ then it follows from (9.2.8) and the discussion in Section 9.1 that

$$P\{(k/V(\alpha))^{1/2}(\hat{\alpha}_k(\log \xi_1, \ldots, \log \xi_k) - \alpha) \le t\} \to \Phi(t), \qquad n \to \infty. \tag{9.2.18}$$

The rate of convergence in (9.2.18) is again of order $O(k^{-1/2})$. Moreover, the efficiency of $\hat{\alpha}_k(\eta_1, \ldots, \eta_k)$ as an estimator of the scale parameter of the Gumbel distribution carries over to $\hat{\alpha}_k(\log \xi_1, \ldots, \log \xi_k)$ as an estimator of the shape parameter of the Fréchet distribution.

9.3. Semiparametric Models for Sample Maxima

The parametric models as studied in Section 9.2 reflect the ideal world where we are allowed to replace the actual distributions of sample maxima by the limiting ones. By stating that the parametric model is an approximation to the real world one acknowledges that the parametric model is incorrect although in many cases the error of the approximation can be neglected.

In the present section we shall study a nonparametric approach, give some bounds for the error of the parametric approximation and discuss the meaning of a statistical decision within the parametric model for the nonparametric model.

Fréchet Type Model

We observe the sample maximum $X_{m:m}$ of m i.i.d. random variables with common d.f. F belonging to the domain of attraction of a Fréchet d.f. $G_{1,1/\alpha}$. Our aim is to find an estimator of the shape parameter α.

More precisely, we assume that F is close to a Pareto d.f. $W_{1,1/\alpha}^{(0,\sigma)}$ (with unknown scale parameter σ) in the following sense: F has a density f satisfying the condition

$$f(x) = (\sigma\alpha)^{-1}(x/\sigma)^{-(1+1/\alpha)}e^{h(x/\sigma)} \quad \text{for } x \geq (x_0\sigma)^{-\alpha} \tag{9.3.1}$$

where $x_0 > 0$ is fixed and h is a (measurable) function such that

$$|h(x)| \leq L|x|^{-\delta/\alpha}$$

for some constants $L > 0$ and $\delta > 0$.

Condition (9.3.1) is formulated in such a way that the results will hold uniformly over σ and α. It is apparent that the Pareto and Fréchet densities satisfy this condition with $h = 0$ and, respectively, $h(x) = -x^{-1/\alpha}$.

The present model can be classified as a semiparametric (in other words, semi-nonparametric) model where the shape parameter α and (or) the scale parameter σ have to be evaluated and the function h is a nonparametric nuisance parameter which satisfies certain side conditions.

Let $X_{m:m}^{(1)}, \ldots, X_{m:m}^{(k)}$ be independent repetitions of $X_{m:m}$. The joint distribution of $X_{m:m}^{(1)}, \ldots, X_{m:m}^{(k)}$ will heavily depend on the parameters σ and α whereas the dependence on h, x_0, and L can be neglected if m is sufficiently large and k is small compared to m.

Let $\xi_1, \ldots, \xi_k$ be i.i.d. random variables with common d.f. $G_{1,1/\alpha}$. From (3.3.12) and Corollary 5.2.7 it follows that

$$\sup_B |P\{(X_{m:m}^{(1)}, \ldots, X_{m:m}^{(k)}) \in B\} - P\{(\sigma m^{\alpha}\xi_1, \ldots, \sigma m^{\alpha}\xi_k) \in B\}|$$
$$= O(k^{1/2}(m^{-\delta} + m^{-1})) \tag{9.3.2}$$

uniformly over k, m and densities f which satisfy (9.3.1) for some fixed values x_0, L and δ. Notice that $\sigma m^{\alpha}\xi_i$ has the d.f. $G_{1,1/\alpha}^{(0,\sigma m^{\alpha})}$.

Let again $\hat{\alpha}_k$ be the solution of the m.l. equation (9.2.12). Combining (9.2.18) and (9.3.2) we get

Theorem 9.3.1.

$$P\{(k/V(\alpha))^{1/2}[\hat{\alpha}_k(\log X_{m:m}^{(1)}, \ldots, \log X_{m:m}^{(k)}) - \alpha] \leq t\}$$
$$= \Phi(t) + O(k^{1/2}(m^{-\delta} + m^{-1}) + k^{-1/2}) \tag{9.3.3}$$

uniformly over t, k, m and densities f which satisfy condition (9.3.1) for some fixed constants x_0, L, and δ. Moreover, $V(\alpha) = 6\,\alpha^2/\pi^2$.

The properties of the m.l. estimator carry over from the parametric to the nonparametric framework.

Sample Maxima within a Fixed Period

If the practitioner insists on observing the data within a fixed period, then it is necessary to modify the results above since now the sample size is random.

This situation e.g. occurs in insurance mathematics. So let us speak for a while in terms of claims and claim sizes.

Assume that the claims come in according to a Poisson process $N(s)$, $s \geq 0$, and that independently the claim sizes $\eta_1, \ldots, \eta_k$ have the common density f which satisfies condition (9.3.1). Thus, the number of claims within a period of length s will be $N(s)$. The claims will be arranged in k groups. Write

$$M \equiv M(s, k) = [N(s)/k]. \tag{9.3.4}$$

Denote by $X_{M:M}^{(i)}$ the maximum claim size of the r.v.'s $\eta_{(i-1)M+1}, \ldots, \eta_{iM}$. Thus, using the notation of (1.1.4) we get the representation

$$X_{M:M}^{(i)} = Z_{M:M}(\eta_{(i-1)M+1}, \ldots, \eta_{iM}). \tag{9.3.5}$$

In analogy to Theorem 9.3.1 we get

Theorem 9.3.2.

$$P\{(k/V(\alpha))^{1/2}[\hat{\alpha}_k(\log X_{M:M}^{(1)}, \ldots, \log X_{M:M}^{(k)}) - \alpha] \leq t\}$$

$$= \Phi(t) + O\left[k^{1/2}\left[\sum_{m=0}^{\infty}(m^{-\delta} + m^{-1})P\{M = m\}\right] + k^{-1/2}\right] \tag{9.3.6}$$

uniformly over t, k, m and densities f which satisfy condition (9.3.1) for some fixed constants x_0, L, and δ. Moreover, $V(\alpha) = 6\,\alpha^2/\pi^2$.

PROOF. Writing $\alpha_k^* = \hat{\alpha}_k(\log X_{M:M}^{(1)}, \ldots, \log X_{M:M}^{(k)})$ and conditioning on M we get

$$P\{\alpha_k^* \leq t\} = \sum_{m=0}^{\infty} P(\alpha_k^* \leq t | M = m)P\{M = m\}$$

$$= \sum_{m=0}^{\infty} P\{\hat{\alpha}_k(\log X_{m:m}^{(1)}, \ldots, \log X_{m:m}^{(k)}) \leq t\}P\{M = m\}$$

with $X_{m:m}^{(i)}$ as in Theorem 9.3.1. Now the assertion is immediate since Theorem 9.3.1 holds uniformly over m. $\qquad\square$

If the distribution of M is highly concentrated about a fixed value, say m, then it is apparent that the right-hand side of (9.3.6) is again that of (9.3.3).

Another interesting problem arises if k periods of length $t_i - t_{i-1}$ are fixed. Notice that the claim numbers $N(t_1)$, $N(t_2) - N(t_3)$, $\ldots$, $N(t_k) - N(t_{k-1})$ of the k periods are independent. Again the statistical inference can be based on the maximum claim sizes of each period. After conditioning on the claim numbers the maximum claim sizes can again approximately be represented by independent Gumbel r.v.'s which, however, are not identically distributed.

9.4. Parametric Models Belonging to Upper Extremes

In Section 9.2 we studied the classical problem of evaluating the unknown parameter in the extreme value model by means of estimators based on i.i.d. random variables. A model of a different kind arises in connection with the limiting joint distributions $G_{i,\alpha,\mathbf{k}}$ of the k largest extremes of a sample of size n as introduced in Section 5.3. More precisely, one has to speak of approximate distributions when the number $k = k(n)$ of extremes goes to infinity as n goes to infinity.

Now the statistical procedures will be based on k r.v.'s which are dependent. However, we shall only study certain sub-models which can be transformed to models involving i.i.d. random variables.

Fréchet Type Model

First we examine a model that corresponds to that in (9.2.1), namely,

$$\{G^{(0,\sigma)}_{1,1/\alpha,\mathbf{k}}\colon \sigma > 0, \alpha > 0\} \tag{9.4.1}$$

with location parameter 0 and scale parameter σ. This model arises out of the Fréchet distributions $G_{1,1/\alpha}$. The model in (9.4.1) can be transformed to the model

$$\{Q^{k-1}_\alpha \times G^{(0,\sigma)}_{1,1/\alpha,k}\colon \sigma > 0, \alpha > 0\} \tag{9.4.2}$$

where Q_α is the exponential distribution with scale parameter α and $G^{(0;\sigma)}_{1,1/\alpha,k}$ is the kth marginal distribution of $G^{(0,\sigma)}_{1,1/\alpha,\mathbf{k}}$.

More precisely, if $(\xi_1, \ldots, \xi_k)$ is a random vector with distribution $G^{(0;\sigma)}_{1,1/\alpha,\mathbf{k}}$ then according to (5.3.3), (1.6.14), and Corollary 1.6.11(iii) the random vector

$$(\eta_1, \ldots, \eta_k) := (\log(\xi_1/\xi_2), 2\log(\xi_2/\xi_3), \ldots, (k-1)\log(\xi_{k-1}/\xi_k), \xi_k) \tag{9.4.3}$$

has the distribution $Q^{k-1}_\alpha \times G^{(0,\sigma)}_{1,1/\alpha,k}$.

Exponential Model

The statistical inference is particularly simple in the exponential model

$$\{Q^{k-1}_\alpha\colon \alpha > 0\}. \tag{9.4.4}$$

Asymptotically, one does not lose information by restricting model (9.4.2) to model (9.4.4) as far as the evaluation of the parameter α is concerned (proof!).

The m.l. estimator

$$\hat{\alpha}_{k-1}(\eta_1, \ldots, \eta_{k-1}) = (k-1)^{-1} \sum_{i=1}^{k-1} \eta_i \tag{9.4.5}$$

is an (asymptotically) efficient estimator of α. This estimator is expectation unbiased and has the variance

$$\text{Var}(\hat{\alpha}_k(\eta_1, \ldots, \eta_k)) = \alpha^2/k. \tag{9.4.6}$$

Moreover, the Fisher information $I(\alpha)$ is given by

$$I(\alpha) = \int \left[\frac{\partial}{\partial \alpha} \log[\exp(-x/\alpha)/\alpha] \right]^2 dQ_\alpha(x) = \alpha^{-2}, \qquad (9.4.7)$$

thus, $\hat{\alpha}_k(\eta_1, \dots, \eta_k)$ attains the Cramér–Rao bound $(kI(\alpha))^{-1}$. The central limit theorem yields the asymptotic normality of $\hat{\alpha}_k(\eta_1, \dots, \eta_k)$. We have

$$P\{(kI(\alpha))^{1/2}(\hat{\alpha}_k(\eta_1, \dots, \eta_k) - \alpha) \le t\} = \Phi(t) + o(1). \qquad (9.4.8)$$

Moreover, (9.4.8) holds with $O(k^{-1/2})$ in place of $o(1)$ according to the Berry–Esséen theorem.

Corresponding to the results of Section 9.2, the m.l. estimator is asymptotically efficient within the class of all locally uniformly asymptotically median unbiased estimators $\alpha_k^*(\eta_1, \dots, \eta_k)$. For $t', t'' > 0$ we get

$$P\{-t'k^{-1/2} \le \alpha_k^*(\eta_1, \dots, \eta_k) - \alpha \le t''k^{-1/2}\}$$
$$\le P\{-t'k^{-1/2} \le \hat{\alpha}_k(\eta_1, \dots, \eta_k) - \alpha \le t''k^{-1/2}\} + o(1). \qquad (9.4.9)$$

9.5. Inference Based on Upper Extremes

In analogy to the investigations in Section 9.3 we are going to examine the relation between the actual model of distributions of upper extremes and the model built by limiting distributions

$$\{G_{1,1/\alpha,\mathbf{k}}^{(0,\sigma)} : \sigma > 0, \alpha > 0\}$$

as introduced in Section 9.4. Let f be a density which satisfies condition (9.3.1), that is,

$$f(x) = (\sigma\alpha)^{-1}(x/\sigma)^{-(1+1/\alpha)}e^{h(x/\sigma)} \quad \text{for } x \ge (x_0\sigma)^{-\alpha} \qquad (9.5.1)$$

where x_0 is fixed and h is a (measurable) function such that

$$|h(x)| \le L|x|^{-\delta/\alpha}$$

for some constants $L > 0$ and $\delta > 0$.

Contrary to Section 9.3, the statistical inference will now be based on the k upper extremes $(X_{n:n}, \dots, X_{n-k+1:n})$ of a sample of n i.i.d. random variables with common density f.

The distribution of $(X_{n:n}, \dots, X_{n-k+1:n})$ will heavily depend on the parameters α and σ whereas the dependence on h, x_0, and L can be neglected if n is sufficiently large and k is small compared to n.

It is immediate from Corollary 5.5.5 that

$$\sup_B |P\{(X_{n:n}, \dots, X_{n-k+1:n}) \in B\} - G_{1,1/\alpha,\mathbf{k}}^{(0,\sigma n^\alpha)}(B)| = O(k/n)^\delta k^{1/2} + k/n)) \qquad (9.5.2)$$

uniformly over n, $k \in \{1, \dots, n\}$ and densities f which satisfy (9.5.1) for some fixed constants δ, L, and x_0.

Thus, the transformation as introduced in (9.4.3) yields

$$\sup_B \left| P\left\{ \left(\log\frac{X_{n:n}}{X_{n-1:n}}, \ldots, (k-1)\log\frac{X_{n-k+2:n}}{X_{n-k+1:n}}, X_{n-k+1:n} \right) \in B \right\} \right.$$

$$\left. - (Q_\alpha^{k-1} \times G_{1,1/\alpha,k}^{(0,\sigma n^\alpha)})(B) \right| = O((k/n)^\delta k^{1/2} + k/n)). \quad (9.5.3)$$

The optimal estimator in the exponential model $\{Q_\alpha^{k-1} : \alpha > 0\}$ with unknown scale parameter α (compare with Section 9.4) is the m.l. estimator $\hat\alpha_k(\eta_1, \ldots, \eta_k) = (k-1)^{-1} \sum_{i=1}^{k-1} \eta_i$ where $\eta_1, \ldots, \eta_k$ are i.i.d. random variables with common distribution Q_α. Thus, within the error bound given in (9.5.3) the estimator

$$\alpha_{k,n}^* = (k-1)^{-1} \sum_{i=1}^{k-1} i \log(X_{n-i+1:n}/X_{n-i:n})$$

$$= \left[(k-1)^{-1} \sum_{i=1}^{k-1} \log X_{n-i+1:n} \right] - \log X_{n-k+1:n} \quad (9.5.4)$$

has the same performance as the m.l. estimator $\hat\alpha_k(\eta_1, \ldots, \eta_k)$ as far as covering probabilities are concerned. We remark that $\alpha_{k,n}^*$ is Hill's (1975) estimator. The optimality property carries over from $\hat\alpha_k(\eta_1, \ldots, \eta_k)$ to $\alpha_{k,n}^*$. From (9.5.3) we get for $t', t'' > 0$,

$$P\{-t'\alpha k^{-1/2} \le \alpha_{k,n}^* - \alpha \le t''\alpha k^{-1/2}\}$$

$$= P\{k(1 - t'k^{-1/2}) \le \gamma_{k-1} \le k(1 + t''k^{-1/2})\} + O((k/n)^\delta k^{1/2} + k/n) \quad (9.5.5)$$

$$= \Phi(t'') - \Phi(-t') + O[(k/n)^\delta k^{1/2} + k/n + k^{-1/2}]$$

where γ_{k-1} is a gamma r.v. with parameter $k-1$.

From (9.5.5) we see that the gamma approximation is preferable to the normal approximation if k is small. From an Edgeworth expansion of length 2 one obtains that the term $k^{-1/2}$ in the 3rd line of (9.5.5) can be replaced by k^{-1} if $t' = t''$.

9.6. Comparison of Different Approaches

In the Sections 9.3 and 9.5 we studied the nonparametric model given by densities f of the form

$$f(x) = (\sigma\alpha)^{-1}(x/\sigma)^{-(1+1/\alpha)} e^{h(x/\sigma)} \quad \text{for } x \ge (x_0\sigma)^{-\alpha} \quad (9.6.1)$$

where h satisfies the condition

$$|h(x) \le L|x|^{-\delta/\alpha}.$$

Let $n = mk$. Given the i.i.d. random variables $\xi_1, \ldots, \xi_n$ with common density f let $X_{m:m}^{(j)}$ be the maximum based on the jth subsample of r.v.'s $\xi_{(j-1)m+1}, \ldots, \xi_{jm}$ for $j = 1, \ldots, k$. Moreover, $X_{n-k+1:n}, \ldots, X_{n:n}$ are the k largest order statistics of $\xi_1, \ldots, \xi_n$. We write

$$\hat{\alpha}_{k,n} = \hat{\alpha}_k(\log X_{m:m}^{(1)}, \ldots, \log X_{m:m}^{(k)}) \tag{9.6.2}$$

where $\hat{\alpha}_k$ is the solution of (9.2.12). From (9.3.3) we know that for every t,

$$P\{(k\pi^2/6)^{1/2}\alpha^{-1}(\hat{\alpha}_{k,n} - \alpha) \le t\}$$
$$= \Phi(t) + O[(k/n)^\delta k^{1/2} + k^{3/2}/n + k^{-1/2}]. \tag{9.6.3}$$

Recall from (9.5.6) that Hill's estimator $\alpha_{k,n}^*$, which is based on the k largest order statistics, has the following property:

$$P\{k^{1/2}\alpha^{-1}(\alpha_{k,n}^* - \alpha) \le t\}$$
$$= \Phi(t) + O[(k/n)^\delta k^{1/2} + k/n + k^{-1/2}] \tag{9.6.4}$$

for every t.

A comparison of (9.6.3) and (9.6.4) shows that the asymptotic relative efficiency of Hill's estimator $\alpha_{k,n}^*$ w.r.t. the estimator $\hat{\alpha}_{k,n}$, based on the sample maxima of subsamples, is given by

$$\text{ARE}(\alpha_{k,n}^*, \hat{\alpha}_{k,n}) = 0.6079 \ldots \tag{9.6.5}$$

Thus, Hill's estimator is asymptotically inefficient if both estimators are based on the same number $k \equiv k(n)$ of observations (where, of course, the error bound in (9.6.3) and (9.6.4) has to go to zero as $n \to \infty$). Notice that the error bounds in (9.6.3) and (9.6.4) are of the same order if $\delta \le 1$ which is perhaps the most interesting case. A numerical comparison of both estimators for small sample sizes showed an excellent agreement to the asymptotic results.

The crucial point is the choice of the number k. This problem is similar to that of choosing the bandwidth in the context of kernel density estimators as discussed in Section 8.2.

The above results are applicable if $(k/n)^\delta k^{1/2}$ is sufficiently small where for the sake of simplicity it is assumed that $\delta \le 1$. On the other hand, the relations (9.6.3) and (9.6.4) show that k should be large to obtain estimators of a good performance. This leads to the proposal to take

$$k = cn^{2\delta/(2\delta+1)} \tag{9.6.6}$$

for some appropriate choice of the constant c. If δ is known to satisfy a condition $0 < \delta_0 \le \delta \le 1$, where δ_0 is known, then one may take k as in (9.6.6) with δ replaced by δ_0.

Within a smaller model, that is, the densities f satisfy a stronger regularity condition, it was proved by Hall and Welsh (1985) that δ can consistently be estimated from the data obtaining in this way an adaptive version of Hill's estimator. S. Csörgő et al. (1985) were able to show that the bias term of Hill's estimator (and of related estimators) restricts the choice of the number k; the balance between the variance and the bias determines the performance of the estimator and the optimal choice of k. These results are proved under conditions weaker than that given in (5.2.18). By using (5.2.18), thus strengthening

(9.6.1), we may suppose that the density f satisfies the condition

$$f(x) = (\sigma\alpha)^{-1}(x/\sigma)^{-(1+1/\alpha)}(1 - K(x/\sigma)^{-\rho/\alpha} + h(x/\sigma)), \qquad x \geq (x_0\sigma)^{-\alpha}, \quad (9.6.7)$$

where

$$|h(x)| \leq L|x|^{-\delta/\alpha}$$

and $0 < \rho \leq \delta \leq 1$. According to the results of Section 5.2, the expansion of length 2 of the form

$$G_{1,1/\alpha}(x/\sigma)\left(1 + m^{-\rho}\frac{K}{1 + \rho}(x/\sigma)^{-(1+\rho)\alpha}\right) \qquad (9.6.8)$$

provides a better approximation to the normalized d.f. of the maximum $X_{m:m}^{(j)}$ than the Fréchet d.f. $G_{1,1/\alpha}$.

The d.f.'s in (9.6.8) define an extended extreme value model that contains the classical one for $K = 0$. Notice that the restricted original model of distributions of sample maxima is approximated by the extended extreme value model with a higher accuracy.

The approach, developed in this chapter, is again applicable. By constructing an estimator of α in the extended model one is able to find an estimator of α in the original model of densities satisfying condition (9.6.7). The details are carried out in Reiss (1989).

It is needless to say that our approach also helps to solve various other problems. We mention two-sample problems or, more general, m-sample problems. If every sample consists of the k largest order statistics with $k \geq 2$ and m tends to infinity then one needs modified versions of the results of Section 5.5, namely, a formulation w.r.t. the Hellinger distance instead of the variational distance, to obtain sharp bounds for the remainder terms of the approximations. Such situations are discussed in articles by R.L. Smith (1986), testing the trend of the Venice sea-level, and I. Gomes (1981).

9.7. Estimating the Quantile Function Near the Endpoints

Let us recall the basic idea standing behind the method adopted in Section 8.2 to estimate the underlying q.f. F^{-1}. Under the condition that F^{-1} has bounded derivatives it is plausible to use an estimator which also has bounded derivatives. Thus, the sample q.f. F_n^{-1} has been smoothened by means of an appropriate kernel. One has to choose a bandwidth which controls to some extent the degree of smoothness of the resulting kernel estimator $F_{n,0}^{-1}$.

For q being close to 0 or 1 the required smoothness condition imposed on F^{-1} will only hold for exceptional cases. So if no further information about F^{-1} is available it is advisable to reduce the degree of smoothing when q approaches 0 or 1 (as it was done in Section 8.2).

However, for q close to 0 or 1 we are in the realm of extreme value theory. In many situations the statistician will accept the condition that the underlying d.f. F belongs to the domain of attraction of an extreme value distribution. As pointed out in Section 5.1 this condition can be interpreted in the way that the tail of F lies in a neighborhood of a generalized Pareto distribution $W_{i,\alpha}$ with shape parameter α.

This suggests to estimate the unknown q.f. F^{-1} near the endpoints by means of the q.f. of a generalized Pareto q.f. where the unknown parameters are replaced by estimates.

When treating the full extreme value model then it is advisable to make use of the von Mises parametrization of generalized Pareto distributions as given in Section 5.1. Then, in a first step, one has to estimate the unknown parameters. As already pointed out the full 3-parameter model contains regular as well as non-regular sub-models so that a satisfactory treatment of this problem seems to be quite challenging from the mathematical point of view.

In practice the statistician will often be able to specify a certain submodel. We shall confine ourselves to the treatment of the upper tails of d.f.'s F which belong to a neighborhood of a Pareto d.f. $W_{1,1/\alpha}^{(0,\sigma)}$ with scale parameter σ. Thus,

$$W_{1,1/\alpha}^{(0,\sigma)}(x) = 1 - (x/\sigma)^{-1/\alpha}, \qquad x > \sigma, \tag{9.7.1}$$

and the q.f. is given by

$$(W_{1,1/\alpha}^{(0,\sigma)})^{-1}(q) = \sigma(1-q)^{-\alpha}, \qquad 0 < q < 1. \tag{9.7.2}$$

The estimator G_n^{-1} is defined by

$$G_n^{-1}(q) = \begin{cases} F_{n,0}^{-1}(q) & q \le x_0 \\ F_{n,0}^{-1}(x_0)\left(\dfrac{1-q}{1-x_0}\right)^{-\alpha_n^*} & \text{if} \quad x_0 < q \end{cases} \tag{9.7.3}$$

where $F_{n,0}^{-1}$ is the kernel q.f. as defined in Section 8.2 and α_n^* is the Hill estimator defined in (9.5.4).

In Figures 9.7.1 and 9.7.2, $n = 100$ pseudo-random numbers were drawn according to the standard Fréchet d.f. $G_{1,1}$. The point x_0 was chosen to be equal to 0.9; the estimate of α is equal to 1.012.

In Figure 9.7.1 the inverse $(F_{n,0})^{-1}$ of the kernel estimator of the d.f. cannot visually be distinguished from the sample q.f. F_n^{-1} (compare this with the remarks to Figure 8.2.6).

As indicated above, the philosophy behind this procedure is the following: Up to some point x_0 we only have information that the underlying q.f. is smooth, thus, the kernel method is applicable. Beyond the point x_0 we are in the realm of extreme value theory, and hence, the use of a Pareto tail with estimated parameters may be appropriate. The choice of the point x_0 is crucial. There seems to be some relationship to the well-known problem of estimating

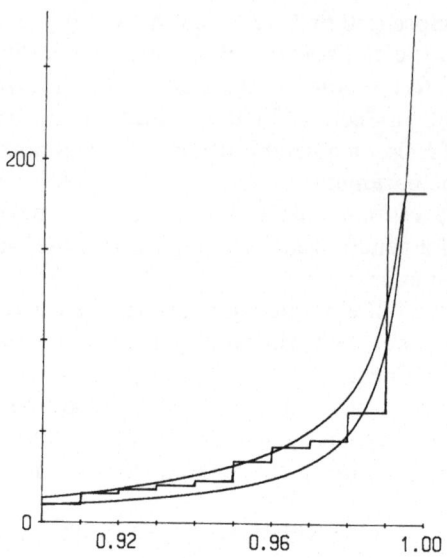

Figure 9.7.1. $G_{1,1}^{-1}$, F_n^{-1}, $F_{n,0}^{-1}$, $(F_{n,0})^{-1}$ with $\beta = 0.08$.

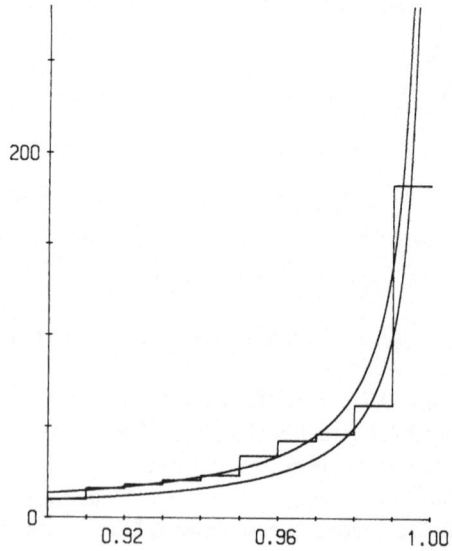

Figure 9.7.2. $G_{1,1}^{-1}$, F_n^{-1}, and estimated Pareto tail G_n^{-1}.

a change point of a sequence of r.v.'s where the underlying distributions changes the parameter after an unknown time point.

P.9. Problems and Supplements

1. Prove that there exists a unique solution of the log-likelihood-equations (9.2.9) and (9.2.10) provided the values $x_1, \ldots, x_k$ are not identical.

2. (Estimators based on sample mean and sample deviation)
   Let $\eta_1, \ldots, \eta_k$ be i.i.d. random variables with mean μ and variance σ^2. Denote by μ_i the ith central moment, by m_k the sample mean and by s_k^2 the sample variance.
   (i) Prove that $k^{1/2}(m_k - \mu, s_k^2 - \sigma^2)$ is asymptotically normal with mean vector zero and covariance matrix given by $\sigma_{1,1} = \sigma^2$, $\sigma_{1,2} = \sigma_{2,1} = \mu_3$, $\sigma_{2,2} = \mu_4 - \sigma^4$.

   (see Serfling, 1980, page 114)

   (ii) Prove the corresponding result for the sample mean m_k and the sample standard deviation s_k.

   [Hint: Apply (i) and Theorem A, Serfling, 1980, page 122.]

   (iii) Let $\eta_1, \ldots, \eta_k$ be i.i.d. random variables with common Gumbel d.f. $G_3^{(\theta,\alpha)}$ where θ and α denote the location and scale parameters. Define

   $$\theta_k^* = m_k - \gamma \alpha_k^* \quad \text{and} \quad \alpha_k^* = (6^{1/2}/\pi)s_k$$

   where γ is Euler's constant. Prove that $k^{1/2}(\theta_k^* - \theta, \alpha_k^* - \alpha)$ is asymptotically normal with mean vector zero and covariance matrix given by

   $$\sigma_{1,1} = [\pi^2/6 + \gamma^2(\beta_2 - 1)/4 - \gamma\pi\beta_1/6^{1/2}]/\alpha^2,$$

   $$\sigma_{1,2} = \sigma_{2,1} = [\beta_1 - 6^{1/2}\gamma(\beta_2 - 1)/2\pi]\pi^2\alpha^2/12,$$

   $$\sigma_{2,2} = (\beta_2 - 1)\alpha^2/4,$$

   where $\beta_1 = \mu_3/\mu_2^{3/2}$ and $\beta_2 = \mu_4/\mu_2^2$.
   (see Tiago de Oliveira, 1963, and Johnson and Kotz, 1970)

3. (Estimators based on order statistics)
   Prove a result corresponding to that in P.9.2(iii) by using estimators as given in (6.2.6) and (6.2.7).

4. In Figure 9.7.2 we see that the second largest and largest observations are about 60 and 180.
   (i) Let $X_{r:n}$ be the rth order statistic of n i.i.d. random variables with common Pareto d.f. $W_{1,\alpha}$. Then, for $u \geq 1$,

   $$P\{X_{n:n} > uX_{n-1:n}\} = u^{-\alpha}.$$

   [Hint: Apply Corollary 1.6.12(ii).]

   (ii) Let $\alpha = 1$ as in Figure 9.7.2. Notice that

   $$P\{X_{n:n} > 3X_{n-1:n}\} = 1/3.$$

Bibliographical Notes

In this book we primarily explore the distributional properties of order statistics and relations between models of actual distributions of order statistics and approximate, parametric models. Statistical procedures are studied as examples to show in which way parametric statistical procedures become relevant within the nonparametric context.

A proper place for an exhaustive list of a greater number of parametric statistical procedures in extreme value models is a book like that of Johnson and Kotz (1970): Chapters 18, 20, and 21 deal with exponential, Weibull, and Gumbel models. We will only give a summary by using keywords out of these chapters: Maximum likelihood (m.l.), minimum variance unbiased, Bayesian, censoring, quick estimators, method of moments, best linear unbiased.

One might add (compare with Herbach (1984) and Mann (1984)) the additional keywords: Best linear invariant, unbiased nearly best linear, simplified linear.

Further references may be found in the following articles. Smith (1985a) studied the asymptotic behavior of m.l. estimators in nonregular models like the Weibull model; see also Polfeldt (1970). By the way, see Reiss (1973, 1978b) and Pitman (1979) for consistency results concerning m.l. estimators in models of unimodal d.f.'s and, respectively, in models with location and scale parameters. New quick estimators of location and scale parameters in the Gumbel model have been proposed by Hüsler and Schüpbach (1986). Quick tests and a locally most powerful test have been studied by van Montfort and Gomes (1985) for testing Gumbel d.f.'s against Fréchet and Weibull alternatives.

In Section 9.2 we mentioned the asymptotic normality of the m.l. estimator of the location and scale parameter in the Gumbel model. Higher order approximations of the distribution of the m.l. estimator can be obtained by means of expansions. These expansions may e.g. be applied to establish asymptotic median unbiasedness of a higher order. We refer to R. Michel (1975) for expansions in the case of vector parameters and to Miebach (1977) for a specialization of these results to families with location and scale parameters.

Next, we make some further comments about the estimation of the tail index and related problems. The statistical extreme value theory is based on the idea that the parametric extreme value model is an approximation of the model of actual distributions of maxima. This idea was made rigorous by Weiss (1971) in a particular case by treating a model of densities in a neighborhood of Weibull densities. Weiss constructed quick estimators of the location, scale, and shape ($\equiv$ tail index) parameter based on extreme and intermediate order statistics. The estimator of the tail index is based on two intermediate order statistics. This is of interest because an alternative approach, namely, the use of the k largest order statistics, with k being fixed, fails to entail consistent estimators. The article of Hill (1975) attracted more attention than that of Weiss. Presumably, the reason for this is that Hill's estimator is efficient and, moreover, is related to the m.l. estimator of the scale parameter in the exponential model. Notice that the estimation of the tail index based on the k largest order statistics, with k fixed and $n \to \infty$, is equivalent to estimating the scale parameter in the exponential model for the fixed sample size k. Hill's estimator and related estimators were extensively studied in literature [e.g. de Haan and Resnick (1980), Hall (1982b), Hall and Welsh (1984), Häusler and Teugels (1985), and Smith (1987)]. The estimation of the endpoint of d.f.'s in the Weibull case was treated by Hall (1982a).

Falk (1985b) took up Weiss' approach of approximating models and derived the properties, as essentially known in literature (Hall (1982b), Häusler and Teugels (1985)), of Hill's estimator by using the properties of the m.l. estimator in the exponential model (compare with Sections 9.4 and 9.5).

The method of taking maxima of subsamples is due to Gumbel; a typical example is to take annual maxima. The results of Sections 9.5 and 9.6 are partly taken from Reiss (1987). A comparison of the two different methods, namely, to base the inference on the k largest order statistics and, respectively, to use the subsample method, was also carried out in the paper by Hüsler and Tiago de Oliveira (1988) within a parametric framework.

The estimation of the parameters of extreme value d.f.'s is related to the estimation of the q.f. near to the ends of the support (see Section 9.7). This subject was dealt with in the articles by Weissman (1978), Boos (1984), Joe (1987), Smith (1987), and Smith and Weissman (1987), among others. In this context another interesting paper is that of Heidelberger and Lewis (1984) who suggested applying the subsample method to reduce the possible correlation of the r.v.'s and to reduce the problem of estimating extreme quantiles to that of estimating the median; moreover, it may have computational advantages to reduce the sample size in certain simulations by applying the subsample method.

The statistical procedures in Sections 9.2–9.6 are either based on the k largest order statistics or on k subsamples. The choice of the number k is crucial for the performance of the statistical procedures. The optimal choice heavily depends on the given model as is pointed out in this chapter. Some work has been done concerning the selection of the model; we refer to Pickands (1975), Hall and Welsh (1985), and to Section 9.5 for some results. The advice of Du Mouchel (1983) to take the upper 10 per cent of the sample might be valuable for practitioners. The visual comparison between sample and extreme value d.f.'s gives further insight into the problem.

CHAPTER 10

Approximate Sufficiency of Sparse Order Statistics

This chapter starts with an introduction to "comparison of statistical models" where in addition to Section 9.1 we also make use of Markov kernels.

In Section 10.2 it is shown that sparse order statistics $X_{r_1:n}, X_{r_2:n}, \dots, X_{r_k:n}$ are approximately sufficient over a nonparametric neighborhood of a fixed d.f. F_0. This result will be proved under particularly weak conditions.

In Section 10.3 the fixed d.f. F_0 will be replaced by a parametric family of d.f.'s. In the case of the location and scale parameter family of uniform distributions, the extended result follows immediately from Section 10.2. In other cases, one has to include an auxiliary estimator of the unknown parameter into the considerations.

Since sparse order statistics are asymptotically jointly normal one obtains a normal approximation of the nonparametric model of distributions of $(X_{r:n}, X_{r+1:n}, \dots, X_{s:n})$ or $(X_{1:n}, \dots, X_{n:n})$. The usefulness of this approach will be demonstrated in Section 10.4 by considering a nonparametric testing problem.

10.1. Comparison of Statistical Models via Markov Kernels

The statistical models $\mathscr{P}$ and $\mathscr{Q}$ we are primarily concerned with are built by the joint distributions of order statistics $X_{1:n}, \dots, X_{n:n}$ and, respectively, $X_{r_1:n}, X_{r_2:n}, \dots, X_{r_k:n}$ where $1 \le r_1 \le \cdots \le r_k \le n$. The order statistics come from n i.i.d. random variables with common d.f. F which belongs to a certain nonparametric family of d.f.'s.

It is obvious that the projection τ defined by

$$\tau(x_1, \ldots, x_n) = (x_{r_1}, \ldots, x_{r_k}) \tag{10.1.1}$$

carries the model $\mathcal{P}$ to the model $\mathcal{Q}$. Notice that the map τ is not one-to-one, and hence to return from $\mathcal{Q}$ to $\mathcal{P}$ one has to make use of a Markov kernel.

Markov Kernels

A Markov kernel K carrying mass from the probability space $(S_1, \mathcal{B}_1, Q)$ to $(S_0, \mathcal{B}_0)$ has the following two properties:

(a) $K(\cdot | y)$ is a probability measure on $\mathcal{B}_0$ for every $y \in S_1$, and
(b) $K(B | \cdot)$ is measurable for every $B \in \mathcal{B}_0$.

Recall that

$$KQ(B) := \int K(B | \cdot) \, dQ \tag{10.1.2}$$

defines a probability measure on $\mathcal{B}_0$. KQ is the distribution of the Markov kernel K (under Q). Thus, the symbol K also denotes a map from the family of probability measures on $\mathcal{B}_1$ into that on $\mathcal{B}_0$.

The reader is reminded of the following interpretation of KQ. First observe y which is an outcome of an experiment governed by Q. Secondly, carry out an experiment governed by $K(\cdot | y)$ and observe x. Then, the 2-step experiment with the final outcome x is governed by KQ.

Note that the distribution of a map T (under Q) can be written as KQ where K is the Markov kernel defined by

$$K(B | y) = 1_B(T(y)) = \varepsilon_{T(y)}(B)$$

with ε_x denoting the Dirac measure with mass 1 at x. In this case, given y the value $T(y)$ is chosen "with probability one."

More Informative and Blackwell-Sufficiency

In this sequel, we are given two models $\mathcal{P} = \{P_\theta : \theta \in \Theta\}$ and $\mathcal{Q} = \{Q_\theta : \theta \in \Theta\}$ such that $TP_\theta = Q_\theta$, $\theta \in \Theta$ (in other words, if ξ is a r.v. with distribution P_θ then $\eta = T(\xi)$ is distributed according to Q_θ).

Notice that the models $\mathcal{P}$ and $\mathcal{Q}$ may be defined on different measurable spaces $(S_0, \mathcal{B}_0)$ and $(S_1, \mathcal{B}_1)$ like Euclidean spaces of different dimensions.

It is desireable to find a Markov kernel K (independent of the parameter θ) such that

$$P_\theta = KQ_\theta, \qquad \theta \in \Theta, \tag{10.1.3}$$

which means that P_θ can be reconstructed from Q_θ by means of the Markov kernel K.

If (10.1.3) holds then $\mathcal{Q}$ is said to be more informative than $\mathcal{P}$. If also $TP_\theta = Q_\theta$, $\theta \in \Theta$, then both models are equivalent, and T is said to be Blackwell-sufficient.

Recall from Section 9.1 that $TP_\theta = Q_\theta$, $\theta \in \Theta$, implies that for every statistical procedure on $\mathcal{Q}$ one finds a procedure on $\mathcal{P}$ of equal performance. Under (10.1.3) also the converse conclusion holds. Let us exemplify this idea in the context of the testing problem.

Let $C \in \mathcal{B}_0$ be a critical region (acting on $\mathcal{P}$). Then, the critical function $K(C|\cdot): S_1 \to [0,1]$ is of equal performance if, as usual, the comparison is based on power functions. This becomes obvious by noting that according to (10.1.2) and (10.1.3),

$$P_\theta(C) = \int K(C|\cdot)\,dQ_\theta, \qquad \theta \in \Theta. \tag{10.1.4}$$

The same conclusion holds if one starts with a critical function ψ defined on S_0. The Fubini theorem for Markov kernels implies that

$$\int \psi\,dP_\theta = \int\left[\int \psi(x)K(dx|y)\right]dQ_\theta(y), \qquad \theta \in \Theta, \tag{10.1.5}$$

and hence the critical functions ψ and $\tilde{\psi} = \int \psi(x)K(dx|\cdot)$ are of equal performance.

Blackwell-Sufficiency and Sufficiency

We continue our discussion of basic statistical concepts being aware that there is a good chance of boring some readers. However, if this is the case, omit the next lines and continue with Example 10.1.2 and the definition of the ε-deficiency for unequal parameter sets.

The classical concept of sufficiency is closely related to that of Blackwell-sufficiency. In fact, under mild regularity conditions, which are always satisfied in our context, Blackwell-sufficiency and sufficiency are equivalent [see e.g. Heyer (1982, Theorem 22.12)].

Recall that $T: S_0 \to S_1$ is sufficient if for every critical function ψ defined on S_0 there exists a version $E(\psi|T)$ of the conditional expectation w.r.t. T which does not depend on the parameter θ. Then, the Blackwell-sufficiency holds with a Markov kernel defined by

$$K(B|y) = Q(B|T = y)$$

where $Q(B|T = y)$ are appropriate versions of the conditional probability of B given $T = y$ (in other words, K is the factorization of the conditional distribution of the identity on S_0 given T). Check that

$$E(\psi|T) = \int \psi(x)K(dx|T) \qquad \text{w.p. 1.}$$

Recall that the Neyman criterion provides a powerful tool for the verification of sufficiency of T. The sufficiency holds if the density p_θ of P_θ (w.r.t. some dominating measure) can be factorized in the form $p_\theta = r(h_\theta \circ T)$.

EXAMPLES 10.1.1.
(i) Let $\mathscr{P}$ be a family of uniform distributions with unknown location parameter. Then, $(X_{1:n}, X_{n:n})$ is sufficient.
(ii) Let $\mathscr{P}$ be a family of exponential distributions with unknown location parameter. Then, $X_{1:n}$ is sufficient.

The concept of Blackwell-sufficiency will be extended in two steps. First we consider the situation where (10.1.3) holds with a remainder term. The second extension also includes the case where the parameter sets of $\mathscr{P}$ and $\mathscr{Q}$ are unequal.

Approximate Sufficiency and ε-Deficiency

If (10.1.3) does not hold for any Markov kernel then one may try to find a Markov kernel K such that the variational distances $\sup_B |P_\theta(B) - KQ_\theta(B)|$, $\theta \in \Theta$, are small. We say that $\mathscr{Q}$ is ε-deficient w.r.t. $\mathscr{P}$ if

$$\sup_B |P_\theta(B) - KQ_\theta(B)| \leq \varepsilon(\theta), \qquad \theta \in \Theta$$

for some Markov kernel K.

In this context, the map T may be called approximately sufficient if $\varepsilon(\theta)$ is small. Define the one-sided deficiency $\delta(\mathscr{Q}, \mathscr{P})$ of $\mathscr{Q}$ w.r.t. $\mathscr{P}$ by

$$\delta(\mathscr{Q}, \mathscr{P}) := \inf_K \sup_{\theta \in \Theta} \sup_B |P_\theta(B) - KQ_\theta(B)| \qquad (10.1.6)$$

where K ranges over all Markov kernels from $(S_1, \mathscr{B}_1)$ to $(S_0, \mathscr{B}_0)$. The deficiency $\delta(\mathscr{Q}, \mathscr{P})$ of $\mathscr{Q}$ w.r.t. $\mathscr{P}$ measures the amount of information which is needed so that $\mathscr{Q}$ is more informative than $\mathscr{P}$. If $TP_\theta = Q_\theta$, $\theta \in \Theta$, then $\delta(\mathscr{P}, \mathscr{Q}) = 0$.

Notice that $\delta(\mathscr{Q}, \mathscr{P})$ is not symmetric. To obtain a symmetric distance between $\mathscr{Q}$ and $\mathscr{P}$ define the symmetric deficiency

$$\Delta(\mathscr{Q}, \mathscr{P}) = \max(\delta(\mathscr{Q}, \mathscr{P}), \delta(\mathscr{P}, \mathscr{Q})). \qquad (10.1.7)$$

The arguments in (10.1.4) and (10.1.5) carry over to the present situation; now, we have to include some remainder term into our consideration. Let again K be a Markov kernel carrying mass from $(S_1, \mathscr{B}_1, \mathscr{Q})$ to $(S_0, \mathscr{B}_0)$. If ψ^* is an optimal critical function acting on $\mathscr{Q}$ then $\psi^{**} = \psi^*(T)$ is optimal on $\mathscr{P}$ within the error bound $\delta(\mathscr{Q}, \mathscr{P})$. To prove this, consider a critical function ψ on S_1. We have

$$\int \psi^{**} \, dP_\theta = \int \psi^* \, dQ_\theta \geq \int \left[\int \psi(x) K(dx|y) \right] dQ_\theta(y) = \int \psi \, dK Q_\theta$$

(10.1.8)

$$\geq \int \psi \, dP_\theta - \sup_B |P_\theta(B) - K Q_\theta(B)|$$

for every Markov kernel K, and hence

$$\int \psi^{**} \, dP_\theta \geq \int \psi \, dP_\theta - \delta(\mathcal{Q}, \mathcal{P})$$

(10.1.9)

showing the desired conclusion.

Next we shall give a simple, (as we hope) illuminating example in order not to remain too theoretical. The technical details are omitted in order not to disturb the flow of the main ideas.

EXAMPLE 10.1.2. Consider the location parameter model $\mathcal{P}_{0,n} = \{P_{0,n,\theta}\}$ of a sample of size n arising out of the densities

$$x \to f(x - \theta)$$

(10.1.10)

with f being fixed. Assume that $f(x) > 0$, $a \leq x \leq b$, and $= 0$, otherwise.

A typical example is given by the uniform density

$$f = 2^{-1} 1_{[-1,1]}.$$

(10.1.11)

Denote by $\mathcal{P}_{0,n}^*$ the special model under condition (10.1.11). Recall from Example 10.1.1 that $(X_{1:n}, X_{n:n})$ is a sufficient statistic in this case.

Step 1 (Approximate Sufficiency of $(X_{1:n}, X_{n:n})$). Under weak regularity conditions it can be shown that $(X_{1:n}, X_{n:n})$ is still approximately sufficient for the location model $\mathcal{P}_{0,n}$. We refer to Weiss (1979b) for a global treatment and to Janssen and Reiss (1988) for a local "one-sided" treatment of this problem. The technique for proving such a result will be developed in the next section. Regularity conditions have to guarantee that no further jumps of the density occur besides those at the points a, b.

Let $\mathcal{P}_{1,n} = \{P_{1,n,\theta}\}$ denote the model of distributions of $(X_{1:n}, X_{n:n})$ under the parameter θ.

Approximate sufficiency means that there exists a Markov kernel K_1 such that $P_{0,n,\theta}$ can approximately be rebuilt by $K_1 P_{1,n,\theta}$. In terms of ε-deficiency we have

$$\Delta(\mathcal{P}_{0,n}, \mathcal{P}_{1,n}) \leq \varepsilon(n)$$

(10.1.12)

where $\varepsilon(n) \to 0$, $n \to \infty$. We remark that $\varepsilon(n) = O(n^{-1})$ under certain regularity conditions.

In the special case of (10.1.11), obviously,

$$\Delta(\mathcal{P}_{0,n}^*, \mathcal{P}_{1,n}) = 0.$$

(10.1.13)

Notice that $\mathcal{P}_{1,n}$ is again a location parameter model.

Step 2 (Asymptotic Independence of $X_{1:n}$ and $X_{n:n}$). Next $X_{1:n}$ and $X_{n:n}$ will be replaced by independent versions $Y_{1:n}$ and $Y_{n:n}$, that is,

$$Y_{i:n} \overset{d}{=} X_{i:n}, \qquad i = 1, n, \tag{10.1.14}$$

and $Y_{1:n}$, $Y_{n:n}$ are independent.

From (4.2.10) we know that the variational distance between the distributions of $(X_{1:n}, X_{n:n})$ and $(Y_{1:n}, Y_{n:n})$ is of order $O(n^{-1})$.

In this case, the Markov kernel which carries one model to the other is simply represented by the identity. Denote by $\mathscr{P}_{2,n} = \{P_{2,n,\theta}\}$ the location parameter model which consists of the distributions of $(Y_{1:n}, Y_{n:n})$. Then,

$$\Delta(\mathscr{P}_{1,n}, \mathscr{P}_{2,n}) = O(n^{-1}). \tag{10.1.15}$$

Step 3 (Limiting Distributions of Extremes). Our journey through several models is not yet finished. Under mild conditions (see Section 5.2), the extremes $Y_{1:n}$ and $Y_{n:n}$ have an exponential distribution with remainder term of order $O(n^{-1})$. More precisely, if the extremes $Y_{i:n}$, $i = 1, n$, are generated under the parameter θ then

$$\sup_B |P_\theta\{(Y_{1:n} - a) \in B\} - Q_{1,n,\theta}(B)| = O(n^{-1})$$

and $\qquad\qquad\qquad\qquad\qquad\qquad\qquad\qquad\qquad\qquad$ (10.1.16)

$$\sup_B |P_\theta\{(Y_{n:n} - b) \in B\} - Q_{2,n,\theta}(B)| = O(n^{-1})$$

where the $Q_{i,n,\theta}$ have the densities $q_{i,n}(\cdot - \theta)$ defined by

$$q_{1,n}(x) = \begin{cases} nf(a)\exp[-nf(a)x] \\ 0 \end{cases} \text{if} \quad \begin{matrix} x \geq 0 \\ x < 0 \end{matrix}$$

and $\qquad\qquad\qquad\qquad\qquad\qquad\qquad\qquad\qquad\qquad$ (10.1.17)

$$q_{2,n}(y) = \begin{cases} nf(b)\exp[nf(b)y] \\ 0 \end{cases} \text{if} \quad \begin{matrix} y \leq 0 \\ y > 0 \end{matrix}.$$

We introduce the ultimate model $\mathscr{P}_{3,n} = \{P_{3,n,\theta}\}$ where

$$P_{3,n,\theta} = Q_{1,n,\theta} \times Q_{n,n,\theta}.$$

Note that $\mathscr{P}_{3,n}$ is again a location parameter model.

Summarizing the steps 1–3 we get

$$\Delta(\mathscr{P}_{0,n}, \mathscr{P}_{3,n}) = O(\varepsilon(n) + n^{-1}). \tag{10.1.18}$$

One may obtain a fixed asymptotic model by starting with the model of distributions of $n(Y_{1:n} - a)$ and $n(Y_{n:n} - b)$ under local parameters $n\theta$ in place of θ.

Step 4 (Estimation of the Location Parameter). In a location parameter model it makes sense to choose an optimal estimator out of the class of estimators that are equivariant under translations; that is, given the model $\mathscr{P}_{3,n}$ the estimator has the property

$$\theta_n(x + \theta, y + \theta) = \theta_n(x, y) + \theta. \tag{10.1.19}$$

If θ_n is an optimal equivariant estimator on $\mathscr{P}_{3,n}$ then

$$\theta_n(X_{1:n} - a, X_{n:n} - b) \tag{10.1.20}$$

is an equivariant estimator operating on $\mathscr{P}_{0,n}$ having the same performance as θ_n besides of a remainder term of order $O(\varepsilon(n) + n^{-1})$.

We remark that in order to show that $\theta_n(X_{1:n} - a, X_{n:n} - b)$ is the optimal estimator on $\mathscr{P}_{0,n}$ one has to verify that θ_n is optimal within the class of all randomized equivariant estimators operating on $\mathscr{P}_{3,n}$.

Let us examine the special case of uniform densities as given in (10.1.11). A moment's reflection shows that necessarily

$$X_{n:n} - 1 \le \theta \le X_{1:n} + 1 \tag{10.1.21}$$

so that any reasonable estimator has to lie between $X_{n:n} - 1$ and $X_{1:n} + 1$.

One could try to adopt the maximum likelihood (m.l.) principle for finding an optimal estimator. However, the likelihood function

$$\theta \to 2^{-n} \prod_{i=1}^{n} 1_{[\theta-1,\theta+1]}(X_{i:n})$$

has its maximum at any θ between $X_{n:n} - 1$ and $X_{1:n} + 1$. Hence, the m.l. principle does not lead to a reasonable solution of the problem.

For location parameter models it is well known that Pitman estimators are optimal within the class of equivariant estimators (see e.g. Ibragimov and Has'minskii (1981), page 22, lines 1–9).

It is a simple exercise to verify that

$$(X_{1:n} + X_{n:n})/2 \tag{10.1.22}$$

is a Pitman estimator w.r.t. any sub-convex loss function $L(\cdot - \cdot)$. Note that $L(\cdot - \cdot)$ is sub-convex if L is symmetric about zero and $L|[0, \infty)$ is nondecreasing.

If L is strictly increasing then the Pitman estimator is uniquely determined.

Let us return to the ultimate model $\mathscr{P}_{3,n}$. A Pitman estimate $\theta_n(x, y)$ w.r.t. the loss function $L(\cdot - \cdot)$ minimizes

$$\int L(\theta - u)g_{1,n}(x - u)g_{2,n}(y - u)\, du \tag{10.1.23}$$

in θ. [Recall that the Pitman estimator is a generalized Bayes estimator with the Lebesgue measure being the prior "distribution."]

Check that (10.1.23) is equivalent to solving the problem

$$\int_y^x L(\theta - u)\exp[n(f(a) - f(b))u]\, du = \min_{\theta}!. \tag{10.1.24}$$

If $f(a) = f(b)$ then for sub-convex loss functions,

$$\theta_n(x, y) = (x + y)/2$$

is a solution of (10.1.24). Moreover, this is the unique solution if L is strictly increasing on $[0, \infty)$.

Thus,

$$[X_{1:n} + X_{n:n} - (a + b)]/2 \tag{10.1.25}$$

is an "approximate" Pitman estimator in the original model $\mathscr{P}_{0,n}$.

The finding of explicit solutions of (10.1.24) for $f(a) \neq f(b)$ is an open problem.

Unequal Parameter Sets

Corresponding to (9.1.7) and (9.1.8) we introduce models

$$\mathscr{P} = \{P_{\theta,g} : \theta \in \Theta, g \in G(\theta)\}$$

and

$$\mathscr{Q} = \{Q_{\theta,h} : \theta \in \Theta, h \in H(\theta)\}$$

where g and h may be regarded as nuisance parameters. The notion and the results above carry over to the present framework.

$\mathscr{Q}$ is said to be ε-deficient w.r.t. $\mathscr{P}$ if

$$\sup_B |P_{\theta,g}(B) - KQ_{\theta,h}(B)| \leq \varepsilon(\theta, g, h), \quad \theta \in \Theta, g \in G(\theta), h \in H(\theta), \tag{10.1.26}$$

for some Markov kernel K.

Define the "one-sided" deficiency $\delta(\mathscr{Q}, \mathscr{P})$ of $\mathscr{Q}$ w.r.t. $\mathscr{P}$ by

$$\delta(\mathscr{Q}, \mathscr{P}) := \inf_K \sup_{\theta,g,h} \sup_B |P_{\theta,g}(B) - KQ_{\theta,h}(B)| \tag{10.1.27}$$

where K ranges over all Markov kernels from $(S_1, \mathscr{B}_1)$ to $(S_0, \mathscr{B}_0)$. Moreover, the symmetric deficiency of $\mathscr{Q}$ and $\mathscr{P}$ is again defined by

$$\Delta(\mathscr{Q}, \mathscr{P}) = \max(\delta(\mathscr{Q}, \mathscr{P}), \delta(\mathscr{P}, \mathscr{Q})). \tag{10.1.28}$$

10.2. Approximate Sufficiency over a Neighborhood of a Fixed Distribution

In this section we compute an upper bound for the deficiency (in the sense of (10.1.7)) of a model defined by the distributions of the order statistic and a second model defined by the joint distribution, say, P_n of sparse order statistics $X_{r_1:n} \leq X_{r_2:n} \leq \cdots \leq X_{r_k:n}$ [suppressing the dependence on $r_1, \ldots, r_k$]. To prove such a result one has to construct an appropriate Markov kernel which carries the second model back to the original model.

Let $X_{1:n} \leq \cdots \leq X_{n:n}$ be the order statistics of n i.i.d. random variables with common d.f. F which is assumed to be continuous. Theorem 1.8.1 provides the conditional distribution

$$K_n(\cdot \,|\, \mathbf{x}) = P((X_{1:n}, \ldots, X_{n:n}) \in \cdot \,|\, (X_{r_1:n}, X_{r_2:n}, \ldots, X_{r_k:n}) = \mathbf{x}) \qquad (10.2.1)$$

of the order statistic $(X_{1:n}, \ldots, X_{n:n})$ conditioned on $(X_{r_1:n}, X_{r_2:n}, \ldots, X_{r_k:n}) = \mathbf{x}$. Recall that K_n is a Markov kernel having the "reproducing" property

$$K_n P_n(B) = \int K_n(B|\cdot)\, dP_n = P\{(X_{1:n}, X_{2:n}, \ldots, X_{n:n}) \in B\} \qquad (10.2.2)$$

for every Borel set B.

Let K_n^* denote the special Markov kernel which is obtained if F is the uniform d.f. on $(0, 1)$, say, F_0. Thus, we have

$$K_n^*(\cdot \,|\, \mathbf{x}) = P((U_{1:n}, \ldots, U_{n:n}) \in \cdot \,|\, (U_{r_1:n}, U_{r_2:n}, \ldots, U_{r_k:n}) = \mathbf{x}).$$

If F is close to F_0—in a sense to be described later—then one can hope that (10.2.2) approximately holds when K_n is replaced by K_n^*. The decisive point is that K_n^* does not depend on the d.f. F.

In light of the foregoing remark the k order statistics $X_{r_1:n}, \ldots, X_{r_k:n}$ carry approximately as much information about F as the full order statistic.

The Main Results

We shall prove a bound for the accuracy of the approximation introduced above under particularly weak conditions on the underlying d.f. F.

Theorem 10.2.1. *Let $1 \leq k \leq n$ and $0 = r_0 < r_1 < \cdots < r_k < r_{k+1} = n + 1$. Denote again by P_n the joint distribution of order statistics $X_{r_1:n}, X_{r_2:n}, \ldots, X_{r_k:n}$ [of n i.i.d. random variables with common d.f. F and density f].*

Assume that $\alpha(F) = 0$ and $\omega(F) = 1$, and that f has a derivative on $(0, 1)$. Then,

$$\sup_B |P\{(X_{1:n}, X_{2:n}, \ldots, X_{n:n}) \in B\} - K_n^* P_n(B)|$$

$$\leq \delta(F) \left[\sum_{j=1}^{k+1} (r_j - r_{j-1} - 1) \left(\frac{r_j - r_{j-1} + 1}{n + 1} \right)^2 \right]^{1/2} \qquad (10.2.3)$$

where

$$\delta(F) = \sup_{y \in (0,1)} |f'(y)| / \inf_{y \in (0,1)} f^2(y). \qquad (10.2.4)$$

PROOF. Let K_n denote the Markov kernel in (10.2.1) given the d.f. F. Applying Theorem 1.8.1 we obtain

$$\sup_{B} |P\{(X_{1:n}, \ldots, X_{n:n}) \in B\} - K_n^* P_n(B)|$$

$$\leq \int \sup_{B} |K_n(B|\cdot) - K_n^*(B|\cdot)| \, dP_n \tag{1}$$

$$\leq \int \sup_{B} \left| \left(\bigotimes_{j=1}^{n} P_{j,x} \right)(B) - \left(\bigotimes_{j=1}^{n} Q_{j,x} \right)(B) \right| dP_n(\mathbf{x})$$

where $P_{r_i,x} = Q_{r_i,x}$ are the Dirac-measures at x_i for $i = 1, \ldots, k$; moreover for $i = 1, \ldots, k + 1$ and $j = r_{i-1} + 1, \ldots, r_i - 1$ the probability measures $P_{j,x}$ and $Q_{j,x}$ are defined by the densities:

$$p_{i,x} = f \, 1_{(x_{i-1}, x_i)} / (F(x_i) - F(x_{i-1}))$$

and

$$q_{i,x} = 1_{(x_{i-1}, x_i)} / (x_i - x_{i-1})$$

[with the convention that $x_0 = 0$ and $x_{n+1} = 1$].

Writing

$$g \equiv g_{j,x} = (p_{j,x}/q_{j,x}) - 1$$

we obtain from (3.3.10) that for every $\mathbf{x}$ with $x_j - x_{j-1} > 0$, $j = 1, \ldots, k + 1$,

$$\sup_{B} \left| \left(\bigotimes_{j=1}^{n} P_{j,x} \right)(B) - \left(\bigotimes_{j=1}^{n} Q_{j,x} \right)(B) \right|$$

$$\leq \left[\sum_{j=1}^{k+1} (r_j - r_{j-1} - 1) \int g_{j,x}^2 \, dQ_{r_j,x} \right]^{1/2} \tag{2}$$

$$\leq \left[\sum_{j=1}^{k+1} (r_j - r_{j-1} - 1) \rho(F)(x_j - x_{j-1})^2 \right]^{1/2}$$

where

$$\rho(F) = \left[\sup_{y \in (0,1)} |f'(y)| \middle/ \inf_{y \in (0,1)} f(y) \right]^2.$$

The second inequality in (2) can easily be verified by using the representation

$$g_{j,x}(y) = \frac{f'(v)}{f(u)} (y - u), \qquad x_{j-1} < y < x_j, \tag{3}$$

with u and v strictly between x_{j-1} and x_j and u not depending on y.

Combining (1) and (2) and applying the Schwarz inequality we obtain

$$\sup_{B} |P\{(X_{1:n}, X_{2:n}, \ldots, X_{n:n}) \in B\} - K_n^* P_n(B)|$$

$$\leq \left[\sum_{j=1}^{k+1} (r_j - r_{j-1} - 1) \rho(F) E(X_{r_j:n} - X_{r_{j-1}:n})^2 \right]^{1/2} \tag{4}$$

where $X_{0:n} = 0$ and $X_{n+1:n} = 1$. Finally, Theorem 1.2.5(i) and (1.7.4) yield

$$E(X_{r_j:n} - X_{r_{j-1}:n})^2 = E[F^{-1}(U_{r_j:n}) - F^{-1}(U_{r_{j-1}:n})]^2$$

$$\leq E[U_{r_j-r_{j-1}:n}^2] \Big/ \inf_{y \in (0,1)} f^2(y) \leq \left(\frac{r_j - r_{j-1} + 1}{n+1}\right)^2 \Big/ \inf_{y \in (0,1)} f^2(y) \qquad (5)$$

thus, (10.2.3) is immediate from (3)–(5).   □

Notice that $\delta(F)$ can be regarded as a distance between F and the uniform d.f. F_0.

EXAMPLE 10.2.2. If the differences $r_i - r_{i-1}$ are of order $O(m(n))$ and $k \equiv k(n)$ is of order $O(n/m(n))$ which means that, roughly speaking, the indices r_i are equi-distant, then the right-hand side of (10.2.3) is of order

$$O(\delta(F)m(n)/n^{1/2}).$$

Thus, if $m(n) = o(n^{1/2})$ (entailing that the number k of order statistics has to be larger than $n^{1/2}$) then the right-hand side of (10.2.3) goes to zero as n goes to infinity even if $\delta(F)$ is bounded away from zero.

If $n^{1/2} = O(m(n))$ then F should also depend on n. In the statistical context this means that our model has to shrink towards the uniform d.f. as n goes to infinity.

A typical situation for such a dependence on the sample size n occurs in the context of a goodness-of-fit test when one is testing the uniform d.f. F_0 against an alternative F_n having a density f_n given by

$$f_n(x) = 1 + \beta(n)n^{-1/2}h(x).$$

Note that $\beta(n)$ is a fixed constant in classical test problems. In Example 10.4.1 we shall study the situation where the dimension of the alternative increases as the sample size increases. Then, $\beta(n)$ has to go to infinity as $n \to \infty$ in order to attain rejection probabilities bounded away from the level α under alternatives of the above form.

If h and h' are bounded then $\delta(F_n) = O(\beta(n)n^{-1/2})$ and, therefore, the right-hand side of (10.2.3) is of order $O[\beta(n)/k(n)]$.

Local Formulation

Theorem 10.2.1 may be extended in various directions. In cases where one is only interested in local properties of F, our considerations will be based on a statistic only depending on certain extreme or central order statistics, say, $X_{r:n} \leq X_{r+1:n} \leq \cdots \leq X_{s:n}$ where $1 \leq r \leq s \leq n$. Again the number of order statistics may be reduced. If $r_1 = r$ and $r_k = s$ then, in contrary to the conditions of Theorem 10.2.1, it suffices to assume that $0 \leq \alpha(F) < \omega(F) \leq 1$.

For the formulation of Addendum 10.2.3 we introduce the projection $\tau \equiv \tau(r,s)$ defined by

$$\tau(x_1, x_2, \ldots, x_n) = (x_r, x_{r+1}, \ldots, x_s).$$

Note that in the following context, Markov kernels will rebuild the joint distribution of $X_{r:n}, X_{r+1:n}, \ldots, X_s$. Define a Markov kernel adjusted to the present problem, namely,

$$K_{n,\tau}^*(\cdot \,|\mathbf{x}) = \tau K_n^*(\cdot \,|\mathbf{x}).$$

Note that $K_{n,\tau}^*(\cdot \,|\mathbf{x})$ is a marginal distribution of $K_n^*(\cdot \,|\mathbf{x})$. Check that $K_{n,\tau}^*(\cdot \,|\mathbf{x})$ is the conditional distribution of $(U_{r:n}, U_{r+1:n}, \ldots, U_{s:n})$ given $(U_{r_1:n}, U_{r_2:n}, \ldots, U_{r_k:n}) = \mathbf{x}$.

Addendum 10.2.3. *Assume that* $1 \leq r = r_1 < r_2 < \cdots < r_k = s \leq n$. *Denote again by* P_n *the joint distribution of the order statistics* $X_{r_1:n}, X_{r_2:n}, \ldots, X_{r_k:n}$. *Assume that* $0 \leq \alpha(F) < \omega(F) \leq 1$, *and that* f *has a derivative on* $(\alpha(F), \omega(F))$. *Then,*

$$\sup_B |P\{(X_{r:n}, X_{r+1:n}, \ldots, X_{s:n}) \in B\} - K_{n,\tau}^* P_n(B)|$$

$$\leq \tilde{\delta}(F) \left[\sum_{j=2}^k (r_j - r_{j-1} - 1)\left(\frac{r_j - r_{j-1} + 1}{n+1}\right)^2 \right]^{1/2} \tag{10.2.5}$$

where

$$\tilde{\delta}(F) = \sup_{y \in (\alpha(F), \omega(F))} |f'(y)| \Big/ \inf_{y \in (\alpha(F), \omega(F))} f^2(y).$$

The proof of (10.2.5) is an almost verbatim repetition of that of (10.2.3) and can be left to the reader. We remark that Addendum 10.2.3 is an immediate consequence of Theorem 10.2.1 if again $\alpha(F) = 0$ and $\omega(F) = 1$.

Transformed Models

The results until now are concerned with d.f.'s F close to the uniform d.f. F_0 on $(0, 1)$. If we fix some other continuous d.f., say G_0 in place of F_0 then the probability integral transformation may be applied to reduce the problem again to the former case.

The d.f.'s G close to G_0 have to be of the form $G = F \circ G_0$ where F (being equal to $G \circ G_0^{-1}$) has to fulfill the conditions of Theorem 10.2.1. If $Y_{i:n}$ are the order statistics of r.v.'s with common d.f. G then $X_{i:n} = G_0^{-1}(Y_{i:n})$ are the order statistics of r.v.'s with common d.f. F. Thus, Theorem 10.2.1 applies to $X_{i:n}$.

In order to formulate the problem for the original order statistics $Y_{i:n}$ we introduce the Markov kernel M_n^* where $M_n^*(\cdot \,|\mathbf{y})$ is the conditional distribution of $(Y_{1:n}, Y_{2:n}, \ldots, Y_{n:n})$ given $(Y_{r_1:n}, Y_{r_2:n}, \ldots, Y_{r_k:n}) = \mathbf{y}$ in the special case of $G = G_0$ (where again the dependence of M_n^* on $r_1, \ldots, r_k$ will be suppressed).

Theorem 10.2.4. *Let $1 \leq k \leq n$ and $0 = r_0 < r_1 < \cdots < r_k < r_{k+1} = n + 1$. Let F be a continuous d.f. with $\alpha(F) = 0$ and $\omega(F) = 1$. Assume that F has two derivatives on $(0, 1)$. Put $f = F'$.*

Denote by Q_n the joint distribution of the order statistics $Y_{r_1:n}, \ldots, Y_{r_k:n}$ where the $Y_{i:n}$ are the order statistics of n i.i.d. random variables with common d.f. $G_1 = F \circ G_0$. Then,

$$\sup_B |P\{(Y_{1:n}, Y_{2:n}, \ldots, Y_{n:n}) \in B\} - M_n^* Q_n(B)|$$

$$\leq \delta(F) \left[\sum_{j=1}^{k+1} (r_j - r_{j-1} - 1)\left(\frac{r_j - r_{j-1} + 1}{n + 1}\right)^2 \right]^{1/2} \tag{10.2.6}$$

where again

$$\delta(F) = \sup_{y \in (0,1)} |f'(y)| \Big/ \inf_{y \in (0,1)} f^2(y).$$

Theorem 10.2.4 was stated in such a way that it can easily be deduced from Theorem 10.2.1, however, this formulation looks rather artificial. Further insight in the nature of the term $\delta(F)$ can be obtained by means of a different representation of the density f.

From P.1.5 and Remark 1.5.3 we conclude that G_1 has the G_0-density $g = f \circ G_0$. Hence, $f = g \circ G_0^{-1}$, according to Criterion 1.2.3. Thus, the conditions of Theorem 10.2.4 can be reformulated in the following way:

Assume that G_1 has the G_0-density g so that $g \circ G_0^{-1}$ is differentiable.

Moreover, the term $\delta(F)$ can be replaced by

$$\sup_{y \in (0,1)} |(g \circ G_0^{-1})'(y)| \Big/ \inf_{y \in (0,1)} (g \circ G_0^{-1})^2(y).$$

Theorem 10.2.4 is an immediate consequence of Theorem 10.2.1 and the following lemma which may also be applied to prove extensions of Addendum 10.2.3.

Lemma 10.2.5. *Let $1 \leq r_1 < r_2 < \cdots < r_k \leq n$. Let $X_{i:n}$ and $Y_{i:n}$ be the order statistics of n i.i.d. random variables with common d.f. F and, respectively, d.f. $G_1 = F \circ G_0$. The d.f.'s F and G_0 are assumed to be continuous, and $0 \leq \alpha(F) < \omega(F) \leq 1$.*

Denote by P_n and Q_n the joint distributions of $X_{r_1:n}, \ldots, X_{r_k:n}$ and $Y_{r_1:n}, \ldots, Y_{r_k:n}$. Then, with M_n^* and K_n^* as defined above, we have

$$\sup_B |P\{(Y_{1:n}, Y_{2:n}, \ldots, Y_{n:n}) \in B\} - M_n^* Q_n(B)|$$

$$= \sup_B |P\{(X_{1:n}, X_{2:n}, \ldots, X_{n:n}) \in B\} - K_n^* P_n(B)|. \tag{10.2.7}$$

PROOF. From P.1.5 we know that $G_0^{-1}(\eta)$ is a r.v. with d.f. $F \circ G_0$ if η is a r.v. with d.f. F. This implies that

$$(Y_{1:n}, Y_{2:n}, \ldots, Y_{n:n}) \stackrel{d}{=} (G_0^{-1}(X_{1:n}), G_0^{-1}(X_{2:n}), \ldots, G_0^{-1}(X_{n:n})). \tag{1}$$

From (1) we know that

$$P\{(Y_{1:n}, Y_{2:n}, \ldots, Y_{n:n}) \in B\} = P\{(X_{1:n}, X_{2:n}, \ldots, X_{n:n}) \in \tilde{B}\} \qquad (2)$$

where $\tilde{B} = \{\mathbf{x}: (G_0^{-1}(x_1), G_0^{-1}(x_2), \ldots, G_0^{-1}(x_n)) \in B\}$. If, moreover,

$$M_n^* Q_n(B) = K_n^* P_n(\tilde{B}) \qquad (3)$$

then it is apparent that (10.2.7) holds. From (1) we also know that

$$M_n^* Q_n(B) = E M_n^*(B | G_0^{-1}(X_{r_1:n}), G_0^{-1}(X_{r_2:n}), \ldots, G_0^{-1}(X_{r_k:n})).$$

Thus, in view of (3) it remains to prove that

$$K_n^*(\tilde{B} | x_1, \ldots, x_k) = M_n^*(B | y_1, y_2, \ldots, y_k) \qquad (4)$$

whenever $\alpha(F) < x_1 < x_2 < \cdots < x_k < \omega(F)$ with y_i denoting $G_0^{-1}(x_i)$.

Since G_0 is continuous we know that G_0^{-1} is strictly increasing thus, $\alpha(G_0) =: y_0 < y_1 < \cdots < y_k < y_{k+1} := \omega(G_0)$.

Put $x_0 = 0$ and $x_{k+1} = 1$. Let ε_x denote the Dirac-measure at x (with mass 1). Moreover, $\tilde{Q}$ denotes the probability measure corresponding to G_0, and Q is the uniform distribution on $(0, 1)$.

It is obvious that ε_{y_i} is induced by ε_{x_i} and G_0^{-1}. Moreover, from P.1.6 we know that the truncation of $\tilde{Q}$ to the interval (y_{i-1}, y_i), say, $\tilde{Q}_{y_{i-1}, y_i}$ is induced by Q_{x_{i-1}, x_i} and G_0^{-1} for $i = 1, \ldots, k+1$. Thus, Theorem 1.2.5(i) yields that $M_n^*(\cdot | y_1, \ldots, y_k)$ is induced by $K_n^*(\cdot | x_1, \ldots, x_k)$ and the map

$$(u_1, u_2, \ldots, u_n) \to (G_0^{-1}(u_1), G_0^{-1}(u_2), \ldots, G_0^{-1}(u_n)).$$

This implies (4). The proof is complete. □

10.3. Approximate Sufficiency over a Neighborhood of a Family of Distributions

In the preceding section it was proved that a small number of order statistics carries nearly all the information about the underlying d.f. G if this d.f. is close to a fixed d.f. G_0. Now we start with a family of d.f.'s $G(\cdot, \theta)$, $\theta \in \Theta$, and build a model containing joint distributions of order statistics under a d.f. G close to one of the d.f.'s $G(\cdot, \theta)$, $\theta \in \Theta$.

Near to Uniform Distributions

The location and scale parameter family of uniform distributions provides an exceptional case. Here the approximate sufficiency of sparse order statistics can directly be proved by means of the results in Section 10.2. The uniform d.f.'s

$$G(\cdot, \theta) \equiv G(\cdot, (\mu, \sigma))$$

are given by $G(x, (\mu, \sigma)) = (x - \mu)/\sigma$ for $\mu < x < \mu + \sigma$.

Corollary 10.3.1. *Let $X_{i:n}$ be the ith order statistic of n i.i.d. random variables with d.f. G given by $G(x) = F((x - \mu)/\sigma)$ for $\mu < x < \mu + \sigma$ with $-\infty < \mu < \infty$ and $\sigma > 0$. It is assumed that F is continuous and has two derivatives on $(0, 1) = (\alpha(F), \omega(F))$. Put $f = F'$.*

Let $1 \le r = r_1 < r_2 < \cdots < r_k = s \le n$. Let $K_{n,\tau}^*$ denote the Markov kernel defined in Addendum 10.2.3, and let again P_n denote the joint distribution of the sparse order statistics $X_{r_1:n}, X_{r_2:n}, \ldots, X_{r_k:n}$. Then,

$$\sup_B |P\{(X_{r:n}, X_{r+1:n}, \ldots, X_{s:n}) \in B\} - K_{n,\tau}^* P_n(B)|$$

$$\le \delta(F) \left[\sum_{j=2}^{k} (r_j - r_{j-1} - 1) \left(\frac{r_j - r_{j-1} + 1}{n+1} \right)^2 \right]^{1/2}$$

where again

$$\delta(F) = \sup_{y \in (0,1)} |f'(y)| \Big/ \inf_{y \in (0,1)} f^2(y).$$

PROOF. Immediate by applying Theorem 10.2.4 to $G_0 = G(\cdot, (\mu, \sigma))$ and noting that $M_{n,\tau}^* = K_{n,\tau}^*$. □

In Corollary 10.3.1 it may as well be assumed that F has two derivatives on $(\alpha(F), \omega(F))$ with $0 \le \alpha(F) < \omega(F) \le 1$. However, this would not yield an extension of Corollary 10.3.1 since the d.f. G can be represented in the former way by choosing different parameters μ and σ. It is also of importance to take $r_1 = r$ and $r_k = s$ since, otherwise, the identity $M_{n,\tau}^* = K_{n,\tau}^*$ does not hold.

Corollary 10.3.1 was immediate from the results of Section 10.2 since the conditional distribution of $(Y_{r:n}, Y_{r+1:n}, \ldots, Y_{s:n})$ given $(Y_{r_1:n}, Y_{r_2:n}, \ldots, Y_{r_k:n})$ is independent of the parameter (μ, σ) where the $Y_{i:n}$ are the order statistics under the uniform d.f. $G(\cdot, (\mu, \sigma))$. This property is not shared by other parametric families of d.f.'s (proof!), and so we need a modification of the concept.

The Main Results

In this sequel, we shall assume that $r = r_1 < \cdots < r_k = s$, and the parameter space Θ is a subset of the Euclidean d-space equipped with the Euclidean norm $\| \|_2$.

For every vector $\theta = (\theta_1, \theta_2, \ldots, \theta_d) \in \Theta$ let $M_{n,\theta}^*$ be the conditional distribution of $(Y_{r:n}, Y_{r+1:n}, \ldots, Y_{s:n})$ given $(Y_{r_1:n}, Y_{r_2:n}, \ldots, Y_{r_k:n}) = \mathbf{x}$ where the $Y_{i:n}$ are the order statistics under the d.f. $G(\cdot, \theta)$. Notice that the Markov kernel $M_{n,\theta}^*$ also depends on $\mathbf{r} = (r_1, r_2, \ldots, r_k)$.

In the next step the unknown parameter θ will be replaced by an estimator

$\hat{\theta}_n$ based on the order statistics $X_{r_1:n}$, $X_{r_2:n}$, ..., $X_{r_k:n}$ under a d.f. G which not necessarily belongs to the parametric family $\{G(\cdot, \theta): \theta \in \Theta\}$. Thus, a new problem arises, namely, one has to estimate the parameter θ under a model which is incorrect.

The conditions in Theorem 10.3.2 will guarantee that M_n^* defined by

$$M_n^*(\cdot | \mathbf{x}) = M_{n,\hat{\theta}_n(x)}^*(\cdot | \mathbf{x}) \tag{10.3.1}$$

is a Markov kernel.

Let again P_n denote the joint distribution of the sparse order statistics $X_{r_1:n}$, $X_{r_2:n}$, ..., $X_{r_k:n}$. We shall use $M_n^* P_n$ as an approximation to the joint distribution of $X_{r:n}$, $X_{r+1:n}$, ..., $X_{s:n}$. The accuracy of this approximation will depend on the performance of the estimator $\hat{\theta}_n$ and the distance of G from the parametric family $\{G(\cdot, \theta): \theta \in \Theta\}$. We assume that the d.f.'s $G(\cdot, \theta)$ have densities, say, $g(\cdot, \theta)$.

Theorem 10.3.2 will be proved under a local Lipschitz condition. Given a fixed parameter $\theta_0 \in \Theta$ assume that

$$\left| \frac{(\partial/\partial y)G(G^{-1}(y_1, \theta_0), \theta)}{(\partial/\partial y)G(G^{-1}(y_2, \theta_0), \theta)} - 1 \right| \le C \|\theta - \theta_0\|_2 |y_1 - y_2| \tag{10.3.2}$$

for every $\theta \in \Theta$ with $\|\theta - \theta_0\|_2 \le \varepsilon$, $C \ge 0$ and y_i with $0 < q_1 < y_i < q_2 < 1$ for $i = 1, 2$.

In (10.3.2) it is implicitely assumed that $g(x, \theta) > 0$ for every x with $G^{-1}(q_1, \theta_0) < x < G^{-1}(q_2, \theta_0)$ and θ with $\|\theta - \theta_0\|_2 \le \varepsilon$.

If $\Theta = \{\theta_0\}$—that is the problem of Section 10.2—then (10.3.2) holds with $C = 0$.

Another set of conditions involving the partial derivatives

$$(\partial^2/\partial\theta_i \partial y)\log g$$

will be examined in Criterion 10.3.3.

Theorem 10.3.2. *Let* $1 \le k \le n$ *and* $1 \le r = r_1 < r_2 < \cdots < r_k = s \le n$. *Let* $X_{i:n}$ *be the* ith *order statistic of* n *i.i.d. random variables with common d.f.* $G = F \circ G(\cdot, \theta_0)$ *where* $\theta_0 \in \Theta$ *and* F *is a d.f. with* $\alpha(F) = 0$ *and* $\omega(F) = 1$. *Moreover, suppose that* F *has two derivatives on* $(0, 1)$. *Put again* $f = F'$.

Suppose that the d.f.'s $G(\cdot, \theta)$ *fulfill condition* (10.3.2) *for some constants* ε, $C > 0$ *and* $0 < q_1 < q_2 < 1$.

Then for every measurable and Θ-*valued estimator* $\hat{\theta}_n$ *we have, with* M_n^* *defined as in* (10.3.1),

$$\sup_B |P\{(X_{r:n}, X_{r+1:n}, \ldots, X_{s:n}) \in B\} - M_n^* P_n(B)|$$

$$\le [\delta(F) + \rho(F, \hat{\theta}_n, C, \varepsilon)] \left[\sum_{j=2}^k (r_j - r_{j-1} - 1)\left(\frac{r_j - r_{j-1} + 3}{n+1}\right)^2 \right]^{1/2}$$

$$+ P\{\|\hat{\theta}_n(X_{r_1:n}, X_{r_2:n}, \ldots, X_{r_k:n}) - \theta_0\|_2 > \varepsilon\}$$

$$+ P\{X_{r:n} \le G^{-1}(q_1, \theta_0)\} + P\{X_{s:n} \ge G^{-1}(q_2, \theta_0)\}$$

with $\delta(F)$ as in (10.2.4) and

$$\rho(F, \hat{\boldsymbol{\theta}}_n, C, \varepsilon)$$

$$= \left(C \middle/ \inf_{y \in (0,1)} f(y) \right) \min(\varepsilon, [E\|\hat{\boldsymbol{\theta}}_n(X_{r_1:n}, X_{r_2:n}, \ldots, X_{r_k:n}) - \boldsymbol{\theta}_0\|_2^4]^{1/4}).$$

PROOF. Our first aim is to prove that

$$\sup_B |P\{(X_{r:n}, X_{r+1:n}, \ldots, X_{s:n}) \in B\} - M_n^* P_n(B)|$$

$$\leq \delta(F) \left[\sum_{j=2}^k (r_j - r_{j-1} - 1) \left(\frac{r_j - r_{j-1} + 1}{n+1} \right)^2 \right]^{1/2}$$

$$+ P\{\|\hat{\boldsymbol{\theta}}_n(X_{r_1:n}, X_{r_2:n}, \ldots, X_{r_k:n}) - \boldsymbol{\theta}_0\|_2 > \varepsilon\} \qquad (1)$$

$$+ P\{X_{r:n} \leq G^{-1}(q_1, \boldsymbol{\theta}_0)\} + P\{X_{s:n} \geq G^{-1}(q_2, \boldsymbol{\theta}_0)\}$$

$$+ \left[\sum_{j=2}^k (r_j - r_{j-1} - 1) \int_A \psi_j(\mathbf{x}) \, dP_n(\mathbf{x}) \right]^{1/2}$$

where, with $\hat{\boldsymbol{\theta}} \equiv \hat{\boldsymbol{\theta}}_n$,

$$A = \{\mathbf{x}: G^{-1}(q_1, \boldsymbol{\theta}_0) < x_1 < x_2 < \cdots < x_k < G^{-1}(q_2, \boldsymbol{\theta}_0) \text{ and } \|\boldsymbol{\theta}(\mathbf{x}) - \boldsymbol{\theta}_0\|_2 < \varepsilon\},$$

$$\psi_j(\mathbf{x}) = \int (h_{j,\mathbf{x}}(y, \hat{\boldsymbol{\theta}}(\mathbf{x}))/h_{j,\mathbf{x}}(y, \boldsymbol{\theta}_0) - 1)^2 h_{j,\mathbf{x}}(y, \boldsymbol{\theta}_0) \, dy,$$

$$h_{j,\mathbf{x}}(y, \theta) = g(y, \theta) 1_{(x_{j-1}, x_j)}(y)/[G(x_j, \theta) - G(x_{j-1}, \theta)].$$

Applying the triangle inequality and Theorem 10.2.4 one obtains

$$\sup_B |P\{(X_{r:n}, X_{r+1:n}, \ldots, X_{s:n}) \in B\} - M_n^* P_n(B)|$$

$$\leq \sup_B |P\{(X_{r:n}, X_{r+1:n}, \ldots, X_{s:n}) \in B\} - M_{n,\boldsymbol{\theta}_0}^* P_n(B)|$$

$$+ \sup_B |M_{n,\boldsymbol{\theta}_0}^* P_n(B) - M_n^* P_n(B)|$$

$$\leq \delta(F) \left[\sum_{j=2}^k (r_j - r_{j-1} - 1) \left(\frac{r_j - r_{j-1} + 1}{n+1} \right)^2 \right]^{1/2}$$

$$+ \int \sup_B \left| \left(\bigtimes_{j=1}^n P_{j,\mathbf{x}} \right)(B) - \left(\bigtimes_{j=1}^n Q_{j,\mathbf{x}} \right)(B) \right| dP_n(\mathbf{x})$$

where $P_{r_i,\mathbf{x}}$ and $Q_{r_i,\mathbf{x}}$ are the Dirac-measures at x_i, and for $i = 2, \ldots, k$ and $j = r_{i-1} + 1, \ldots, r_i - 1$ the probability measures $P_{j,\mathbf{x}}$ and $Q_{j,\mathbf{x}}$ are defined by the densities $h_{j,\mathbf{x}}(\cdot, \hat{\boldsymbol{\theta}}(\mathbf{x}))$ and $h_{j,\mathbf{x}}(\cdot, \boldsymbol{\theta}_0)$. Now (1) is immediate from inequality (3.3.10) and the Schwarz inequality.

For every $\mathbf{x} \in A$ we obtain, with $z_j = G(x_j, \boldsymbol{\theta}_0)$, that

$$\psi_j(\mathbf{x}) \leq C^2 \|\hat{\boldsymbol{\theta}}(\mathbf{x}) - \boldsymbol{\theta}_0\|_2^2 |z_j - z_{j-1}|^2. \qquad (2)$$

From the mean value theorem and substituting y by $G^{-1}(z, \boldsymbol{\theta}_0)$ we obtain for some u_j between z_{j-1} and z_j that

$$\psi_j(\mathbf{x}) = \frac{1}{z_j - z_{j-1}} \int_{G^{-1}(z_{j-1}, \theta_0)}^{G^{-1}(z_j, \theta_0)} \left[\frac{g(y, \hat{\theta}(\mathbf{x}))}{g(y, \theta_0)} \right.$$

$$\left. \times \frac{z_j - z_{j-1}}{G(G^{-1}(z_j, \theta_0), \hat{\theta}(\mathbf{x})) - G(G^{-1}(z_{j-1}, \theta_0), \hat{\theta}(\mathbf{x}))} - 1 \right]^2 g(y, \theta_0)\, dy$$

$$= \frac{1}{z_j - z_{j-1}} \int_{z_{j-1}}^{z_j} \left(\frac{\partial/\partial z \, G(G^{-1}(z, \theta_0), \hat{\theta}(\mathbf{x}))}{\partial/\partial z \, G(G^{-1}(u_j, \theta_0), \hat{\theta}(\mathbf{x}))} - 1 \right)^2 dz$$

and hence (2) follows at once from condition (10.3.2) by noting that $q_1 < z_1 < z_2 < \cdots < z_k < q_2$.

It is immediate from (2) and the Schwarz inequality that

$$\int_A \psi_j(\mathbf{x})\, dP_n(\mathbf{x}) \le C^2 \min\{\varepsilon^2, (E\|\hat{\theta}(X_{r_1:n}, X_{r_2:n}, \ldots, X_{r_k:n}) - \theta_0\|_2^4)^{1/2}\} \qquad (3)$$
$$\times (E(G(X_{r_j:n}, \theta_0) - G(X_{r_{j-1}:n}, \theta_0))^4)^{1/2}.$$

Applying (1.7.4) we obtain (as in the proof of Theorem 10.2.1) that

$$E(G(X_{r_j:n}, \theta_0) - G(X_{r_{j-1}:n}, \theta_0))^4 = E(F^{-1}(U_{r_j:n}) - F^{-1}(U_{r_{j-1}:n}))^4$$

$$\le \left(\inf_{y \in (0,1)} f(y) \right)^{-4} EU_{r_j - r_{j-1}:n}^4 \qquad (4)$$

$$\le \left(\inf_{y \in (0,1)} f(y) \right)^{-4} ((r_j - r_{j-1} + 3)/(n+1))^4.$$

Combining (1), (3), and (4) the proof is complete. $\square$

Condition (10.3.2) holds—as already mentioned—in the degenerate case where $\Theta = \{\theta_0\}$. Another special case will be studied in the following.

Criterion 10.3.3. *Assume that Θ is an open and convex subset of the Euclidean d-space. Assume that the partial derivatives $(\partial^2/\partial\theta_i \partial y)\log g$ exist.*

Then condition (10.3.2) holds with

$$C = \exp[\varepsilon|q_2 - q_1|\kappa(g)]\kappa(g)$$

where

$$\kappa(g) = \sup\|((\partial^2/\partial\theta_i \partial y)\log g(G^{-1}(y, \theta_0), \theta))_{i=1}^m\|_2$$

with the supremum ranging over all (y, θ) with $q_1 < y < q_2$ and $\|\theta - \theta_0\|_2 \le \varepsilon$.

PROOF. Applying the mean value theorem we get

$$\left| \log\frac{\partial}{\partial y} G(G^{-1}(y_1, \theta_0), \theta) - \log\frac{\partial}{\partial y} G(G^{-1}(y_2, \theta_0)\theta) \right|$$

$$= \left| \frac{\partial}{\partial y}\log\frac{\partial}{\partial y} G(G^{-1}(\tilde{y}, \theta_0), \theta))(y_1 - y_2) \right|$$

$$= \left| \frac{\partial}{\partial y}\log g(G^{-1}(\tilde{y}, \theta_0), \theta) - \frac{\partial}{\partial y}\log g(G^{-1}(\tilde{y}, \theta_0), \theta_0) \right| |y_1 - y_2|$$

$$\le \kappa(g)\|\theta - \theta_0\|_2 |y_1 - y_2|$$

with $\bar{y}$ between y_1 and y_2. Since $z_1/z_2 = \exp(\log z_1 - \log z_2)$ and $|\exp(z) - 1| \leq \exp(z)z$ for $z, z_1, z_2 > 0$ the proof can easily be completed.   □

Final Remarks

Let us examine the problem of testing the parametric null-hypothesis $\{G(\cdot, \theta): \theta \in \Theta\}$ against certain nonparametric alternatives G_n.

It is easy to see that G_n is of the form $F_n \circ G(\cdot, \theta_0)$ where F_n has the density $f_n(y) = 1 + h(G^{-1}(y, \theta_0))\alpha(n)$ if, and only if, G_n has the density

$$g_n(x) = g(x, \theta_0)(1 + h(x)\alpha(n))$$

where $\int h(x)g(x, \theta_0)\,dx = 0$. In this case if h and $h'(G^{-1}(\cdot, \theta_0))/g(G^{-1}(\cdot, \theta_0))$ are bounded we have $\delta(F_n) = O(\alpha(n))$ and $\inf_{y \in (0,1)} f_n(y) \leq 1 - O(\alpha(n))$.

Within the present framework one has to find an appropriate estimator of θ. The problem of constructing estimators which are optimal in the sense of minimizing the upper bound in Theorem 10.3.2 is also connected to the problem of finding an "optimal" parameter θ_0 which makes $\delta(F) = \delta(G \circ G^{-1}(\cdot, \theta_0))$ small.

Given a functional T on the family of all q.f.'s so that $T(G^{-1}(\cdot, \theta)) = \theta$, the statistical functional $T(F_n^{-1})$ is an appropriate estimator of $T(G^{-1})$ and thus of θ_0 if G^{-1} is close to $G^{-1}(\cdot, \theta_0)$. Since the estimator $\hat{\theta}_n$ is only allowed to depend on the sparse order statistics $X_{r_1:n}, X_{r_2:n}, \ldots, X_{r_k:n}$ one has to take a statistical functional w.r.t. a version of the sample q.f. which is based on these sparse order statistics.

10.4. Local Comparison of a Nonparametric Model and a Normal Model

Let us summarize the results of Sections 10.2 and 10.3 without going into the technical details. The nucleus of our model is a parametric family $G(\cdot, \theta)$, $\theta \in \Theta$, of d.f.'s. In Section 10.2 we studied the particular case where Θ consists of one parameter. In Section 10.3 the model is built by d.f.'s G close to $G(\cdot, \theta)$ for some $\theta \in \Theta$. Under appropriate conditions on $\mathbf{r} = (r_1, \ldots, r_k)$ and G we find a Markov kernel M_n^* such that

$$\sup_B |P\{(X_{r:n}, X_{r+1:n}, \ldots, X_{s:n}) \in B\} - M_n^* P_n(B)| \leq \varepsilon_0(G, \mathbf{r}, n) \quad (10.4.1)$$

where $X_{1:n} \leq \cdots \leq X_{n:n}$ are the order statistics of n i.i.d. random variables with common d.f. G, and P_n is the joint distribution of $X_{r_1:n}, X_{r_2:n}, \ldots, X_{r_k:n}$. The decisive point in (10.4.1) is that the Markov kernel M_n^* is independent of G.

Let us also apply the result of Section 4.5, namely, that central order statistics $X_{r_1:n}, X_{r_2:n}, \ldots, X_{r_k:n}$ are approximately normally distributed.

Denote by g the density of G. We have

$$\sup_B |P\{(X_{r_1:n}, X_{r_2:n}, \ldots, X_{r_k:n}) \in B\} - P\{(Y_1', Y_2', \ldots, Y_k') \in B\}| \le \varepsilon_1(G, \mathbf{r}, n)$$
(10.4.2)

where the explicit form of $\varepsilon_1(G, \mathbf{r}, n)$ is given in Theorem 4.5.3, and $(Y_1', Y_2', \ldots, Y_k')$ is a normal random vector with mean vector

$$\boldsymbol{\mu}(G) = \left(G^{-1}\left(\frac{r_1}{n+1}\right), \ldots, G^{-1}\left(\frac{r_k}{n+1}\right) \right)$$
(10.4.3)

and covariance matrix $\Sigma(G) = (\sigma_{i,j})$ given by

$$\sigma_{i,j} = \frac{r_i}{n+1}\left(1 - \frac{r_j}{n+1}\right) \bigg/ \left[(n+1)g\left(G^{-1}\left(\frac{r_i}{n+1}\right)\right) g\left(G^{-1}\left(\frac{r_j}{n+1}\right)\right) \right]$$
(10.4.4)

for $1 \le i \le j \le k$.

Since (10.4.2) can be extended to $[0, 1]$-valued measurable functions (see P.3.5) we obtain

$$\sup_B |M_n^* P_n(B) - M_n^* N_{(\boldsymbol{\mu}(G), \Sigma(G))}(B)| \le \varepsilon_1(G, \mathbf{r}, n).$$
(10.4.5)

Combining (10.4.1) and (10.4.5) we have

$$\sup_B |P\{(X_{r:n}, X_{r+1:n}, \ldots, X_{s:n}) \in B\} - M_n^* N_{(\boldsymbol{\mu}(G), \Sigma(G))}(B)|$$
$$\le \varepsilon(G, \mathbf{r}, n) := \varepsilon_0(G, \mathbf{r}, n) + \varepsilon_1(G, \mathbf{r}, n).$$
(10.4.6)

(10.4.6) connects the following two models. The first one is given by joint distributions of order statistics $X_{r:n}, \ldots, X_{s:n}$ with "parameter" G; the second one is a family of k-dimensional normal distributions with parameters $(\boldsymbol{\mu}(G), \Sigma(G))$. In the sense of (10.1.26), the model, given by normal distributions $N_{(\boldsymbol{\mu}(G), \Sigma(G))}$, is $\varepsilon(G, \mathbf{r}, n)$-deficient w.r.t. the model determined by the order statistics $X_{r:n}, X_{r+1:n}, \ldots, X_{s:n}$.

If (10.4.6) holds for $r = 1$ and $s = n$ then the following result also holds: Let $\xi_1, \xi_2, \ldots, \xi_n$ be the original i.i.d. random variables. Since the order statistic is sufficient we find a Markov kernel M_n^{**} (see also P.1.29) such that

$$\sup_B |P\{(\xi_1, \xi_2, \ldots, \xi_n) \in B\} - M_n^{**} N_{(\boldsymbol{\mu}(G), \Sigma(G))}(B)| \le \varepsilon(G, \mathbf{r}, n).$$ (10.4.7)

Next we present the main ideas of an example due to Weiss (1974, 1977) where the approximating normal distribution depends on the original d.f. F only through the mean vector. Moreover, we indicate the possibility of calculating a bound of the remainder term of the approximation.

EXAMPLE 10.4.1. As a continuation of Example 10.2.2, the uniform d.f. F_0 on $(0, 1)$ will be tested against a composite alternative of d.f.'s F_n having densities f_n given by

$$f_n(x) = 1 + \beta(n)n^{-1/2}h(x), \qquad 0 \le x \le 1,$$

and $= 0$, otherwise, where $\int_0^1 h(x)\,dx = 0$. The term $\beta(n)$ will be specified later.

Part 1 (Asymptotic Sufficiency). Recall from Example 10.2.2 that sparse order statistics

$$X_{r_1:n} < X_{r_2:n} < \cdots < X_{r_k:n}$$

are asymptotically sufficient under weak conditions.

Part 2 (Asymptotic Normality). Put again

$$\lambda_i = r_i/(n + 1).$$

Let $\beta_{i,i}$ and $\beta_{i,i-1}$ be given as in the proof of Lemma 4.4.2. Recall that the $\beta_{i,j}$ define a map S such that $SN_{(0,\Sigma)} = N_{(0,I)}$ where $\Sigma = (\sigma_{i,j})$ and $\sigma_{i,j} = \lambda_i(1 - \lambda_j)$, $i \le j$. The decisive point is that these values do not depend on F. Define

$$Z_i = (n + 1)^{1/2}(\beta_{i,i}(X_{r_i:n} - \lambda_i) + \beta_{i,i-1}(X_{r_{i-1}:n} - \lambda_{i-1})) \qquad (10.4.8)$$

for $i = 1, \ldots, k$ where $\beta_{1,0} = 0$. Notice that $Z_1, \ldots, Z_k$ are known to the statistician, and hence tests may be based on these r.v.'s. The Z_i are closely related to spacings, however, the use of spacings would not lead to asymptotically independent r.v.'s (compare with P.4.4).

Applying (10.4.2) we obtain that $Z_i, \ldots, Z_k$ can be replaced by independent normal r.v.'s $Y_1, \ldots, Y_k$ with unit variances and expectations equal to

$$\mu_i = -(\lambda_i - \lambda_{i-1})^{1/2}\beta(n)h(\lambda_i), \qquad i = 1, \ldots, k. \qquad (10.4.9)$$

Thus, we have

$$\sup_B |P\{(Z_1, \ldots, Z_k) \in B\} - P\{(Y_1, \ldots, Y_k) \in B\}| = o(1). \qquad (10.4.10)$$

A bound for the remainder term in (10.4.10) may be proved by means of P.4.2(i) and P.4.2(v) [see also P.10.7].

Thus, the original testing problem has become a problem of testing, within a model of normal distributions $N_{(\mu, I)}$, the null-hypothesis

$$\mathbf{\mu} = (\mu_1, \ldots, \mu_k) = \mathbf{0}$$

against   (10.4.11)

$$\mu_i = -(\lambda_i - \lambda_{i-1})^{1/2}\beta(n)h(\lambda_i), \qquad i = 1, \ldots, k,$$

where the alternative has to be specified more precisely.

Part 3 (Discussion). The above considerations enable us to apply the non-asymptotic theory of linear models to the original problem of testing the uniform distribution against a parametric or nonparametric alternative. By finding an optimum procedure within the linear model one gets an approximately optimum procedure for the original model.

Recall from P.3.8 that the most powerful level-α-test of a sample of size n, for testing the uniform density against the density $1 + \beta(n)n^{-1/2}h$, rejects the null-hypothesis with probability

$$\Phi\left(\Phi^{-1}(\alpha) + \beta(n)\left(\int_0^1 h^2(x)\,dx\right)^{1/2}\right) + O(n^{-1/2}) \qquad (10.4.12)$$

under appropriate regularity conditions. However, in general, this power cannot be attained uniformly over a composite alternative. It is well known that test procedures with high efficiency w.r.t. one "direction" h have a bad efficiency w.r.t. other directions. The Kolmogorov–Smirnov test provides a typical example of a test having such a behavior.

In view of (10.4.12) a plausible requirement is that a test in the original model should be of equal performance under every alternative $1 + \beta(n)n^{-1/2}h$ satisfying the condition

$$\left(\int_0^1 h^2(x)\,dx\right)^{1/2} = \delta \qquad (10.4.13)$$

for fixed $\delta > 0$.

Let again $Y_1, \ldots, Y_k$ be i.i.d. normal r.v.'s with unit variance and mean vector $\mu = (\mu_1, \ldots, \mu_k)$ as given in (10.4.11). Denote again by $\|\cdot\|_2$ the Euclidean norm. Notice that $\sum_{i=1}^k (\lambda_i - \lambda_{i-1})h(\lambda_i)^2$ is an approximation to δ^2 and hence $\|\mu\|_2$ is an approximation to $\beta(n)\delta$.

Thus, within the normal model, one has to test the null-hypothesis $\mathcal{H}_0 = \{0\} = \{\mu: \|\mu\|_2 = 0\}$ against an alternative

$$\mathcal{H}_1 \subset \{\mu: \|\mu\|_2 > 0\} \qquad (10.4.14)$$

under the additional requirement that the performance of the test procedure depends on the underlying parameter μ through $\|\mu\|_2$ only; thus, the test is invariant under orthogonal transformations. In Parts 4 and 5 we shall recall some basic facts from classical, parametric statistics.

Part 4 (A χ^2-Test). Let us first consider the case where $\mathcal{H}_1 = \{\mu: \|\mu\|_2 > 0\}$ without taking into account that h has to satisfy a certain smoothness condition that also restricts the choice of the parameters μ. The uniformly most powerful, invariant test of level α is given by the critical region

$$C_k = \{T_k > \chi^2_{k,\alpha}\} \qquad (10.4.15)$$

where

$$T_k = \sum_{i=1}^k Y_i^2 \qquad (10.4.16)$$

and $\chi^2_{k,\alpha}$ is the $(1 - \alpha)$-quantile of the central χ^2-distribution with k degrees of freedom. According to Weiss (1977) the critical region C_k is also a Bayes test for testing $\|\mu\|_2 = 0$ against $\|\mu\|_2 = \delta$ with prior probability uniformly distributed over the sphere $\{\boldsymbol{\mu}: \|\boldsymbol{\mu}\|_2 = \delta\}$ (proof!). Moreover, C_k is minimax for this testing problem.

Since $\mathbf{Y}_k = (Y_1, \ldots, Y_k)$ is a vector of normal r.v.'s with unit variance and mean vector $\boldsymbol{\mu}$ we know that T_k is distributed according to a noncentral χ^2-distribution with k degrees of freedom and noncentrality parameter $\|\boldsymbol{\mu}\|_2^2$.

If $k \equiv k(n)$ tends to infinity as $n \to \infty$, the central limit theorem implies that

$$(2k + 4\|\boldsymbol{\mu}\|_2^2)^{-1/2} \left(\sum_{i=1}^{k} (Y_i^2 - 1) - \|\boldsymbol{\mu}\|_2^2 \right) \tag{10.4.17}$$

is asymptotically standard normal. Consequently, C_k has the asymptotic power function

$$\boldsymbol{\mu} \to \Phi(\Phi^{-1}(\alpha) + \|\boldsymbol{\mu}\|_2^2/(2k)^{1/2}) + o(k^0). \tag{10.4.18}$$

This yields that asymptotically the rejection probability is strictly larger than α if $\|\boldsymbol{\mu}\|_2^2/k^{1/2}$ is bounded away from zero.

In the original model, the critical region

$$\tilde{C}_k = \left\{ \sum_{i=1}^{k} Z_i^2 > \chi_{k,\alpha}^2 \right\}, \tag{10.4.19}$$

with Z_i defined in (10.4.8), attains the rejection probability

$$\Phi\left(\Phi^{-1}(\alpha) + \left(\int_0^1 h^2(x)\,dx \right) \right) + o(k^0) \tag{10.4.20}$$

under alternatives $1 + [(2k)^{1/2}/n]^{1/2} h$.

The critical region $\tilde{C}_k$ is closely related to a χ^2-test based on a random partition of the interval $[0, 1]$.

Part 5 (Linear Regression). We indicate a natural generalization of Part 4 that also takes into account the required smoothness condition imposed on h. Assume that

$$\mathscr{H}_{1,s} \subset \left\{ \sum_{j=1}^{s} \theta_j \mathbf{v}_j \right\}, \quad s < k, \tag{10.4.21}$$

where $\mathbf{v}_j = (v_j(1), \ldots, v_j(k))$, $j = 1, \ldots, s$, are orthonormal vectors w.r.t. the inner product $(\mathbf{x}, \mathbf{y}) = \sum_{i=1}^{k} x_i y_i$. The well-known solution of the problem is to take the critical region

$$C_s = \{T_s > \chi_{s,\alpha}^2\} \tag{10.4.22}$$

where

$$T_s = \sum_{j=1}^{s} (\mathbf{v}_j, \mathbf{Y}_k)^2 = \sum_{j=1}^{s} \left(\sum_{i=1}^{k} v_j(i) Y_i \right)^2. \tag{10.4.23}$$

Notice that $T_s = \|\hat{\mathbf{Y}}_k\|_2$ where $\hat{\mathbf{Y}}_k = \sum_{j=1}^{s} (\mathbf{v}_j, \mathbf{Y}_k)^2 \mathbf{v}_j$ is the orthogonal projection of $\mathbf{Y}_k$ onto the s-dimensional linear sub-space. The statistic T_s is again distributed according to a noncentral χ^2-distribution with s degrees of freedom and non-centrality parameter $\|\boldsymbol{\mu}\|_2^2$. We refer to Witting and Nölle (1970) or Lehmann (1986) for the details. Now the remarks made above concerning the asymptotic performance of the critical regions C_k and $\tilde{C}_k$ carry over with k replaced by s.

Part 6 (Parametric and Nonparametric Statistics). If s is fixed as $n \to \infty$ then, obviously, our asymptotic considerations belong to parametric statistics. If $s \equiv s(n) \to \infty$ as $n \to \infty$ then, e.g. in view of the Fourier expansion of square integrable functions, the sequence of original models approaches the space of square integrable densities close to the uniform density showing that the testing problem is of a nonparametric nature.

The foregoing remarks seem to be of some importance for nonparametric density testing (and estimation). Note that the functions h may belong to the linear space spanned by the trigonometric functions $e_1, \ldots, e_s$ (see P.8.5(i)). So there is some relationship to the orthogonal series method adopted in nonparametric density estimation. The crucial problem in nonparametric density estimation is to find a certain balance between the variance and the bias of estimation procedures. Our present point of view differs from that taken up in literature. First, we deduce the asymptotically optimum procedure w.r.t. the $s(n)$-dimensional model. These considerations belong to classical statistics. In a second step, we may examine the performance of the test procedure if the $s(n)$-dimensional model is incorrect.

P.10. Problems and Supplements

1. Let $\xi_1, \ldots, \xi_n$ and, respectively, $\eta_1, \ldots, \eta_n$ be i.i.d. random variables and denote by $X_{1:n} \le \cdots \le X_{n:n}$ and $Y_{1:n} \le \cdots \le Y_{n:n}$ the corresponding order statistics. Prove that

$$\sup_B |P\{(\xi_1, \ldots, \xi_n) \in B\} - P\{(\eta_1, \ldots, \eta_n) \in B\}|$$
$$= \sup_B |P\{(X_{1:n}, \ldots, X_{n:n}) \in B\} - P\{(Y_{1:n}, \ldots, Y_{n:n}) \in B\}|.$$

2. Prove that Theorem 10.2.1 holds with

$$\delta(F) = \exp\left(\sup_{y \in (0,1)} |f'(y)/f(y)| \right) \sup_{y \in (0,1)} |f'(y)/f(y)| \Big/ \inf_{y \in (0,1)} f(y).$$

[Hint: Use the fact that $f(y)/f(x) = \exp[(f'(z)/f(z))(y - x)]$ with z between x and y.]

3. Theorem 10.2.1 holds with the upper bound replaced by

$$\left(C \Big/ \inf_{y \in (0,1)} f^2(y) \right) \left[\sum_{j=1}^{k+1} (r_j - r_{j-1} - 1) \left(\frac{r_j - r_{j-1} + 1}{n + 1} \right)^{2\alpha} \right]^{1/2}$$

if the density f statisfies a Lipschitz condition of order $\alpha \in (0, 1]$ on $(0, 1)$.

4. (i) If $0 = r_0 < r_1 < r_2 < \cdots < r_k = s$, $r = 1$ and $\alpha(F) = 0$ then (10.2.5) holds with $\sum_{j=2}^{k}$ replaced by $\sum_{j=1}^{k}$.

   (ii) If $r = r_1 < r_2 < \cdots < r_k < r_{k+1} = n + 1$, $s = n$ and $\omega(F) = 1$ then (10.2.5) holds with $\sum_{j=2}^{k}$ replaced by $\sum_{j=2}^{k+1}$.

5. Let $\tau(x_1, \ldots, x_n) = (x_{n-k+1}, \ldots, x_n)$. Under the conditions of Addendum 10.2.3, if $\alpha(F) \geq 0$ and $\omega(F) = 1$,

$$\sup_{B} |P\{(X_{n-k+1:n}, \ldots, X_{n:n}) \in B\} - K_{n,\tau}^* P_n(B)|$$

$$\leq \left[\sup_{y \in (\alpha(F), 1)} |f'(y)| \bigg/ \inf_{y \in (\alpha(F), 1)} f^2(y) \right] k^{3/2} \bigg/ n.$$

6. (i) Verify condition (10.3.2) with $C = \rho(g)\kappa(g)$ where $\kappa(g)$ is given as in Criterion 10.3.3 and

$$\rho(g) = \sup_{\|\theta - \theta_0\|_2 \leq \varepsilon} \left(\sup_{q_1 < y < q_2} G(G^{-1}(y, \theta_0), \theta) \bigg/ \inf_{q_1 < y < q_2} G(G^{-1}(y, \theta_0), \theta) \right).$$

   (ii) Prove a modified version of Theorem 10.3.2 under the condition

$$\left| \frac{(\partial/\partial y)G(G^{-1}(y_1, \theta_0), \theta)}{(\partial/\partial y)G(G^{-1}(y_2, \theta_0), \theta)} - 1 \right| \leq C_1 \|\theta - \theta_0\|_2 |y_1 - y_2|$$

$$+ C_2(\|\theta - \theta_0\|_2 |y_1 - y_2|)^2$$

   for every $\theta \in \Theta$ with $\|\theta - \theta_0\|_2 \leq \varepsilon$, and $q_1 < y_1$, $y_2 < q_2$. Here, $C_1 = \kappa(g)$ and $C_2 = \rho(g)\kappa^2(g)$.

7. In analogy to (10.4.8) define

$$Z_i' = (n + 1)^{1/2}(\beta_{i,i}(X_{r_i:n} - G^{-1}(\lambda_i)) + \beta_{i,i-1}(X_{r_{i-1}:n} - G^{-1}(\lambda_i))).$$

   Denote by P_n the joint distribution of

$$(n + 1)^{1/2} g_i(X_{r_i:n} - G^{-1}(\lambda_i)), \qquad i = 1, \ldots, k,$$

   where $g_i = g(G^{-1}(\lambda_i))$. $N_{(0,\Sigma)}$ again denotes the k-variate normal distribution with mean vector zero and covariances $\sigma_{i,j} = \lambda_i(1 - \lambda_j)$, $1 \leq i \leq j \leq k$. Prove that

$$\sup_{B} |P\{(Z_1', \ldots, Z_k') \in B\} - N_{(0,1)}^k(B)|$$

$$\leq \|P_n - N_{(0,\Sigma)}\| + 2^{-1/2} \left[\sum_{i=1}^{k} (\sigma_{i,i}' - 1 + 2 \log g_i) \right]^{1/2}$$

   where $\sigma_{1,1}' = g_1^{-2}$ and

$$\sigma_{i,i}' = g_i^{-2} + \frac{\lambda_{i-1}(1 - \lambda_i)}{g_i^2(\lambda_i - \lambda_{i-1})} \left(\frac{g_i}{g_{i-1}} - 1 \right)^2, \qquad i = 2, \ldots, k.$$

   [Hint: Let H be the diagonal matrix with diagonal elements $\eta_{i,i} = 1/g_i$. Let $\Sigma = B \circ H \circ \Sigma \circ H^t \circ B^t$ where B is defined as in the proof of Lemma 4.4.2. Notice that $\det(\Sigma') = (\det(H))^2$.]

8. Specialize Example 10.4.1, Part 5, to trigonometric functions (see P.8.5).

9. Extend Example 10.4.1 to the composite null-hypothesis of uniform distributions.

Bibliographical Notes

The reader who is interested in the theoretical background concerning the comparison of experiments is referred to Torgersen (1976), Strasser (1985), and Le Cam (1986). The article of Torgersen gives a short, illuminating introduction to this subject.

The magnificent idea to study a construction like that in Theorem 10.2.1 is due to L. Weiss (1974) who also gave some asymptotic results. The extension of the problem from a single d.f. to a parametric family of d.f.'s was suggested by Weiss (1980). Weiss carried out a detailed study in the location and scale parameter case. Further insight into the problem of comparing models based on order statistics was obtained by Reiss et al. (1984) where a sharp bound of the remainder term of the approximation was also established. The present approach is taken from Reiss (1986). Some results concerning the sufficiency of extremes within a parametric framework can be found in the articles by Weiss (1979b) and Janssen and Reiss (1988). In the second article the location model of a Weibull sample is locally compared with location models defined by

$$(S_m^{1/\alpha} + \theta)_{m \leq k} \quad \text{and, respectively,} \quad (S_m^{1/\alpha} + \theta)_{m=1,2,3,\ldots}$$

where θ is the location parameter, and S_m is the sum of m i.i.d. standard exponential r.v.'s.

The optimum test procedure described in Example 10.4.1, (10.4.22), depends on the special choice of the set of alternatives. Weiss (1977) also describes a Bayes test that has the properties of an "all purpose" test. Moreover, it is apparent that, by using the approach of Section 10.4, larger parts of the theory of linear models can be made applicable to nonparametric statistics. A similar procedure based on spacings is dealt with by Weiss (1965).

APPENDIX 1

The Generalized Inverse

Extending the definition of a q.f. (see (1.1.10)) we define the inverse ψ^* of a real-valued, nondecreasing and right continuous function ψ with domain (α, ω) by setting

$$\psi^*(y) = \inf\{t \in (\alpha, \omega): \psi(t) \geq y\} \qquad (A.1.1)$$

for $-\infty < y < \infty$ (with the convention that $\inf \varnothing = \omega$). Moreover, we define

$$\psi^{-1} = \psi^*|(\inf \psi(s), \sup \psi(s)); \qquad (A.1.2)$$

that is, ψ^{-1} is the restriction of ψ^* to the interval $(\inf \psi(s), \sup \psi(s))$.

Thus, in the particular case of the q.f. we have $\psi = F$, $(\alpha, \omega) \equiv$ real line, $(\inf \psi(s), \sup \psi(s)) = (0, 1)$, and $\psi^{-1} = F^{-1}$. From the definitions of ψ^* and ψ^{-1} one can easily conclude that ψ^* is $[\alpha, \omega]$-valued and ψ^{-1} is (α, ω)-valued.

Lemma A.1.1. *For ψ as above, if $\alpha < x < \beta$ then for every real y,*

$$y \leq \psi(x) \quad \text{iff} \quad \psi^*(y) \leq x. \qquad (A.1.3)$$

PROOF. Since $\psi^*(y)$ is the inf of all $t \in (\alpha, \omega)$ such that $\psi(t) \geq y$ it is clear that $\psi(x) \geq y$ implies $x \geq \psi^*(y)$. Conversely, for every $z > x \geq \psi^*(y)$ we have $\psi(z) \geq y$, and thus, $y \leq \lim_{z \downarrow x} \psi(z) = \psi(x)$ since ψ is right continuous. $\quad\square$

It is clear that (A.1.3) also holds for ψ^{-1} and $\inf \psi(s) \leq y \leq \sup \psi(s)$ in place of ψ^* and $-\infty < y < \infty$. Thus (1.2.9) is a special case of (A.1.3).

We already know that ψ^{-1} is a (α, ω)-valued function with domain $(\inf \psi(s), \sup \psi(s))$. More precisely, one can easily check that ψ^{-1} is an $(\alpha(\psi), \omega(\psi))$-valued function where

$$\alpha(\psi) = \inf\{t \in (\alpha, \omega): \psi(t) > \inf \psi(s)\} \qquad (A.1.4)$$

and

$$\omega(\psi) = \sup\{t \in (\alpha, \omega): \psi(t) > \sup \psi(s)\}. \tag{A.1.5}$$

It is clear that $\alpha \leq \alpha(\psi) \leq \omega(\psi) \leq \omega$. Notice that in the particular case of a d.f. F we have

$$\alpha(F) = \inf\{t: F(t) > 0\} \tag{A.1.6}$$

and

$$\omega(F) = \sup\{t: F(t) < 1\}. \tag{A.1.7}$$

For the proof of Theorem 1.2.8 we also need the following auxiliary result.

Lemma A.1.2. *If ψ is as above then ψ^* is nondecreasing and left continuous. Moreover,*

$$\lim_{y \to -\infty} \psi^*(y) = \alpha, \qquad \lim_{y \to \infty} \psi^*(y) = \omega,$$

and

$$\lim_{y \to \inf \psi(s)} \psi^{-1}(y) = \alpha(\psi), \qquad \lim_{y \to \sup \psi(s)} \psi^{-1}(y) = \omega(\psi).$$

PROOF. From the definition of ψ^* it is obvious that ψ^* is nondecreasing. Moreover, ψ^* is left continuous if $y_n \uparrow y$ implies $\psi^*(y_n) > t$, eventually, whenever $\alpha < t < \psi^*(y)$. Lemma A.1.1 implies that $\psi(t) < y$. Consequently, $\psi(t) < y_n$ and thus, by Lemma A.1.1 again $t < \psi^*(y_n)$, eventually. By similar arguments one can verify the other assertions. $\square$

In analogy to the inverse of a nondecreasing, right continuous function ψ one can define the inverse of a nondecreasing left continuous function, say, φ with domain (α, ω). Put

$$\varphi^{**}(y) = \sup\{t \in (\alpha, \omega): \varphi(t) \leq y\} \tag{A.1.8}$$

for $-\infty < y < \infty$ (with the convention that $\sup \varnothing = \alpha$). An application of Lemma A.1.1 to the nondecreasing, right continuous function defined by $\psi(x) = -\varphi(-x)$ leads to

Lemma A.1.3. *For φ as above, if $\alpha < x < \omega$, then for every y,*

$$y \leq \varphi(x) \quad \textit{iff} \quad \varphi^{**}(y) \geq x.$$

PROOF. Verify that $\varphi^{**}(y) = -\psi(-y)$. $\square$

From Lemma A.1.2 we conclude

Lemma A.1.4. (i) *φ^{**} is nondecreasing and right continuous.*
(ii) *Moreover,*

$$\lim_{y \to -\infty} \varphi^{**}(y) = \alpha, \quad \text{and} \quad \lim_{y \to \infty} \varphi^{**}(y) = \beta.$$

Now we are in the proper position to carry out the

PROOF OF THEOREM 1.2.8. (i) is immediate from Lemma A.1.2.
(ii) Put $F = G^{**}$. From Lemma A.1.4 it is clear that F is a d.f. To prove
that $G = F^{-1}$ we apply Lemma A.1.1 and Lemma A.1.3. For $q \in (0,1)$ and
$-\infty < x < \infty$ we have $G(q) \le x$ iff $q \le F(x)$, and this holds iff $F^{-1}(q) \le x$.
This equivalence implies that $G = F^{-1}$.

Finally we show that for d.f.'s F_1 and F_2 with $F_1^{-1} = F_2^{-1} = G$ we have
$F_1 = F_2$. Suppose that $F_1^{-1} = F_2^{-1}$ and $F_1(x) \ne F_2(x)$ for some x. W.l.g. we can
assume that $F_1(x) < q < F_2(x)$ for some $q \in (0,1)$. Lemma A.1.1 implies
that $F_2^{-1}(q) \le x < F_1^{-1}(q)$ which is a contradiction to $F_1^{-1} = F_2^{-1}$. Thus,
$F_1 = F_2$.   □

From the proof to Theorem 1.2.8 we also know that $(F^{-1})^{**} = F$. Thus F
is the "generalized inverse" of F^{-1} which does, however, not imply that
$F \circ F^{-1}$ is the identity function as we already know from Criterion 1.2.3.
In analogy to Criterion 1.2.3 we obtain

Criterion A.1.5. *The d.f. F^{-1} is continuous if, and only if, $F^{-1}(F(x)) = x$ for
every x with $0 < F(x) < 1$.*

Two Technical Lemmas on Expansions

The results below will provide us with the basic tools for proving asymptotic expansions for distributions of extreme and central order statistics.

Expansion of $(1 + x/n)^n$

When studying extreme order statistics we are interested in an expansion of finite length of $e^{-x}(1 + x/n)^n$ where n is a positive integer and $x > 0$. We remark that $e^{-x}(1 + x/n)^n$ can easily be written as an infinite series by multiplying the absolutely convergent series

$$\sum_{i=0}^{\infty} \binom{n}{i}(x/n)^i \quad \text{and} \quad \sum_{i=0}^{\infty} (-x)^i/i!.$$

We have

$$e^{-x}(1 + x/n)^n = \sum_{i=0}^{\infty} \beta(i, n)x^i \tag{A.2.1}$$

where

$$\beta(i, n) = \sum_{j=1}^{i} \frac{(-1)^j}{j!} \binom{n}{i - j} n^{j-1}. \tag{A.2.2}$$

We will prove that also an expansion of finite length arranged in powers of n^{-1} holds. This result will be proved for real numbers $\alpha \geq 1$ instead of positive integers n.

If $k = 1, 3, 5, \ldots$ and $\alpha \geq 1$ then by writing $(1 + x/\alpha)^{\alpha}$ as $\exp[\alpha \log(1 + x/\alpha)]$ and by using a Taylor expansion of log about 1 it is immediate that

$$\exp\left[-\frac{(2x)^{k+1}}{(k+1)\alpha^k}\right] \le e^{-x}(1+x/\alpha)^\alpha \exp\left[-\sum_{i=2}^{k}\frac{(-1)^{i+1}x^i}{i\alpha^{i-1}}\right] \le 1 \qquad \text{(A.2.3)}$$

for $x \ge -\alpha/2$. Moreover, the upper bound still holds for $x \ge -\alpha$. The inequalities are strict for $x \ne 0$. Since $\exp(x) \ge 1 + x$ we obtain from (A.2.3), applied to $k = 1$, that

$$|e^{-x}(1+x/\alpha)^\alpha - 1| \le 2x^2/\alpha \quad \text{for } x \ge -\alpha/2. \qquad \text{(A.2.4)}$$

For $k = 3, 5, 7, \ldots$ the term $\exp[\sum_{i=2}^{k}(-1)^{i+1}x^i/(i\alpha^{i-1})]$ is a higher order approximation to $e^{-x}(1+x/\alpha)^\alpha$ but this approximation is not an expansion of the type as discussed in Section 3.2. However, a Taylor expansion of exp about zero yields the following result.

Lemma A.2.1. *For every positive integer m there exists a constant $C_m > 0$ such that for every $\alpha \ge 1$ and x with $-\alpha/2 \le x \le \alpha^{2/3}$ the following inequality holds:*

$$\left|e^{-x}(1+x/\alpha)^\alpha - \left[1 + \sum_{i=2}^{2(m-1)}\beta(i,\alpha)x^i\right]\right| \le C_m\alpha^{-m}(|x|^{2m-1} + |x|^{2m})$$

$$\text{(A.2.5)}$$

where $\beta(i,\alpha)$ are real numbers which have the property $\max\{|\beta(2k-1,\alpha)|,$ $|\beta(2k,\alpha)|\} \le C_m\alpha^{-k}$ for $k = 1, \ldots, m-1$.

Moreover, we have

$$\beta(2,\alpha) = -1/(2\alpha), \qquad \beta(3,\alpha) = 1/(3\alpha^2), \qquad \beta(4,\alpha) = 1/(8\alpha^2) - 1/(4\alpha^3).$$

PROOF. We have

$$\left|\exp\left[\sum_{i=2}^{2m-1}\frac{(-1)^{i+1}x^i}{i!\alpha^{i-1}}\right] - \sum_{j=0}^{m-1}\frac{1}{j!}\left[\sum_{i=1}^{2(m-1)}\frac{(-1)^i x^{i+1}}{(i+1)!\alpha^i}\right]^j\right| \le C\alpha^{-m}x^{2m}$$

$$\text{(A.2.6)}$$

where C will be used as a generic constant which only depends on m. By some tedious (however straightforward) computations one can prove that

$$\left|\sum_{j=0}^{m-1}\frac{1}{j!}\left[\sum_{i=1}^{2(m-1)}\frac{(-1)^i x^{i+1}}{(i+1)!\alpha^i}\right]^j - \sum_{i=2}^{2(m-1)}\beta(i,\alpha)x^i\right| \le C\alpha^{-m}(|x^{2m-1}| + |x^{2m}|)$$

where the values $\beta(i,\alpha)$ have the desired property. This together with (A.2.3) and (A.2.6) implies (A.2.5).   □

By writing down the proof of Lemma A.2.1 in detail one realizes that the upper bound in A.2.5 still holds for values x with $-\alpha \le x \le \alpha^{2/3}$.

For every positive integer n the terms $\beta(i,\alpha)$ in (A.2.5) are identical to the corresponding values, say, $\beta^*(i,\alpha)$ in (A.2.2). This becomes obvious by noting that there exists $A > 0$ and $B > 0$ such that

$$\left|e^{-x}(1+x/n)^n - \left(1 + \sum_{i=2}^{2(m-1)}\beta^*(i,n)x^i\right)\right| \le B|x|^{2m-1}$$

for every $|x| \leq A$. Now a comparison of this inequality to (A.2.5) leads to the desired identification.

The Second Lemma

The next lemma will provide us with an expansion of the function

$$g_{\alpha,\beta}(x) = e^{-x^2/2}\left[1 + \left(\frac{\beta}{(\alpha+\beta)\alpha}\right)^{1/2} x\right]^{\alpha}\left[1 - \left(\frac{\alpha}{(\alpha+\beta)\beta}\right)^{1/2} x\right]^{\beta} \qquad (A.2.7)$$

where $\alpha \geq 1$ and $\beta \geq 1$. This expansion will be arranged in powers of the terms $((\alpha+\beta)/\alpha\beta)^{1/2}$. As an application of this result one obtains expansions of densities and, in a second step, of distributions of central order statistics (see Section 4.2).

Lemma A.2.2. *For every positive integer m there exists a constant $C_m > 0$ such that for every $\alpha \geq 1$, $\beta \geq 1$ and $|x| \leq (\alpha\beta/(\alpha+\beta))^{1/6}$ the following inequality holds:*

$$\left|g_{\alpha,\beta}(x) - \left(1 + \sum_{i=1}^{m-1} g_{i,\alpha,\beta}(x)\right)\right| \leq C_m\left(\frac{\alpha+\beta}{\alpha\beta}\right)^{m/2}(|x|^{m+2} + |x|^{3m})$$

where $g_{i,\alpha,\beta}$ is a polynomial of degree $\leq 3i$ and the coefficients of $g_{i,\alpha,\beta}$ are smaller than $C_m((\alpha+\beta)/\alpha\beta)^{i/2}$ for $i = 1,\ldots, m-1$.

In the proof of Lemma A.2.2 one has to choose the polynomials $g_{i,\alpha,\beta}$ in such a way that

$$\left|\sum_{j=1}^{m-1}\frac{1}{j!}\left(\sum_{i=1}^{m-1} a_{i,\alpha,\beta}x^{i+2}\right)^j - \sum_{i=1}^{m-1} g_{i,\alpha,\beta}(x)\right| \leq C_m\left(\left(\frac{\alpha+\beta}{\alpha\beta}\right)^{1/2}|x|^3\right)^m \qquad (A.2.8)$$

where for $i = 1,\ldots, m-1$

$$a_{i,\alpha,\beta} = \frac{1}{i+2}\left[(-1)^{i+1}\alpha\left(\frac{\beta}{(\alpha+\beta)\alpha}\right)^{(i+2)/2} - \beta\left(\frac{\alpha}{(\alpha+\beta)\beta}\right)^{(i+2)/2}\right].$$

Particularly, for $i = 1, 2, 3$

$$g_{1,\alpha,\beta}(x) = a_{1,\alpha,\beta}x^3,$$
$$g_{2,\alpha,\beta}(x) = a_{2,\alpha,\beta}x^4 + a_{1,\alpha,\beta}^2 x^6/2, \qquad (A.2.9)$$
$$g_{3,\alpha,\beta}(x) = a_{3,\alpha,\beta}x^5 + a_{1,\alpha,\beta}a_{2,\alpha,\beta}x^7 + a_{1,\alpha,\beta}^3 x^9/6.$$

PROOF OF LEMMA A.2.2. Starting as in the proof of Lemma A.2.1 we obtain, by using a Taylor expansion of $\log(1 + x)$ of length $m + 1$, that for every $|x| \leq (\alpha\beta/(\alpha+\beta))^{1/6}$:

$$g_{\alpha,\beta}\begin{Bmatrix}\leq\\\geq\end{Bmatrix}\exp\left[\sum_{i=3}^{m+1}a_{i-2,\alpha,\beta}x^i \begin{matrix}+\\-\end{matrix} C\left(\frac{\alpha+\beta}{\alpha\beta}\right)^{m/2}|x|^{m+2}\right]$$

$$\begin{Bmatrix}\leq\\\geq\end{Bmatrix}\exp\left[\sum_{j=0}^{m-1}\frac{1}{j!}\left(\sum_{i=1}^{m-1}a_{i,\alpha,\beta}x^{i+2}\right)^j\right]\begin{matrix}+\\-\end{matrix} C_m\left(\frac{\alpha+\beta}{\alpha\beta}\right)^{m/2}(|x|^{m+2}+|x|^{3m}).$$

Now the proof can easily be completed by choosing the polynomials $g_{i,\alpha,\beta}$ as indicated in (A.2.8). □

If $\alpha=\beta$ then Lemma A.2.1 and Lemma A.2.2 roughly coincide for non-negative x.

We believe that an expansion of the function $g_{\alpha,\beta}$ is of some interest in its own right, however, this function is not properly adjusted to the particular problem of computing an expansion of the distribution of a central order statistic. For this purpose one has to deal with functions $h_{\alpha,\beta}$ defined by

$$h_{\alpha,\beta}(x)=e^{-x^2/2}\left[1+\left(\frac{\beta}{(\alpha+\beta)\alpha}\right)^{1/2}x\right]^{\alpha-1}\left[1-\left(\frac{\alpha}{(\alpha+\beta)\beta}\right)^{1/2}x\right]^{\beta-1}\quad\text{(A.2.10)}$$

or with some other variation of the function $g_{\alpha,\beta}$ according to the standardization of the distribution of the order statistic. By using Taylor expansions of

$$\left[1+\left(\frac{\beta}{(\alpha+\beta)\alpha}\right)^{1/2}x\right]^{-1}\quad\text{and}\quad\left[1-\left(\frac{\alpha}{(\alpha+\beta)\beta}\right)^{1/2}x\right]^{-1}$$

about 1, one can easily deduce from Lemma A.2.2 the following:

Corollary A.2.3. *For $\alpha,\beta\geq 1$ let $h_{\alpha,\beta}$ be defined as in (A.2.10).*

Then, Lemma A.2.2 holds true for $h_{\alpha,\beta}$, $h_{i,\alpha,\beta}$ and the term $(|x|^m+|x|^{3m})$ in place of $g_{\alpha,\beta}$, $g_{i,\alpha,\beta}$ and $(|x|^{m+2}+|x|^{3m})$ where $h_{i,\alpha,\beta}$ is a polynomial which has the same properties as $g_{i,\alpha,\beta}$ for $i=1,\ldots,m-1$.

The polynomials $h_{1,\alpha,\beta}$ and $h_{2,\alpha,\beta}$ are given by

$$h_{1,\alpha,\beta}(x)=g_{1,\alpha,\beta}(x)-\left[\left(\frac{\beta}{(\alpha+\beta)\alpha}\right)^{1/2}-\left(\frac{\alpha}{(\alpha+\beta)\beta}\right)^{1/2}\right]x$$

$$h_{2,\alpha,\beta}(x)=g_{2,\alpha,\beta}(x)-\left[\left(\frac{\beta}{(\alpha+\beta)\alpha}\right)^{1/2}-\left(\frac{\alpha}{(\alpha+\beta)\beta}\right)^{1/2}\right]xg_{1,\alpha,\beta}(x)$$

$$+\left[\frac{\beta}{(\alpha+\beta)\alpha}-\frac{1}{\alpha+\beta}+\frac{\alpha}{(\alpha+\beta)\beta}\right]x^2.$$

Further Results on Distances of Measures

The aim of the following lines is to extend some of the results of Section 3.3 to finite signed measures. Moreover, we prove some highly technical inequalities which do not belong to the necessary prerequisites for the understanding of the main ideas of this volume. However, these inequalities are useful for certain computations.

In the sequel, let v_i be a finite signed measure (on a measurable space $(S, \mathscr{B})$) represented by the density f_i w.r.t. a dominating measure μ.

A Further Remark about the Scheffé Lemma

An extension of Lemma 3.3.4 to finite signed measures is easily obtained by splitting the measure v_i to the positive and negative part v_i^+ and v_i^- with the respective densities f_i^+ and f_i^- (the positive and negative part of f). Check that

$$|f_0 - f_n| = |f_0^+ - f_n^+| - |f_0^- - f_n^-|.$$

Now, if the conditions of Lemma 3.3.4 are satisfied by f_i^+ and f_i^- then again

$$\lim_{n \to \infty} \int |f_0 - f_n| \, d\mu = 0. \tag{A.3.1}$$

The Variational Distance and the L_1-Distance

Define again

$$\|v_0 - v_1\| = \sup_B |v_0(B) - v_1(B)| \tag{A.3.2}$$

as the variational distance between v_0 and v_1. As an extension to Lemma 3.3.1 (the proof can be left to the reader) we get

Lemma A.3.1. (i) $\|v_0 - v_1\| \leq \int |f_0 - f_1| \, d\mu \leq 2\|v_0 - v_1\|$.
(ii) *If* $v_0(S) = v_1(S)$ *then*

$$\|v_0 - v_1\| = 2^{-1} \int |f_0 - f_1| \, d\mu.$$

We note that under the condition that $v_0(S) = v_1(S)$ we have again

$$\|v_0 - v_1\| = v_0\{f_0 > f_1\} - v_1\{f_0 > f_1\}.$$

The following modification of Lemma A.3.1(i) will be useful when the error term of an approximation has to be computed. In our applications no estimate of the term $\int g \, d\mu$ has to be computed.

Lemma A.3.2. *Let* f *and* g *be* μ-*integrable functions with* $g \geq 0$, $\int g \, d\mu > 0$, *and* $\int f \, d\mu > 0$. *Denote by* Q *the probability measure with* μ-*density* $g_0 = g/\int g \, d\mu$, *and by* v *the signed measure with* μ-*density* $f_0 = f/\int f \, d\mu$.
Then for every $B \in \mathcal{B}$,

$$\|Q - v\| \leq \left(\int f \, d\mu \right)^{-1} \int_B |g - f| \, d\mu + \int_{B^c} |g_0 - f_0| \, d\mu$$

where B^c *denotes the complement of* B.

PROOF. From Lemma A.3.1(ii) and the triangle inequality we get

$$\|Q - v\| \leq \left(2 \int f \, d\mu \right)^{-1} \int_B |g - f| \, d\mu + 2^{-1} \int_B \left| g_0 - g \middle/ \int f \, d\mu \right| d\mu \tag{1}$$
$$+ 2^{-1} \int_{B^c} |g_0 - f_0| \, d\mu.$$

Moreover, since $g \geq 0$ and $\int g_0 \, d\mu = \int f_0 \, d\mu = 1$ we have

$$\int_B \left| g_0 - g \middle/ \int f \, d\mu \right| d\mu = \left| \int_B \left(g_0 - g \middle/ \int f \, d\mu \right) d\mu \right|$$
$$= \left| - \int_B \left(g \middle/ \int f \, d\mu \right) d\mu - \int_{B^c} g_0 \, d\mu + \int f_0 \, d\mu \right|$$
$$\leq \left(\int f \, d\mu \right)^{-1} \int_B |g - f| \, d\mu + \int_{B^c} |g_0 - f_0| \, d\mu.$$

This together with (1) implies the asserted inequality. $\qquad \square$

In the applications the set B in Lemma A.3.2 is an exceptional set (with $Q(B)$ and $v(B)$ close to one) on which the integrand $|g - f|$ can easily be computed.

The Variational Distance between Product Measures

Given finite signed measures v_i with μ_i-density f_i put $|v_i| \equiv |v_i(S_i)| = \int |f_i| \, d\mu_i$ where S_i denotes the underlying space. Fubini's theorem implies that $|v_1 \times v_2| = |v_1||v_2|$ where the product measure $v_1 \times v_2$ is defined by

$$(v_1 \times v_2)(B) = \int_B f_1(x_1) f_2(x_2) \, d(\mu_1 \times \mu_2)(x_1, x_2).$$

Lemma A.3.3. *Let v_i and λ_i be finite signed measures for $i = 1, \ldots, k$. Then,*

$$\left\| \underset{i=1}{\overset{k}{\times}} v_i - \underset{i=1}{\overset{k}{\times}} \lambda_i \right\| \leq \sum_{i=1}^{k} \left| \underset{j=1}{\overset{i-1}{\times}} v_j \right| \left| \underset{j=i+1}{\overset{k}{\times}} \lambda_j \right| \|v_i - \lambda_i\|.$$

PROOF. For notational simplicity we will prove the assertion for $k = 2$ only. The general case can easily be proved by induction over k.

For measurable sets A in the product space we get by Fubini's theorem

$$|(v_1 \times \lambda_2)(A) - (\lambda_1 \times \lambda_2)(A)| = \left| \int (v_1(A_{x_2}) - \lambda_1(A_{x_2})) \, d\lambda_2(x_2) \right| \tag{1}$$

$$\leq |\lambda_2| \sup_{x_2} |v_1(A_{x_2}) - \lambda_1(A_{x_2})|$$

where A_{x_2} is the x_2-section of A. In analogy to (1) we get

$$|(v_1 \times v_2)(A) - (v_1 \times \lambda_2)(A)| \leq |v_1| \sup_{x_1} |v_2(A_{x_1}) - \lambda_2(A_{x_1})|. \tag{2}$$

Thus, combining (1) and (2) we get the desired inequality in the case of $k = 2$. The proof is complete. $\qquad\square$

Notice that Lemma 3.3.7 is an immediate consequence of Lemma A.3.3. Moreover, Lemma A.3.3 is an extension of the following well-known formula

$$\left| \prod_{i=1}^{k} a_i - \prod_{i=1}^{k} b_i \right| \leq \sum_{i=1}^{k} \left(\prod_{j=1}^{i-1} |a_j| \right) \left(\prod_{j=i+1}^{k} |b_j| \right) |a_i - b_i|$$

which holds for all real (as well as complex) numbers a_i and b_i.

Corollary A.3.4. *For probability measures Q_i and finite signed measures λ_i with $\lambda_i(S_i) = 1$ we have*

$$\left\| \underset{i=1}{\overset{k}{\times}} Q_i - \underset{i=1}{\overset{k}{\times}} \lambda_i \right\| \leq \exp\left[2 \sum_{i=1}^{k} \|Q_i - \lambda_i\| \right] \sum_{i=1}^{k} \|Q_i - \lambda_i\|. \tag{A.3.3}$$

PROOF. Check that

$$\left| \underset{j=i+1}{\overset{k}{\times}} \lambda_j \right| = \underset{j=i+1}{\overset{k}{\times}} |\lambda_j| \leq \prod_{j=i+1}^{k} (1 + 2\|Q_j - \lambda_j\|) \leq \exp\left[2 \sum_{i=1}^{k} \|Q_i - \lambda_i\| \right]. \qquad\square$$

The proof of Lemma A.3.3 gives a little bit more than stated there. For every measurable set A we obtain

$$\left|\left(\bigtimes_{i=1}^{k} \nu_i\right)(A) - \left(\bigtimes_{i=1}^{k} \lambda_i\right)(A)\right| \leq \sum_{i=1}^{k} \left|\bigtimes_{j=1}^{i-1} \nu_j\right| \left|\bigtimes_{j=i+1}^{k} \lambda_j\right| \sup_{\mathbf{x}_i} |\nu_i(A_{\mathbf{x}_i}) - \lambda_i(A_{\mathbf{x}_i})|$$

where $\mathbf{x}_i = (x_1, \ldots, x_{i-1}, x_{i+1}, \ldots, x_k)$ and $A_{\mathbf{x}_i}$ is the x_i-section of A; that is, $A_{\mathbf{x}_i} = \{x_i : (x_1, \ldots, x_k) \in A\}$. Thus, if e.g. A is a convex set then $A_{\mathbf{x}_i}$ is an interval.

The Hellinger Distance and the Kullback–Leibler Distance

Next we give the proof of the inequality (3.3.9) where it was stated that

$$H(Q_0, Q_1) \leq K(Q_0, Q_1)^{1/2}.$$

This is immediate from inequality (A.3.4) applied to $B = S$.

Lemma A.3.5. Let Q_0 and Q_1 be probability measures with μ-densities f_0 and f_1. Then, for every measurable set B,

$$H(Q_0, Q_1) \leq \left[2Q_0(B^c) + \int_B (-\log(f_1/f_0)) \, dQ_0 \right]^{1/2}. \tag{A.3.4}$$

PROOF. According to (3.3.5) we have to establish a lower bound of $\int (f_1 f_0)^{1/2} \, d\mu$. W.l.g. let $Q_0(B) > 0$. Since $\exp(x) \geq 1 + x$, we obtain from the Jensen inequality that

$$\int (f_1 f_0)^{1/2} \, d\mu \geq Q_0(B) \int_B (f_1/f_0)^{1/2} \, d(Q_0/Q_0(B))$$

$$\geq Q_0(B) \exp\left[(2Q_0(B))^{-1} \int_B \log(f_1/f_0) \, dQ_0 \right]$$

$$\geq Q_0(B) + 2^{-1} \int_B \log(f_1/f_0) \, dQ_0.$$

Now the assertion is immediate from (3.3.5). □

Further Bounds for the Variational Distance of Product Measures

Finally, we establish upper bounds for the variational distance of product measures via the χ^2-distance D. One special case was already proved in (3.3.10), namely, that

$$\left\| \underset{i=1}{\overset{k}{\bigtimes}} Q_i - \underset{i=1}{\overset{k}{\bigtimes}} P_i \right\| \le \left[\sum_{i=1}^{k} D(Q_i, P_i)^2 \right]^{1/2}$$

for probability measures Q_i and P_i where P_i has to be dominated by Q_i.

Next P_i will be replaced by a signed measure v_i with $v_i(S) = 1$. Again one has to assume that v_i is dominated by Q_i. Lemma A.3.6, applied to $m = 0$, yields

$$\left\| \underset{i=1}{\overset{k}{\bigtimes}} Q_i - \underset{i=1}{\overset{k}{\bigtimes}} v_i \right\| \le 2^{-1} \exp\left[2^{-1} \sum_{i=1}^{k} D(Q_i, v_i)^2 \right] \left[\sum_{i=1}^{k} D(Q_i, P_i)^2 \right]^{1/2}. \quad \text{(A.3.5)}$$

At the end of this section we will discuss in detail the special case of $m = 1$.

Lemma A.3.6. *Assume that Q_i and v_i satisfy the conditions above. Let $1 + g_i$ be a Q_i-density of v_i. Then, for every $m \in \{0, \ldots, k\}$,*

$$\sup_B \left| \left(\underset{i=1}{\overset{k}{\bigtimes}} v_i \right)(B) - \int_B h_m \, d \underset{i=1}{\overset{k}{\bigtimes}} Q_i \right|$$

$$\le 2^{-1} \left(\exp\left[((m + 1)!)^{-1} \sum_{i=1}^{k} D(Q_i, v_i)^2 \right] \right)^{1/2} \left[\sum_{i=1}^{k} D(Q_i, P_i)^2 \right]^{(m+1)/2}$$

where

$$h_m(x_1, \ldots, x_k) = 1 + \sum_{i=1}^{m} \sum_{1 \le i_1 < \cdots < i_j \le k} \prod_{r=1}^{j} g_{i_r}(x_{i_r}).$$

PROOF. Notice that

$$\prod_{i=1}^{k} (1 + a_i) = 1 + \sum_{i=1}^{k} \sum_{1 \le i_1 < \cdots < i_j \le k} \prod_{r=1}^{j} a_{i_r},$$

and, therefore, h_k is the $\bigtimes_{i=1}^{k} Q_i$-density of $\bigtimes_{i=1}^{k} v_i$. From the Schwarz inequality and the fact that the functions $(x_1, \ldots, x_k) \to \prod_{r=1}^{j} g_{i_r}(x_{i_r})$ for $1 \le i_1 < \cdots < i_j \le k$ and $j = 1, \ldots, k$ form a multiplicative system w.r.t. $\bigtimes_{i=1}^{k} Q_i$ we obtain

$$\sup_B \left| \left(\underset{i=1}{\overset{k}{\bigtimes}} v_i \right)(B) - \int_B h_m \, d \underset{i=1}{\overset{k}{\bigtimes}} Q_i \right| \le 2^{-1} \int |h_k - h_m| \, d \underset{i=1}{\overset{k}{\bigtimes}} Q_i$$

$$\le 2^{-1} \left[\int (h_k - h_m)^2 \, d \underset{i=1}{\overset{k}{\bigtimes}} Q_i \right]^{1/2}$$

$$= 2^{-1} \left[\sum_{j=m+1}^{k} \sum_{1 \le i_1 < \cdots < i_j \le k} \prod_{r=1}^{j} \int g_{i_r}^2 \, dQ_{i_r} \right]^{1/2}.$$

This implies the asserted inequality since

$$\sum_{1 \le i_1 < \cdots < i_j \le k} \prod_{r=1}^{j} a_{i_r} \le \left(\sum_{i=1}^{k} |a_i| \right)^j \bigg/ j!$$

and

$$\sum_{i=m}^{\infty} z^i/i! \le \exp(z) z^m/m! \qquad \text{for } z \ge 0. \qquad \square$$

In the special case of $m = 1$ we get

$$\sup_{B} \left| \left(\underset{i=1}{\overset{k}{\times}} v_i \right)(B) - \left(\underset{i=1}{\overset{k}{\times}} Q_i \right)(B) - \int_B \left(\sum_{i=1}^{k} g_i(x_i) \right) d\left(\underset{i=1}{\overset{k}{\times}} Q_i \right)(x_1, \ldots, x_k) \right|$$

$$\le 8^{-1} \exp\left[2^{-1} \sum_{i=1}^{k} D(Q_i, v_i)^2 \right] \sum_{i=1}^{k} D(Q_i, P_i)^2 =: R_k \qquad (A.3.6)$$

and hence

$$\left\| \underset{i=1}{\overset{k}{\times}} v_i - \underset{i=1}{\overset{k}{\times}} Q_i \right\| - 2^{-1} \int \left| \sum_{i=1}^{k} g_i(x_i) \right| d\left(\underset{i=1}{\overset{k}{\times}} Q_i \right)(x_1, \ldots, x_k) \le R_k.$$

This shows that for $k \to \infty$ further insight into the variational distance of product measures may be gained by means of the central limit theorem.

Bibliography

Alam, K. (1972). Unimodality of the distribution of an order statistic. Ann. Math. Statist. 43, 2041–2044.

Albers, W., Bickel, P. J. and van Zwet, W.R. (1976). Asymptotic expansions for the power of distribution-free tests in one-sample problem. Ann. Statist. 4, 108–156.

Ali, M.M. and Kuan, K.S. (1977). On the joint asymptotic normality of quantiles. Nanta Math. 10, 161–165.

Anderson, C.W. (1971). Contributions to the Asymptotic Theory of Extreme Values. Ph.D. Thesis, University of London.

Anderson, C.W. (1984). Large deviations of extremes. In: Statistical Extremes and Applications, Ed. J. Tiago de Oliveira, pp. 325–340. Dordrecht: Reidel.

Arnold, B.C., Becker, A., Gather, U. and Zahedi, H. (1984). On the Markov property of order statistics. J. Statist. Plann. Inference 9, 147–154.

Bahadur, R.R. (1966). A note on quantiles in large samples. Ann. Math. Statist. 37, 577–580.

Bain, L.J. (1978). Statistical Analysis of Reliability and Life-Testing Models. New York: Marcel Dekker.

Balkema, A.A. and Haan, L. de (1978a). Limit distributions for order statistics I. Theory Probab. Appl. 23, 77–92.

Balkema, A.A. and Haan, L. de (1978b). Limit distributions for order statistics II. Theory Probab. Appl. 23, 341–358.

Barndorff-Nielsen, O. (1964). On the limit distribution of the maximum of a random number of independent random variables. Acta Math. Acad. Sci. Hungar. 15, 399–403.

Barnett, V. (1975). Probability plotting methods and order statistics. Appl. Statist. 24, 95–108.

Barnett, V. (1976). The ordering of multivariate data. J. Roy. Statist. Soc., Ser. A, 139, 318–344.

Barnett, V. and Lewis, T. (1978). Outliers in Statistical Data. Chichester: Wiley.

Beirlant, J. and Teugels, J.L. (1987). Asymptotics of Hill's estimator. Theory Probab. Appl. 31, 463–469.

Beran, J. (1985). Stochastic procedures: Bootstrap and random search methods in statistics: Proceedings of 45th Session of the ISI, Vol. 4 (Amsterdam), 25.1.

Berman, S.M. (1961). Convergence to bivariate limiting extreme value distributions. Ann. Inst. Statist. Math. 13, 217–223.

Bhattacharya, R.N. and Rao, R.R. (1976). Normal Approximation and Asymptotic Expansion. New York: Wiley.

Bhattacharya, R.N. and Gosh, J.K. (1978). On the validity of the formal Edgeworth expansions. Ann. Statist. 6, 434–451.

Bickel, P.J. (1967). Some contributions to the theory of order statistics. In: Proc. 5th Berkeley Symp. Math. Statistics and Prob., Vol. I., pp. 575–591. Berkeley: Univ. California Press.

Bickel, P.J. and Freedman D.A. (1981). Some asymptotic theory for the bootstrap. Ann. Statist. 9, 1196–1217.

Bickel, P.J. and Rosenblatt, M. (1973). On some global measures of the deviation of density function estimates. Ann. Statist. 1, 1071–1095.

Bickel, P.J. and Rosenblatt, M. (1975). Correction to "On some global measures of the deviation of density function estimates". Ann. Statist. 3, 1370.

Bloch, D.A. and Gastwirth, J.L. (1968). On a simple estimate of the reciprocal of the density function. Ann. Math. Statist. 39, 1083–1085.

Blom, G. (1958). Statistical Estimates and Transformed Beta-Variables. New York: Wiley.

Blum, J.R. and Pathak, P.K. (1972). A note on the zero-one law. Ann. Math. Statist. 43, 1008–1009.

Boos, D.D. (1984). Using extreme value theory to estimate large percentiles. Technometrics 26, 33–39.

Bortkiewicz, L. von (1922). Variationsbreite und mittlere Fehler. Sitzungsberichte Berliner Math. Ges. 21, 3–11.

Brown, B.M. (1981). Symmetric quantile averages and related estimators. Biometrika 68, 235–242.

Brozius, H. and Haan, L. de (1987). On limit laws for the convex hull of a sample. J. Appl. Probab. 24, 863–874.

Chernoff, H., Gastwirth, J.L. and John, M.V. (1967). Asymptotic distribution of linear combinations of functions of order statistics with applications to estimation. Ann. Math. Statist. 38, 52–72.

Chibisov, D.M. (1964). On limit distributions for order statistics. Theory Probab. Appl. 9, 150–165.

Chow, Y.S. and Teicher, H. (1978). Probability Theory. New York: Springer.

Cohen, J.P. (1982a). The penultimate form of approximation to normal extremes. Adv. Appl. Probab. 14, 324–339.

Cohen, J.P. (1982b). Convergence rates for the ultimate and penultimate approximation in extreme-value theory. Adv. Appl. Probab. 14, 833–854.

Cohen, J.P. (1984). The asymptotic behaviour of the maximum likelihood estimates for univariate extremes. In: Statistical Extremes and Applications, Ed. J. Tiago de Oliveira, pp. 435–442. Dordrecht: Reidel.

Cooil, B. (1985). Limiting multivariate distributions of intermediate order statistics. Ann. Probab. 13, 469–477.

Cooil, B. (1988). When are intermediate processes of the same stochastic order? Statist. Probab. Letters 6, 159–162.

Consul, P.C. (1984). On the distributions of order statistics for a random sample size. Statist. Neerlandica 38, 249–256.

Craig, A.T. (1932). On the distribution of certain statistics. Amer. J. Math. 54, 353–366.

Cramér, H. (1946). Mathematical Methods of Statistics. Princeton: Princeton Univ. Press.

Csiszár, I. (1975). I-Divergence geometry of probability distributions and minimization problems. Ann. Probab. 3, 146–158.

Csörgő, M. (1983). Quantile Processes with Statistical Applications. Philadelphia: SIAM.

Csörgő, M., Csörgő, S., Horváth, L. and Mason, D.M. (1986). Normal and stable convergence of integral functions of the empirical distribution function. Ann. Probab. 14, 86–118.

Csörgő, M. and Révész, P. (1981). Strong Approximations in Probability and Statistics. New York: Academic Press.

Csörgő, S., Deheuvels, P. and Mason, D.M. (1985). Kernel estimates of the tail index of a distribution. Ann. Statist. 13, 1050–1078.

Csörgő, S., Horváth, L. and Mason, D.M. (1986). What portion of the sample makes a partial sum asymptotically stable or normal? Probab. Th. Rel. Fields 72, 1–16.

Csörgő, S., and Mason, D.M. (1986). The asymptotic distributions of sums of extreme values from a regularly varying distribution. Ann. Probab. 14, 974–983.

David, F.N. and Johnson, N.L. (1954). Statistical treatment of censored data, Part I, Fundamental formulae. Biometrika 44, 228–240.

David, H.A. (1981). Order Statistics. 2nd ed. New York: Wiley.

Davis, R.A. (1982). The rate of convergence in distribution of the maxima. Statist. Neerlandica 36, 31–35.

Deheuvels, P. and Pfeifer, D. (1988). Poisson approximations of multinomial distributions and point processes. J. Multivariate Anal. 25, 65–89.

Dodd, E.L. (1923). The greatest and the least variate under general laws of error. Trans. Amer. Math. Soc. 25, 525–539.

Dronskers, J.J. (1958). Approximate formulae for the statistical distributions of extreme values. Biometrika 45, 447–470.

Du Mouchel, W. (1983). Estimating the stable index α in order to measure tail thickness. Ann. Statist. 11, 1019–1036.

Dwass, M. (1966). Extremal processes, II. Illinois J. Math. 10, 381–391.

Dziubdziela, W. (1976). A note on the k-th distance random variables. Zastosowania Matematykai 15, 289–291.

Eddy, W.F. and Gale, J.D. (1981). The convex hull of a spherically symmetric sample. Adv. Appl. Probab. 13, 751–763.

Efron, B. (1979). Bootstrap methods: another look at the jackknife. Ann. Statist. 7, 1–26.

Egorov, V.A. and Nevzorov, V.B. (1976). Limit theorems for linear combinations of order statistics. In: Proc. 3rd Japan-USSR Symp. Probab. Theory, Eds. G. Maruyama and J.V. Prokhorov, pp. 63–79. Lecture Notes in Mathematics 550. New York: Springer.

Englund, G. (1980). Remainder term estimates for the asymptotic normality of order statistics. Scand. J. Statist. 7, 197–202.

Erdélyi, A., Magnus, W., Oberhettinger, F. and Tricomi, F.G. (1953). Higher Transcendental Functions, Vol. I. New York: McGraw-Hill.

Falk, M. (1983). Relative efficiency and deficiency of kernel type estimators of smooth distribution functions. Statist. Neerlandica 37, 73–83.

Falk, M. (1984a). Relative deficiency of kernel type estimators of quantiles. Ann. Statist. 12, 261–268.

Falk, M. (1984b). Berry-Esséen theorems for a global measure of performance of kernel density estimators. South African Statist. J. 19, 1–19.

Falk, M. (1985a). Asymptotic normality of the kernel quantile estimator. Ann. Statist. 13, 428–433.

Falk, M. (1985b). Uniform convergence of extreme order statistics. Habilitationsschrift, University of Siegen.

Falk, M. (1986a). Rates of uniform convergence of extreme order statistics. Ann. Inst. Statist. Math., Ser. A, 38, 245–262.

Falk, M. (1986b). On the estimation of the quantile density function. Statist. Probab. Letters 4, 69–73.

Falk, M. (1989a). Best attainable rate of joint convergence of extremes. In: Extreme Value Theory, Eds. J. Hüsler and R.-D. Reiss, pp. 1–9. Lecture Notes in Statistics 51. New York: Springer.

Falk, M. (1989b). A note on uniform asymptotic normality of intermediate order statistics. Ann. Inst. Statist. Math., Ser. A.

Falk, M. and Kohne, W. (1986). On the rate at which the sample extremes become independent. Ann. Probab. 14, 1339–1346.

Falk, M. and Reiss, R.-D. (1988). Independence of order statistics. Ann. Probab. 16, 854–862.

Falk, M. and Reiss, R.-D. (1989). Weak convergence of smoothed and nonsmoothed bootstrap quantile estimates. Ann. Probab. 17.

Feldman, D. and Tucker, H.G. (1966). Estimation of non-unique quantiles. Ann. Math. Statist. 37, 451–457.

Feller, W. (1972). An Introduction to Probability Theory and its Applications. Vol. 2, 2nd ed. New York: Wiley.

Ferguson, T.S. (1967). Mathematical Statistics. New York: Academic Press.

Finkelstein, B.V. (1953). Limiting distribution of extreme terms of a variational series of a two-dimensional random variable. Dokl. Ak. Nauk. S.S.S.R. 91, 209–211 (in Russian).

Fisher, R.A. (1922). On the mathematical foundation of theoretical statistics. Phil. Trans. Roy. Soc. A 222, 309–368. Reprint in: Collected Papers of R.A. Fisher, Vol. I, Ed. J.H. Bennett, pp. 276–335. University of Adelaide.

Fisher, R.A. and Tippett, L.H.C. (1928). Limiting forms of the frequence distribution of the largest or smallest member of a sample. Proc. Camb. Phil. Soc. 24, 180–190.

Floret, K. (1981). Mass- und Integrationstheorie. Stuttgart: Teubner.

Fréchet, M. (1927). Sur la loi de probabilité de l'écart maximum. Ann. de la Soc. Polonaise de Math. 6, 93–116.

Galambos, J. (1975). Order statistics of samples from multivariate distributions. J. Amer. Statist. Assoc. 70, 674–680.

Galambos, J. (1984). Order statistics. In: Handbook of Statistics. Vol. 4, Eds. P.R. Krishnaiah and P.K. Sen, pp. 359–382. Amsterdam: North-Holland.

Galambos, J. (1987). The Asymptotic Theory of Extreme Order Statistics. 2nd ed. Malabar, Florida: Krieger.

Geffroy, J. (1958/59). Contributions à la théorie des valeurs extrêmes. Publ. Inst. Statist. Univ. Paris 7/8, 37–185.

Gini, C. and Galvani, L. (1929). Di talune estensioni dei concetti di media ai caratteri qualitativi. Metron 8. Partial English translation in: J. Amer. Statist. Assoc. 25, 448–450.

Gnedenko, B. (1943). Sur la distribution limit du terme maximum d'une série aléatoire. Ann. Math. 44, 423–453.

Goldie, C.M. and Smith, R.L. (1987). Slow variation with remainder: Theory and applications. Quart. J. Math. Oxford 38, 45–71.

Gomes, M.I. (1978). Some probabilistic and statistical problems in extreme value theory. Ph.D. Thesis, University of Sheffield.

Gomes, M.I. (1981). An i-dimensional limiting distribution function of largest values and its relevance to the statistical theory of extremes. In: Statistical Distribution in Scientific Work, Eds. C. Taillie et al., Vol. 6., pp. 389–410. Dordrecht: Reidel.

Gomes, M.I. (1984). Penultimate limiting forms in extreme value theory. Ann. Inst. Statist. Math., Ser. A, 36, 71–85.

Gross, A.J. (1975). Survival Distributions: Reliability Applications in the Biomedical Sciences. New York: Wiley.

Guilbaud, O. (1982). Functions of non-iid random vectors expressed as functions of iid random vectors. Scand. J. Statist. 9, 229–233.

Gumbel, E.J. (1933). Das Alter des Methusalem. Z. Schweizerische Statistik und Volkswirtschaft 69, 516–530.

Gumbel, E.J. (1946). On the independence of the extremes in a sample. Ann. Math. Statist. 17, 78–81.

Gumbel, E.J. (1958). Statistics of Extremes. New York: Columbia Univ. Press.

Haan, L. de (1970). On Regular Variation and its Application to the Weak Convergence of Sample Extremes. Amsterdam, Math. Centre Tracts 32.

Haan, L. de (1976). Sample extremes: an elementary introduction. Statist. Neerlandica 30, 161–172.

Haan, L. de and Resnick, S.I. (1980). A simple asymptotic estimate for the index of a stable distribution. J. Roy. Statist. Soc., Ser. B., 42, 83–87.

Haan, L. de and Resnick, S.I. (1982). Local limit theorems for sample extremes. Ann. Probab. 10, 396–413.

Häusler, E. and Teugels, J.L. (1985). On asymptotic normality of Hill's estimator for the exponent of regular variation. Ann. Statist. 13, 743–756.

Haldane, J.B.S. and Jayakar, S.G. (1963). The distribution of extremal and nearly extremal values in samples from a normal distribution. Biometrika 50, 89–94.

Hall, P. (1978). Some asymptotic expansions of moments of order statistics. Stoch. Proc. Appl. 7, 265–275.

Hall, P. (1979). On the rate of convergence of normal extremes. J. Appl. Probab. 16, 433–439.

Hall, P. (1982a). On estimating the endpoint of a distribution. Ann. Statist. 10, 556–568.

Hall, P. (1982b). On some simple estimates of an exponent of regular variation. J. Roy. Statist. Soc., Ser. B., 44, 37–42.

Hall, P. (1983). On near neighbour estimates of a multivariate density. J. Multivariate Anal. 13, 24–39.

Hall, P. and Welsh, A.H. (1984). Best attainable rates of convergence for estimates of parameters of regular variation. Ann. Statist. 12, 1079–1084.

Hall, P. and Welsh, A.H. (1985). Adaptive estimates of parameters of regular variation. Ann. Statist. 13, 331–341.

Hall, W.J. and Wellner, J.A. (1979). The rate of convergence in law of the maximum of an exponential sample. Statist. Neerlandica 33, 151–154.

Hájek, J. and Sidák, Z. (1967). Theory of Rank Tests. New York: Academic Press.

Harrel, F.E. and Davis, C.E. (1982). A new distribution-free quantile estimator. Biometrika 69, 635–640.

Harter, H.L. (1983). The chronological annotated bibliography of order statistics. Vol. I: pre-1950. Vol. II: 1950–1959. Columbus, Ohio: American Sciences Press.

Hecker, H. (1976). A characterization of the asymptotic normality of linear combinations of order statistics from the uniform distribution. Ann. Statist. 4, 1244–1246.

Heidelberger, P. and Lewis, P.A.W. (1984). Quantile estimation in dependent sequences. Opns. Res. 32, 185–209.

Helmers, R. (1981). A Berry-Esséen theorem for linear combinations of order statistics. Ann. Probab. 9, 342–347.

Helmers, R. (1982). Edgeworth Expansions for Linear Combinations of Order Statistics. Amsterdam, Math. Centre Tracts 105.

Herbach, L. (1984). Introduction, Gumbel model. In: Statistical Extremes and Applications, Ed. J. Tiago de Oliveira, pp. 49–80. Dordrecht: Reidel.

Hewitt, E. and Stromberg, K. (1975). Real and Abstract Analysis. 3rd ed. New York: Springer.

Heyer, H. (1982). Theory of Statistical Experiments. Springer Series in Statistics. New York: Springer.

Hill, B.M. (1975). A simple approach to inference about the tail of a distribution. Ann. Statist. 3, 1163–1174.

Hillion, A. (1983). On the use of some variation distance inequalities to estimate the difference between sample and perturbed sample. In: Specifying Statistical Models, Eds. J.P. Florens et al., pp. 163–175. Lecture Notes in Statistics 16. New York: Springer.

Hodges, J.L. Jr. and Lehmann, E.L. (1967). On medians and quasi medians. J. Amer. Statist. Assoc. 62, 926–931.

Hodges, J.L. Jr. and Lehmann, E.L. (1970). Deficiency. Ann. Math. Statist. 41, 783–801.

Hoeffding, W. and Wolfowitz, J. (1958). Distinguishability of sets of distributions. Ann. Math. Statist. 29, 700–718.

Hosking, J.R.M. (1985). Maximum-likelihood estimation of the parameter of the generalized extreme-value distribution. Applied Statistics 34, 301–310.

Huang, J.S. and Gosh, M. (1982). A note on the strong unimodality of order statistics. J. Amer. Statist. Soc. 77, 929–930.

Hüsler, J. and Reiss, R.-D. (1989). Maxima of normal random vectors: Between independence and complete dependence. Statist. Probab. Letters 7.

Hüsler, J. and Schüpbach, M. (1988). On simple block estimators for the parameters of the extreme-value distribution. Commun. Statist.-Simula. 15, 61–76.

Hüsler, J. and Tiago de Oliveira, J. (1986). The usage of the largest observations for parameter and quantile estimation for the Gumbel distribution; an efficiency analysis. Publ. Inst. Stat. Univ. 33, 41–56.

Ibragimov, J.A. (1956). On the composition of unimodal distributions. Theory Probab. Appl. 1, 225–260.

Ibragimov, J.A. and Has'minskii, R.Z. (1981). Statistical Estimation. Springer-Verlag, Berlin.

Iglehardt, D.L. (1976). Simulating stable stochastic systems; VI. Quantile estimation. J. Assoc. Comput. Mach. 23, 347–360.

Ikeda, S. (1963). Asymptotic equivalence of probability distributions with applications to some problems of asymptotic independence. Ann. Inst. Statist. Math. 15, 87–116.

Ikeda, S. (1975). Some criteria for uniform asymptotic equivalence of real probability distributions. Ann. Inst. Statist. Math. 27, 421–428.

Ikeda, S. and Matsunawa, T. (1970). On asymptotic independence of order statistics. Ann. Inst. Statist. Math. 22, 435–449.

Ikeda, S. and Matsunawa, T. (1972). On the uniform asymptotic joint normality of sample quantiles. Ann. Inst. Statist. Math. 24, 33–52.

Ikeda, S. and Nonaka, Y. (1983). Uniform asymptotic joint normality of a set of increasing number of sample quantiles. Ann. Inst. Statist. Math. 35, Ser. A, 329–341.

Isogai, T. (1985). Some extensions of Haldane's multivariate median and its applications. Ann. Inst. Statist. Math. 37, Ser. A, 289–301.

Ivchenko, G.I. (1971). On limit distributions for the order statistics of the multinomial distribution. Theory Probab. Appl. 16, 102–115.

Ivchenko, G.I. (1974). On limit distributions for middle order statistics for double sequence. Theory Probab. Appl. 19, 267–277.

Jacod, J. and Shiryaev, A.N. (1987). Limit Theorems for Stochastic Processes. Berlin: Springer.

Janssen, A. (1988). Uniform convergence of sums of order statistics to stable laws. Probab. Th. Rel. Fields 78, 261–272.

Janssen, A. and Reiss, R.-D. (1988). Comparison of location models of Weibull type samples and extreme value processes. Probab. Th. Rel. Fields 78, 273–292.

Joag-Dev, K. (1983). Independence via uncorrelatedness under certain dependence structures. Ann. Probab. 11, 1037–1041.

Joe, H. (1987). Estimation of quantiles of the maximum of N observations. Biometrika 74, 347–354.

Johnson, N.L. and Kotz, S. (1970). Distributions in Statistics: Continuous Univariate Distributions—1. New York: Wiley.

Johnson, N.L. and Kotz, S. (1972). Distributions in Statistics: Continuous Multivariate Distributions. New York: Wiley.

Kabanov, Yu. and Lipster, R.S. (1983). On convergence in variation of the distributions of multivariate point processes. Z. Wahrsch. verw. Geb. 63, 475–485.

Karr, A.F. (1986). Point Processes and their Statistical Inference. New York: Marcel Dekker.

Kendall, M.G. (1940). Note on the distributions of quantiles for large samples. J. Roy. Statist. Soc., Suppl. 7, 83–85.

Kendall, M.G. and Stuart, A. (1958). The Advanced Theory of Statistics. Vol. 1. London: Griffin.

Kiefer, J. (1967). On Bahadur's representation of sample quantiles. Ann. Math. Statist. 38, 1323–1342.

Kiefer, J. (1969a). Deviations between the sample quantile process and the sample df. In: Nonparametric Techniques in Statistical Inference, Ed. M.L. Puri, pp. 299–319. Cambridge: Cambridge Univ. Press.

Kiefer, J. (1969b). Old and new methods for studying order statistics and sample quantiles. In: Nonparametric Techniques in Statistical Inference, Ed. M.L. Puri, pp. 349–357. Cambridge: Cambridge Univ. Press.

Kinnison, R.R. (1985). Applied Extreme Value Statistics. Columbus: Battelle Press.

Klenk, A. and Stute, W. (1987). Bootstrapping of L-estimates. Statist. Decisions 5, 77–87.

Kohne, W. and Reiss, R.-D. (1983). A note on uniform approximation to distributions of extreme order statistics. Ann. Inst. Statist. Math., Ser. A, 35, 343–345.

Kolchin, V.F. (1980). On the limiting behaviour of extreme order statistics in a polynomial scheme. Theory Probab. Appl. 14, 458–469.

Koziol, J.A. (1980). A note on limiting distributions for spacings statistics. Z. Wahrsch. verw. Gebiete 51, 55–62.

Kuan, K.S. and Ali, M.M. (1960). Asymptotic distribution of quantiles from a multivariate distribution. In: Mult. Statist. Analysis, Ed. R.P. Gupta, pp. 109–120. Amsterdam: North-Holland.

Lamperti, J. (1964). On extreme order statistics. Ann. Math. Statist. 35, 1726–1736.

Landers, D. and Rogge, L. (1985). Asymptotic normality of the estimators of the natural median. Statist. Decisions 3, 77–90.

Laplace, P.S. de (1818). Deuxième supplément à la théorie analytique de probabilitiés. Paris: Courcier, Reprint (1886) in: Ouevres complètes de Laplace 7, pp. 531–580. Paris: Gauthier-Villars.

Lawless, J.F. (1982). Statistical Models and Methods for Lifetime Data. New York: Wiley.

Leadbetter, M.R., Lindgren, G. and Rootzén, H. (1983). Extremes and Related Properties of Random Sequences and Processes. Springer Series in Statistics. New York: Springer.

Le Cam, L. (1986). Asymptotic Methods in Statistical Decision Theory. Springer Series in Statistics. New York: Springer.

Lehmann, E.L. (1986). Testing Statistical Hypothesis. 2nd ed. New York: Wiley.

Loève, M. (1963). Probability Theory. 3rd ed. New York: Van Nostrand.

Mack, Y.P. (1984). Remarks on some smoothed empirical distribution functions and processes. Bull. Informatics Cybernetics 21, 29–35.

Malmquist, S. (1950). On a property of order statistics from a rectangular distribution. Skand. Aktuar. 33, 214–222.

Mammitzsch, V. (1984). On the asymptotically optimal solution within a certain class of kernel type estimators. Statist. Decisions 2, 247–255.

Mann, N.R., Schafer, R.E. and Singpurwalla, N.D. (1974). Methods for Statistical Analysis of Reliability and Life Data. New York: Wiley.

Mann, N.R. (1984). Statistical estimation of the Weibull and Fréchet distributions. In: Statistical Extremes and Applications, Ed. J. Tiago de Oliveira, pp. 81–89. Dordrecht: Reidel.

Marshall, A.W. and Olkin, I. (1983). Domains of attraction of multivariate extreme value distributions. Ann. Probab. 11, 168–177.

Matsunawa, T. (1975). On the error evaluation of the joint normal approximation for sample quantiles. Ann. Inst. Statist. Math. 27, 189–199.

Matsunawa, T. and Ikeda, S. (1976). Uniform asymptotic distribution of extremes. In: Essays in Probab. Statist., Eds. S. Ikeda et al., pp. 419–432. Tokyo: Shinko Tsusho.

Michel, R. (1975). An asymptotic expansion for the distribution of asymptotic maximum likelihood estimators of vector parameters. J. Multivariate Anal. 5, 67–85.

Miebach, B. (1977). Asymptotische Theorie für Familien von Maßen mit Lokalisations- und Dispersionsparameter. Diploma Thesis, University of Cologne.

Mises von, R. (1923). Über die Variationsbreite einer Beobachtungsreihe. Sitzungsberichte Berliner Math. Ges. 22, 3–8.

Mises von, R. (1936). La distribution de la plus grande de n valeurs. Rev. Math. Union Interbalcanique 1, 141–160. Reproduced in Selected Papers of Richard von Mises, Amer. Math. Soc. 2 (1964), 271–294.

Miyamoto, Y. (1976). Optimum spacings for goodness of fit tests based on sample quantiles. In: Essays in Probab. Statist., Eds. S. Ikeda et al., pp. 475–483. Tokyo: Shinko Tsusho.

Montfort, M.A.J. van (1982). Modellen voor maximum en minima, schattingen en betrouwbaarheidsintervallen, kreuze tussen modellen, Agricultural University Wageningen, Netherlands, Dept. Math., Statist. Division, Technical Note 82-02.

Montfort, M.A.J. van and Gomes, I.M. (1985). Statistical choice of extremal models for complete and censored data. J. Hydrology 77, 77–87.

Mood, A. (1941). On the joint distribution of the median in sample from a multivariate population. Ann. Math. Statist. 12, 268–278.

Moore, D.S. and Yackel, J.W. (1977). Large sample properties of nearest neighbour density function estimates. In: Statistical Decision Theory and Related Topics, Eds. S.S. Gupta and D.S. Moore, pp. 269–279. New York: Academic Press.

Mosteller, F. (1946). On some useful inefficient statistics. Ann. Math. Statist. 17, 377–408.

Nadaraya, E.A. (1964). Some new estimates for distribution functions. Theory Probab. Appl. 10, 186–190.

Nagaraja, H.N. (1982). On the non-Markovian structure of discrete order statistics. J. Statist. Plann. Inference 7, 29–33.

Nagaraja, H.N. (1986). Structure of discrete order statistics. J. Statist. Plann. Inference 13, 165–177.

Nelson, W. (1982). Applied Life Data Analysis. New York: Wiley.

Nowak, W. and Reiss, R.-D. (1983). Asymptotic expansions of distributions of central order statistics under discrete distributions. Technical Report 101, University of Siegen.

Oja, H. and Niinimaa, A. (1985). Asymptotic properties of the generalized median in the case of multivariate normality. J. Roy. Statist. Soc., Ser. B, 47, 372–377.

O'Reilley, F.J. and Quesenberry, C.P. (1973). The conditional probability integral transformation and applications to obtain composite chi-square goodness-of-fit tests. Ann. Statist. 1, 74–83.

Pantcheva, E.I. (1985). Limit theorems for extreme order statistics under nonlinear normalization. In: Stability Problems for Stochastic Models, Eds. V.V. Kalashnikov and V.M. Zolotarev, pp. 284–309. Lecture Notes in Mathematics 1155, Berlin: Springer.

Parzen, E. (1962) On estimation of a probability density function and mode. Ann. Math. Statist. 33, 1065–1076.

Parzen, E. (1979). Nonparametric statistical data modeling. J. Amer. Statist. Assoc. 74, 105–121.

Pearson, K. (1902). Note on Francis Galton's problem. Biometrika 1, 390–399.

Pearson, K. (1920). On the probable errors of frequency constants. Biometrika 13, 113–132.

Pfanzagl, J. (1973a). Asymptotically optimum estimation and test procedures. In: Proc. Prague Symp. Asymptotic Statistics, Vol. 1, Ed. J. Hájek, pp. 201–272. Prague: Charles University.

Pfanzagl, J. (1973b). The accuracy of the normal approximation for estimates of vector parameters. Z. Wahrsch. verw. Gebiete 25, 171–198.

Pfanzagl, J. (1973c). Asymptotic expansions related to minimum contrast estimators. Ann. Statist. 1, 993–1026.

Pfanzagl, J. (1975). Investigating the quantile of an unknown distribution. In: Statistical Methods in Biometry, Ed. W.J. Ziegler, pp. 111–126. Basel: Birkhäuser.

Pfanzagl, J. (1982). Contributions to a General Asymptotic Statistical Theory. (With the assistence of W. Wefelmeyer). Lecture Notes in Statistics 13. New York: Springer.

Pfanzagl, J. (1985). Asymptotic Expansions for General Statistical Models. (With the assistance of W. Wefelmeyer). Lecture Notes in Statistics 31. New York: Springer.

Pickands, J. (1967). Sample sequences of maxima. Ann. Math. Statist. 38, 1570–1574.

Pickands, J. (1968). Moment convergence of sample extremes. Ann. Math. Statist. 39, 881–889.

Pickands, J. (1975). Statistical inference using extreme order statistics. Ann. Statist. 3, 119–131.

Pickands, J. (1981). Multivariate extreme value distributions. Proc. 43th Session of the ISI (Buenos Aires), 859–878.

Pickands, J. (1986). The continuous and differentiable domains of attractions of the extreme value distributions. Ann. Probab. 14, 996–1004.

Pitman, E.J.G. (1979). Some Basic Theory for Statistical Inference. London: Chapman and Hall.

Plackett, R.L. (1976). In: Discussion of Professor Barnett's Paper. J.R. Statist. Soc., Ser. A, 139, 344–346.

Polfeldt, T. (1970). Asymptotic results in non-regular estimation. Skand. Aktuar., Suppl. 1–2, 2–78.

Prakasa Rao, B.L.S. (1983). Nonparametric Functional Estimation. Orlando: Academic Press.

Puri, M.L. and Ralescu, S.S. (1986). Limit theorems for random central order statistics. In: Adaptive Statistical Procedures and Related Topics, Ed. J. van Ryzin, pp. 447–475. IMS Lecture Notes 8.

Pyke, R. (1965). Spacings. J. Roy. Statist. Soc., Ser. B. 27, 395–436. Discussion: 437–449.

Pyke, R. (1972). Spacings revisited. In: Proc. 6th Berkeley Symp., Math. Statist. Probability, Vol. 1, Eds. L.M. Le Cam et al., pp. 417–427. Berkeley: Univ. California Press.

Radtke, M. (1988). Konvergenzraten und Entwicklungen unter von Mises Bedingungen der Extremwerttheorie. Ph.D. Thesis, University of Siegen.

Ramachandran, G. (1984). Approximate values for the moments of extreme order statistics in large samples. In: Statistical Extremes and Applications, Ed. J. Tiago de Oliveira, pp. 563–578. Dordrecht: Reidel.

Rao, J.S. and Kuo, M. (1984). Asymptotic results on the Greenwood statistic and some of its generalizations. J. Roy. Statist. Soc., Ser. B, 46, 228–237.

Raoult, J.P., Criticou, D. and Terzakis, D. (1983). The probability integral transformation for not necessarily absolutely continuous distribution functions, and its application to goodness-of-fit tests. In: Specifying Statistical Models, Ed. J.P. Florens et al., pp. 36–49. New York: Springer.

Reiss, R.-D. (1973). On the measurability and consistence of maximum likelihood estimates for unimodal densities. Ann. Statist. 1, 888–901.

Reiss, R.-D. (1974a). On the accuracy of the normal approximation for quantiles. Ann. Probab. 2, 741–744.

Reiss, R.-D. (1974b). Asymptotic expansions for sample quantiles. Technical Report 6, University of Cologne.

Reiss, R.-D. (1975a). The asymptotic normality and asymptotic expansions for the joint distribution of several order statistics. In: Limit Theorems of Prob. Theory, Ed. P. Révész, pp. 297–340. Amsterdam: North-Holland.

Reiss, R.-D. (1975b). Consistency of a certain class of empirical density functions. Metrika 22, 189–203.

Reiss, R.-D. (1976). Asymptotic expansions for sample quantiles. Ann. Probab. 4, 249–258.

Reiss, R.-D. (1977a). Asymptotic Theory of Order Statistics. Lecture Notes, University of Freiburg.

Reiss, R.-D. (1977b). Optimum confidence bands for density functions. Studia Sci. Math. Hungar. 12, 207–214.

Reiss, R.-D. (1978a). Approximate distribution of the maximum deviation of histograms. Metrika 25, 9–26.

Reiss, R.-D. (1978b). Consistency of minimum contrast estimators in nonstandard cases. Metrika 25, 129–142.

Reiss, R.-D. (1980). Estimation of quantiles in certain non-parametric models. Ann. Statist. 8, 87–105.

Reiss, R.-D. (1981a). Approximation of product measures with an application to order statistics. Ann. Probab. 9, 335–341.

Reiss, R.-D. (1981b). Asymptotic independence of distributions of normalized order statistics of the underlying probability measure. J. Multivariate Anal. 11, 386–399.

Reiss, R.-D. (1981c). Nonparametric estimation of smooth distribution functions. Scand. J. Statist. 8, 116–119.

Reiss, R.-D. (1981d). Uniform approximation to distributions of extreme order statistics. Adv. Appl. Probab. 13, 533–547.

Reiss, R.-D. (1982). One sided test for quantiles in certain non-parametric models. In: Nonparametric Statistical Inference, Colloq. Math. Soc. János Bolyai 32, Eds. P.V. Gnedenko et al., pp. 759–772. Amsterdam: North Holland.

Reiss, R.-D. (1984). Statistical inference using approximate extreme value models. Technical Report 124, University of Siegen.

Reiss, R.-D. (1985a). Asymptotic expansions of moments of central order statistics. In: Probability and Statistical Decision Theory, Vol. A., Proc. 4th Pann. Symp., Eds. Mogyordi et al., pp. 293–300. Dordrecht: Reidel.

Reiss, R.-D. (1985b). Approximations to the distributions of ordered distance random variables. Ann. Inst. Statist. Math., Ser. A, 37, 529–533.

Reiss, R.-D. (1986). A new proof of the approximate sufficiency of sparse order statistics. Statist. Probab. Letters 4, 233–235.

Reiss, R.-D. (1987). Estimating the tail index of the claim size distribution. Blätter DGVM 18, 21–25.

Reiss, R.-D. (1989). Extended extreme value models and adaptive estimation of the tail index. In: Extreme Value Theory, Eds. J. Hüsler and R.-D. Reiss, pp. 156–165. Lecture Notes in Statistics 51. New York: Springer.

Reiss, R.-D., Falk, M. and Weller, M. (1984). Inequalities for the relative sufficiency between sets of order statistics. In: Statistical Extremes and Applications, Ed. J. Tiago de Oliveira, pp. 597–610. Dordrecht: Reidel.

Rényi, A. (1953). On the theory of order statistics. Acta Math. Acad. Sci. Hungar. 4, 191–231.

Resnick, S.I. (1987). Extreme Values, Regular Variation, and Point Processes. Applied Probability. Vol. 4. New York: Springer.

Rice, J. and Rosenblatt, M. (1976). Estimation of the log survivor function and hazard function. Sankhyā, Ser. A, 38, 60–78.

Rootzén, H. (1984). Attainable rates of convergence of maxima. Statist. Probab. Letters 2, 219–221.

Rootzén, H. (1985). Asymptotic distributions of order statistics from stationary normal sequences. In: Contribution to Probability and Statistics in Honour of Gunnar Blom, Eds. J. Lanke and G. Lindgren, pp. 291–302. University of Lund.

Rosenblatt, M. (1952). Remarks on a multivariate transformation. Ann. Statist. 23, 470–472.

Rosenblatt, M. (1956). Remarks on some nonparametric estimates of a density function. Ann. Math. Statist. 27, 832–837.

Rosengard, A. (1962). Etude des lois-limites jointes et marginales de la moyenne et des valeurs extrêmes d'un échantillon. Publ. Inst. Statist. Univ. Paris 11, 3–53.

Rossberg, H.J. (1965). Die asymptotische Unabhängigkeit der kleinsten und größten Werte einer Stichprobe vom Stichprobenmittel. Math. Nachr. 28, 305–318.

Rossberg, H.J. (1967). Über das asymptotische Verhalten der Rand- und Zentralglieder einer Variationsreihe (II). Publ. Math. Debrecen 14, 83–90.

Rossberg, H.J. (1972). Characterization of the exponential and the Pareto distribution by means of some properties of the distributions which the differences and quotients of order statistics are subject to. Math. Operationsforsch. Statist. 3, 207–316.

Rüschendorf, L. (1985a). Two remarks on order statistics. J. Statist. Plann. Inference 11, 71–74.

Rüschendorf, L. (1985b). The Wasserstein distance and approximation theorems. Z. Wahrsch. verw. Geb. 66, 117–129.

Ryzin, J. van (1973). A histogram method of density estimation. Commun. Statist. 2, 493–506.

Sen, P.K. (1968). Asymptotic normality of sample quantiles for m-dependent processes. Ann. Math. Statist. 39, 1724–1730.

Sen, P.K. (1972). On the Bahadur representation of sample quantiles for sequences of φ-mixing random variables. J. Multivariate Anal. 2, 77–95.

Sendler, W. (1975). A note on the proof of the zero-one law of Blum and Pathak. Ann. Probab. 3, 1055–1058.

Serfling, R.J. (1980). Approximation Theorems of Mathematical Statistical. New York: Wiley.

Shaked, M. and Tong, Y.L. (1984). Stochastic ordering of spacings from dependent random variables. In: Inequalities in Statistics and Probability. IMS Lecture Notes 5, 141–149.

Shorack, G.R. and Wellner, J.A. (1986). Empirical Processes with Applications to Statistics. New York: Wiley.

Sibuya, M. (1960). Bivariate extreme statistics. Ann. Inst. Stat. Math. 19, 195–210.

Siddiqui, M.M. (1960). Distribution of quantiles in samples from a bivariate population. J. Res. Nat. Bureau Standards 64, Ser. B, 124–150.

Singh, K. (1979). Representation of quantile processes with non-uniform bounds. Sankhyā, Ser. A, 41, 271–277.

Singh, K. (1981). On the asymptotic accuracy of Efron's bootstrap. Ann. Statist. 9, 1187–1195.

Smid, B. and Stam, A.J. (1975). Convergence in distribution of quotients of order statistics. Stoch. Proc. Appl. 3, 287–292.

Smirnov, N.V. (1935). Über die Verteilung des allgemeinen Gliedes in der Variationsreihe. Metron 12, 59–81.

Smirnov, N.B. (1944). Approximation of distribution laws of random variables by empirical data. Uspechi Mat. Nauk 10, 179–206 (in Russian).

Smirnov, N.V. (1949). Limit distributions for the term of a variational series. Trudy Mat. Inst. Steklov 25, 1–60. (In Russian). English translation in Amer. Math. Soc. Transl. (1), 11 (1952), 82–143.

Smirnov, N.V. (1967). Some remarks on limit laws for order statistics. Theory Probab. Appl. 12, 337–339.

Smith, R.L. (1982). Uniform rates of convergence in extreme value theory. Adv. Appl. Probab. 14, 600–622.

Smith, R.L. (1984). Threshold methods for sample extremes. In: Statistical Extremes and Applications, Ed. J. Tiago de Oliveira, pp. 621–638. Dordrecht: Reidel.

Smith, R.L. (1985a). Maximum likelihood estimation in a class of non-regular cases. Biometrika 72, 67–92.

Smith, R.L. (1985b). Statistics of extreme values. Proc. 45th Session of the ISI, Vol. 4 (Amsterdam), 26.1.

Smith, R.L. (1986). Extreme value theory based on the r largest annual events. J. Hydrology 86, 27–43.

Smith, R.L. (1987). Estimating tails of probability distributions. Ann. Statist. 15, 1174–1207.

Smith, R.L. and Weissman, I. (1987). Large deviations of tail estimators based on the Pareto approximation. J. Appl. Probab. 24, 619–630.

Smith, R.L., Tawn, J.A. and Yuen, H.K. (1987). Statistics of multivariate extremes. Preprint, University of Surrey.

Sneyers, R. (1984). Extremes in meteorology. In: Statistical Extremes and Applications, Ed. J. Tiago de Oliveira, pp. 235–252. Dortrecht: Reidel.

Stigler, S.M. (1973). Studies in the history of probability and statistics. XXXII. Biometrika 60, 439–445.

Strasser, H. (1985). Mathematical Theory of Statistics. De Gruyter Studies in Math. 7, Berlin: De Gruyter.

Stute, W. (1982). The oscillation behaviour of empirical processes. Ann. Probab. 10, 86–107.

Sukhatme, P.V. (1937). Tests of significance for sample of the χ^2-population with two degrees of freedom. Ann. Eugenics 8, 52–56.

Sweeting, T.J. (1985). On domains of uniform local attraction in extreme value theory. Ann. Probab. 13, 196–205.

Teugels, J.L. (1981). Limit theorems on order statistics. Ann. Probab. 9, 868–880.

Thompson, W.R. (1936). On confidence ranges for the median and other expectation distributions for populations of unknown distribution form. Ann. Math. Statist. 7, 122–128.

Tiago de Oliveira, J. (1958). Extremal distributions. Rev. Fac. Cienc. Univ. Lisboa A 7, 215–227.

Tiago de Oliveira, J. (1961). The asymptotic independence of the sample means and the extremes. Rev. Fac. Cienc. Univ. Lisboa A 8, 299–310.

Tiago de Oliveira, J. (1963). Decision results for the parameters of the extreme value (Gumbel) distribution based on the mean and standard deviation. Trabajos de Estadistica 14, 61–81.

Tiago de Oliveira, J. (1984). Bivariate models for extremes; statistical decisions. In: Statistical Extremes and Applications, Ed. J. Tiago de Oliveira, pp. 131–153. Dordrecht: Reidel.

Tippett, L.H.C. (1925). On the extreme individuals and the range of samples taken from a normal population. Biometrika 17, 364–387.

Torgersen, E.N. (1976). Comparison of statistical experiments. Scand. J. Statist. 3, 186–208.

Tusnády, G. (1974). On testing density functions. Period. Math. Hungar. 5, 161–169.

Umbach, D. (1981). A note on the median of a distribution. Ann. Inst. Statist. Math. 33, Ser. A, 135–140.

Uzgören, N.T. (1954). The asymptotic development of the distribution of the extreme values of a sample. In: Studies in Mathematics and Mechanics. Presented to Richard von Mises, pp. 346–353. New York: Academic Press.

Vaart, H.P. van der (1961). A simple derivation of the limiting distribution function of a sample quantile with increasing sample size. Statist. Neerlandica 15, 239–242.

Walsh, J.E. (1969). Asymptotic independence between largest and smallest of a set of independent observations. Ann. Inst. Statist. Math. 21, 287–289.

Walsh, J.E. (1970). Sample sizes for appropriate independence of largest and smallest order statistic. J. Amer. Statist. Assoc. 65, 860–863.

Watson, G. and Leadbetter, M. (1964a). Hazard analysis I. Biometrika 51, 175–184.

Watson, G. and Leadbetter, M. (1964b). Hazard analysis II. Sankhyā, Ser. A, 26, 101–116.

Watts, V., Rootzén, H. and Leadbetter, M.R. (1982). On limiting distributions of intermediate order statistics from stationary sequences. Ann. Probab. 10, 653–662.

Weinstein, S.B. (1973). Theory and applications of some classical and generalized asymptotic distributions of extreme values. IEEE Trans. Inf. Theory 19, 148–154.

Weiss, L. (1959). The limiting joint distribution of the largest and smallest sample spacings. Ann. Math. Statist. 30, 590–593.

Weiss, L. (1964). On the asymptotic joint normality of quantiles from a multivariate distribution. J. Res. Nat. Bureau Standards 68, Ser. B, 65–66.

Weiss, L. (1965). On asymptotic sampling theory for distributions approaching the uniform distribution. Z. Wahrsch. verw. Gebiete 4, 217–221.

Weiss, L. (1969a). The joint asymptotic distribution of the k-smallest sample spacings. J. Appl. Probab. 6, 442–448.

Weiss, L. (1969b). The asymptotic joint distribution of an increasing number of sample quantiles. Ann. Inst. Statist. Math. 21, 257–263.

Weiss, L. (1969c). Asymptotic distributions of quantiles in some nonstandard cases. In: Nonparametric Techniques in Statistical Inference, Ed. M.L. Puri, pp. 343–348. Cambridge: Cambridge Univ. Press.

Weiss, L. (1971). Asymptotic inference about a density function at an end of its range. Nav. Res. Logist. Quart. 18, 111–114.

Weiss, L. (1973). Statistical procedures based on a gradually increasing number of order statistics. Commun. Statist. 2, 95–114.

Weiss, L. (1974). The asymptotic sufficiency of a relatively small number of order statistics in test of fit. Ann. Statist. 2, 795–802.

Weiss, L. (1976). The normal approximations to the multinomial with an increasing number of classes. Nav. Res. Logist. Quart. 23, 139–149.

Weiss, L. (1977). Asymptotic properties of Bayes tests of nonparametric hypothesis. In: Statistical Decision Theory and Related Topics, II, Eds. D.S. Moore and S.S. Gupta, pp. 439–450. New York: Academic Press.

Weiss, L. (1978). The error in the normal approximation to the multinomial with an increasing number of classes. Nav. Res. Logist. Quart. 25, 257–261.

Weiss, L. (1979a). The asymptotic distribution of order statistics. Nav. Res. Logist. Quart. 26, 437–445.

Weiss, L. (1979b). Asymptotic sufficiency in a class of nonregular cases. Selecta Statistica Canadiana V, 141–150.

Weiss, L. (1980). The asymptotic sufficiency of sparse order statistics in test of fit with nuisance parameters. Nav. Res. Logist. Quart. 27, 397–406.

Weiss, L. (1982). Asymptotic joint normality of an increasing number of multivariate order statistics and associated cell frequencies. Nav. Res. Logist. Quart. 29, 75–96.

Weissman, I. (1975). Multivariate extremal processes generated by independent non-identically distributed random variables. J. Appl. Probab. 12, 477–487.

Weissman, I. (1978). Estimation of parameters and large quantiles based on the k largest observations. J. Amer. Statist. Assoc. 73, 812–815.

Wellner, J.A. (1977). A law of the iterated logarithm for functions of order statistics. Ann. Statist. 5, 481–494.

Wilks, S.S. (1948). Order Statistics. Bull. Amer. Math. Soc. 54, 6–50.

Wilks, S.S. (1962). Mathematical Statistics. New York: Wiley.

Winter, B.B. (1973). Strong uniform consistency of integrals of density estimators. Canad. J. Statist. 1, 247–253.

Witting, H. (1985). Mathematische Statistik I (Parametrische Verfahren bei festem Stichprobenumfang). Stuttgart: Teubner.

Witting, H. and Nölle, G. (1970). Angewandte Mathematische Statistik. Stuttgart: Teubner.

Wu, C.Y. (1966). The types of limit distributions for some terms of variational series. Sci. Sinica 15, 749–762.

Yang, S.-S. (1985). A smooth nonparametric estimator of a quantile function. J. Amer. Statist. Assoc. 80, 1004–1011.

Yamato, H. (1973). Uniform convergence of an estimator of a distribution function. Bull. Math. Statist. 15, 69–78.

Zolotarev, V.M. and Rachev, S.T. (1985). Rate of convergence in limit theorems for the max scheme. In: Stability Problems for Stochastic Models, Eds. V.V. Kalashnikov and V.M. Zolotarev, pp. 415–442. Lecture Notes in Mathematics 1155. Berlin: Springer.

Zwet, W.R. van (1964). Convex Transformations of Random Variables. Amsterdam. Math. Centre Tracts 7.

Zwet, W.R. van (1984). A Berry-Esséen bound for symmetric statistics. Z. Wahrsch. verw. Gebiete 66, 425–440.

Author Index

Subject Index